Henry Shaler Williams

Geological Biology

An Introduction to the geological History of Organisms

Henry Shaler Williams

Geological Biology

An Introduction to the geological History of Organisms

ISBN/EAN: 9783337216832

Printed in Europe, USA, Canada, Australia, Japan

Cover: Foto ©berggeist007 / pixelio.de

More available books at **www.hansebooks.com**

AN INTRODUCTION TO THE

GEOLOGICAL HISTORY OF ORGANISMS

BY

HENRY SHALER WILLIAMS

SILLIMAN PROFESSOR OF GEOLOGY
IN YALE COLLEGE

NEW YORK
HENRY HOLT AND COMPANY
1895

ROBERT DRUMMOND, PRINTER AND ELECTROTYPER, NEW YORK.

PREFACE.

THE following chapters were originally written in the form of lectures, delivered first at Cornell University, where they were supplemented by special laboratory work and illustrated by actual specimens of the organisms or fossils described. The attempt was made to replace the ordinary treatment of the dry statistics of historical geology and paleontology by something which would bring the chief problems of the history of organisms within the comprehension of the ordinary college student, and kindle in the special student enthusiasm for deeper research. In preparing them for publication the lecture form was dropped, such revision of the language and treatment was made as to provide for readers who might not have at hand full museums from which to draw illustrative material, and a few of the more characteristic examples, used in elucidating the principles discussed, were selected and more fully and precisely elaborated, so as to make a text-book for use in earnest and exact study as well as an exposition of general principles.

Two classes of readers were considered in giving the book its present form, viz., students in colleges and universities who have begun to appreciate the importance of understanding the principles of the nature and history of organisms, either as a preparation for further special studies or as a part of a liberal education; and second, the general reader, who is supposed to know something of the present popular theories regarding organic life, and has, perhaps, already become aware of the increasing sense of disappointment which those are meeting who have attempted seriously to apply them to the solution of the problems of human life. It is not assumed that the reader has any special knowledge of biology or geology to start with. On this account some details have been given which would be unnecessary for the specialist, while, on the other hand, many elaborations which would interest him have been omitted in order to bring under discussion as many as possible of the essential problems.

The book is not intended to be a complete treatise upon paleontology, nor a detailed report of the relation of fossils to geological formations or to time. It is rather a reconnaissance of a fascinating region, from which the few explorers who have already penetrated it have brought back accounts of the most remarkable and unexpected discoveries. A reconnaissance aims to discover the characteristic features and the relative importance of the various elements making up the territory traversed, and is merely introductory to a more minute and careful survey; its purpose is to aid the judgment, to direct the course of further research, and when difficulties of travel and distances are great, it is particularly useful in preventing distraction from the most expeditious way to the facts of chief importance.

The tendency of modern science is now, and for more than a quarter of a century has been, so much to specialization, and our minds have become so fascinated by the minute and the particular, that our common judgments of the true proportion of things have become more or less distorted. Theories and ideas which have been drummed into our ears have come to appear the most important truths in the world, and all our thoughts have become colored by them. We cannot read the newspapers or listen to the talk on the street without being convinced that the thought of the people, however little they may know of the sciences involved, is thus biased by current theories about life and organisms. The bearing of biological theories upon our judgments of the rightness or wrongness of conduct, both of ourselves and of society, is too direct to admit of any uncertainty regarding the validity of their foundations or their precise import. While the facts and phenomena upon which some of the theories rest are purely biological, others of them, which concern man most intimately, have their chief evidence in the historical records of geology.

Among the latter none is more important than those gathered about the phenomena of evolution; but it is evident upon reflection that the biologist proper, who deals alone with the organisms now living upon the earth, must rest with a theoretical interpretation of the laws of evolution. To the geologist the records of evolution are open for direct examination, and geological biology is a scientific treatment of the observed facts of evolution.

While there are no end of books on evolution, and modern biologists seem content to assume that some theory of evolution is true, without being able to decide which it shall be; and although the students of sociology, the moralist, and the theologian are basing their theories about man on the "working hypotheses" of the naturalist as if "law and gospel,"—it seems to have escaped serious attention that we have open for study a genuine record of the actual evolution of organisms, extending from near the beginning of life up to the present time. Men have been speculating in all conceivable directions to form some theory as to how evolution ought to work, and as to what the history of organisms ought to be: it is the province of geological biology to tell us what the history of organisms has actually been. The geologist does not ask what is the theory of evolution, but what are the facts of evolution. "The primary and direct evidence in favor of evolution can be furnished only by paleontology. The geological record, so soon as it approaches completeness, must, when properly questioned, yield either an affirmative or a negative answer: if evolution has taken place, there will its mark be left; if it has not taken place, there will lie its refutation." The late Professor Huxley, who framed this most true and pertinent sentence, knew very well the evidence which those records furnish, although he often treated evolution as if it were a doctrine requiring argumentative defense, rather than a science which only needs elucidation.

The treatment which evolution receives in these pages is designed for those who wish to know what the chief facts and factors of evolution are, not those who are looking for further debate of the arguments either for or against a theory of evolution. To the student who approaches the subject from the historical side evolution becomes the very key to the mystery of organic life. The phenomena of growth are fairly well understood, the development of the individual has been systematized into a science of embryology; and as we also discover the grand features of the evolution of species and races and kinds of organisms. life begins to assume the proportions of one of the fundamental forces in the world. When considered from this point of view the question what causes the evolution of organisms seems as impertinent as what causes the motion of the celestial spheres. The answer to both is the same.

That the form and functions of successive organisms should be accurately adjusted to their organic and physical environments is no more surprising than that the size and weight of the revolving planets should be accurately adjusted to the orbits in which they swing; but once grant that the systems are in motion, and it is not reasonable to suppose in either case that at any point in the succession of phenomena misadjustment should occur which would require any hypothetical selective force to put them right again. Evolution thus becomes one of the fundamental expressions of life force, requiring no theory to support it, but calling only for investigation to reveal its laws; and it is in geological biology that we find the direct evidences of the course of its operation. But evolution is not all of biology, and therefore sufficient illustration of their respective phenomena has been borrowed from physiology and embryology to present a comprehensive view of all the three great factors of organic life, viz., growth, development, and evolution.

A few of the chapters are somewhat technical in their language, and deal with particulars of slight interest to those unfamiliar with the nomenclature of natural history. These chapters may be omitted by readers willing to take the author's statements without verification. Such persons may omit the purely geological part of the book by passing directly from Chapter I to Chapter V, where the discussion of the biological problem begins. The more technical passages are Chapter II; Chapter IV, except the summary at the close; pages 98 to 110 of Chapter V; all but the summary of Chapter VII; the latter parts of Chapters XII and XIII; and the fine print of Chapters XVIII and XIX. The remainder of the book, although occasionally expressed in scientific terms, will be found, it is hoped, fully intelligible to the ordinary reader.

Special students of paleontology and geology will miss the expected descriptions of fossils and the means for identifying them and for recognizing the horizons they indicate. To such readers the author has to say that this book is offered only as an *introduction* to the grand field of study open before them, with the hope that it may be useful in guiding and suggesting methods of investigation, and in encouraging that deep research which will be found necessary to interpret the full story of the history of organisms, of which only a glimpse is here attempted.

H. S. W.

NEW HAVEN, October 5, 1895.

CONTENTS.

(The numbers refer to the pages of the text.)

CHAPTER I.

THE HISTORY OF ORGANISMS, ITS SCOPE AND IMPORTANCE.

CHAPTER II.

THE MAKING OF THE GEOLOGICAL TIME-SCALE.

CHAPTER III.

THE DIVISIONS OF THE GEOLOGICAL TIME-SCALE AND THEIR TIME-VALUES.

CHAPTER IV.

STRATIFIED ROCKS—THEIR NATURE, NOMENCLATURE, AND FOSSIL CONTENTS.

CHAPTER V.

FOSSILS—THEIR NATURE AND INTERPRETATION, AND THE GEOLOGICAL RANGE OF ORGANISMS.

CHAPTER VI.

GEOGRAPHICAL DISTRIBUTION—THE GENERAL RELATION OF ORGANISMS TO THE CONDITIONS OF ENVIRONMENT.

CHAPTER VII.

GEOGRAPHICAL DISTRIBUTION: SPECIAL CONSIDERATION: THE ADJUSTMENT OF ORGANISMS TO ENVIRONMENT.

CHAPTER VIII.

WHAT IS A SPECIES?—VARIOUS DEFINITIONS AND OPINIONS.

CHAPTER IX.

WHAT IS AN ORGANISM? THE CHARACTERISTICS OF THE INDIVIDUAL AND ITS MODE OF DEVELOPMENT.

CHAPTER X.

WHAT IS THE ORIGIN OF SPECIES?—THE PROBLEM AND ITS EXPLANATION.

CHAPTER XI.

THE PRINCIPLES OF NATURAL HISTORY CLASSIFICATION; ILLUSTRATED BY A STUDY OF THE CLASSIFICATION OF THE ANIMAL KINGDOM.

CHAPTER XII.

THE TYPES OF CONSTRUCTION IN THE ANIMAL KINGDOM.

CHAPTER XIII.

PHYLOGENESIS IN CLASSIFICATION.

CHAPTER XIV.

THE ACQUIREMENT OF CHARACTERS OF GENERIC, FAMILY OR HIGHER RANK; ILLUSTRATED BY A STUDY OF THE BRACHIOPODS.

CHAPTER XV.

WHAT IS EVOLVED IN EVOLUTION?—INTRINSIC AND EXTRINSIC CHARACTERS.

CHAPTER XVIII.

THE RATE OF MORPHOLOGICAL DIFFERENTIATION IN A GENETIC SERIES, ILLUSTRATED BY A STUDY OF CEPHALOPODS.

CHAPTER XIX.

PROGRESSIVE MODIFICATION OF AN EXTRINSIC CHARACTER; ILLUSTRATED BY THE EVOLUTION OF THE SUTURE-LINES OF AMMONOIDS.

CHAPTER XX.

THE LAWS OF EVOLUTION EMPHASIZED BY STUDY OF THE GEOLOGICAL HISTORY OF ORGANISMS.

CHAPTER XXI.

PHILOSOPHICAL CONCLUSIONS REGARDING THE CAUSES DETERMINING THE COURSE OF EVOLUTION.

GEOLOGICAL BIOLOGY.

CHAPTER I.

THE HISTORY OF ORGANISMS. ITS SCOPE AND IMPORTANCE.

Man an Organism among Organisms.—Man has been very slow to grasp the fact that he is an organism among organisms. Darwin was the first to speak with such loud emphasis as to thoroughly rouse the world to an appreciation of the very intimate relationship man bears to the whole series of organic forms of not only present but all past time. We are apt to be offended by the bold statement that man is descended from the monkeys, but, without insisting upon the truth of this specific statement, the investigations of modern science have demonstrated beyond controversy that the same conditions of affinity and relationship which lead to the classification of animals into species, genera, or classes, and as connected with each other by direct genetic descent, apply to man as one of the organisms.

For want of a better name this relationship of man to other organisms may be called his natural-history relationship. Man is an organism among organisms, and it is this fact that lifts the history of organisms out of the field of simple morphological or physiological sciences into a place of direct human interest. Man's origin and history is intimately associated with the origin and history of other living beings in the world.

Not only is there human interest in the subject of the history of organisms, but because of this interest there is a demand for discussion of the facts themselves from a special point of view.

The naturalist takes interest in the form and functions of individual organisms from a scientific point of view; they are to him objects of interest in themselves. He classifies and arranges them as favorite objects of knowledge. But the general student, the active thinker, the busy worker in human affairs finds the details of such studies irrelevant, and to him the vital interest is in the questions concerning the relations of organisms to the past and to himself.

More than this, the deepest interest of all attaches to the philosophy which is involved in the proposition that man is not so distinct from the dumb organic world around him as was up to a few years ago universally believed to be the case.

History of Organisms and Man's Relationship to Living Things.—If man has arisen from organisms that were not men; if the machinery of his vital organization is represented in less complex form in other animals; if he may find his functions in operation in simpler forms of life, and separated into their elements in lower types, then he has in the organic world a field of study of the greatest interest, which he cannot neglect without ignoring knowledge that is, in a literal sense, vital to his best interests as a man.

The study of the laws of organisms, their relations to each other and to the conditions of environment, their antiquity, their history, and the nature of those laws of adjustment which are suggested by the words heredity and descent, variability, natural and unfavorable habitat, struggle for existence, adaptation to environment, evolution, and many others which have arisen within the last fifty years, is of more importance than we ordinarily attach to the study of the curiosities of natural history.

The Discussion not from the Zoological and Botanical Side.—The approach to the study of organisms, from the zoological or botanical side, presents great difficulty in the very immensity of the subject. When we attempt to analyze the characters of a single animal, to classify animals and describe them, the mere mass of detail—the abundance of the characters to be distinguished—removes the subject from a place in a general course of liberal education. Such a treatment of organisms as may be sufficient for the illustration of their history does not necessarily enter into an analysis of the structural characters of any particular species. Hence, from the point of view of a technical course of study in biology, this treatise will seem quite superficial.

The Geological Aspect of the History of Organisms.—On the other hand, there are characters distinguishing groups of organisms, evidence of which may be preserved in the rocks, which are of far greater importance than the specific details in indicating the relationship organisms bear to each other, to the conditions in which they have lived, and to the place they have occupied in the history of the life of the globe. Such characters are those which will concern us here. In defining our topic as geological biology, we are not proposing to investigate the anatomical organs and tissues of which particular animals are made, but to review the facts and theories which have led to the belief that each living animal and plant is but the last of a long line of organisms whose remains can be recognized in more or less perfect fossils, and whose varying characters can be traced back into the immense antiquity of geological time.

Geological History not a Repetition of Like Events, but a Progressive Change of Phenomena.—If there were only repetition of the same things, this would not constitute history. If different things have succeeded each other, to ascertain the relationship borne by those that follow to those that preceded them becomes an important problem. We do not, at the outset, assume to explain the causes, but geology makes the fact clear that there has been a very elaborate history of the organisms that have lived on the earth. The question we propose to answer is, "What are the prominent laws expressed in this history?"

The geologist observes that there has been a history for the earth itself: the rocks, as geological formations; the lands, as parts of the crust above the surface of the ocean; the surface of the earth, as a whole, in all its complexity—all these have come to be what they are through innumerable changes. The geological conditions in the past have been associated with the history of the organisms. It is proposed to examine and note what have been the relations existing between organic form and geological and geographical conditions and progress.

Investigation of the Laws of Evolution.—Evolution has been discussed and applied in a thousand ways of late years, until the word has become a kind of shibboleth of modern science. It is proposed, in the following chapters, to ascertain what the term really means in the one field in which it may be properly and scientifically applied. For this purpose it is necessary to use the methods of philosophy, as well as those of science; to weigh the arguments and reasonings of naturalists, as well as to examine, analyze, and classify the facts of nature.

Old Notion of an Organism contrasted with the New.—Within the last thirty years very great change has taken place in the general ideas regarding the nature of organisms and their relations to each other. The old idea of an organism perpetuating its kind by generation, in which difference of kind was at once evidence of difference of origin, has of late been almost entirely replaced by the new idea in which there is not only repetition by generation of the characters of its ancestors, but a constant slight and slow divergence from them, resulting, in the course of many generations, in bringing about all the differences of form which distinguish the various species of the world, present and past. The new theory has led to an exhaustive study of the relations which organisms bear to one another and the interrelations existing between geographical and geological conditions on the one hand and the form of organisms on the other.

Work of the Paleontologist.—While embryologists have been tracing out in detail the changes experienced by the individual in passing from the embryonic to the adult stage of growth, and while the zoologist and botanist have been minutely examining and teaching the differences in structure and function of the various parts of each animal and plant, the paleontologist has been accumulating data to show the order of succession of life in the past, and thus has been opening the way for the particular study of organisms in their relations to time and space, their geological sequence, their geographical distribution, and the various laws regulating these modifications and adjustments. The paleontologist is able actually to see the orderly succession of organisms in the past, and he is constantly called upon to note the relation of the several forms under his view to the environing conditions of their life, and thus to interpret the history of the great races of beings that have peopled the world.

Botanists and Zoologists observe Individual Characters.—The development of the individual organism from the embryo to the mature individual is familiar to us all in its general principles. We know how the seed or the acorn grows to become the flowering plant or the oak tree. We know that the egg, by some mysterious process inside the shell, changes so as to become the chick which cracks its way out, breathes and develops into the crowing cock or the egg-laying hen. In each of these cases the history is the history of an individual organism. It is the history of a single organism, and the science teaching about these phenomena is the science of Embryology, and is concerned with the laws of individual development.

Botany and Zoology, too, are mainly concerned with a study of the morphology of the characters of the individual, its form and structure, and particularly the analysis of its organs and their functions, in their morphological relations, the relations of the organs as they are combined for the functions of life of the individual. What there is of history is life-history of the *individual*, and what there is of study of form is of the form of the *parts*, or of the whole as a complex of such parts, of an individual organism. And what there is of classification is classification to bring out the *differences* existing between the component parts of separate individuals. In these studies the individual organism is the highest unit, and the investigations are conducted in each case as if there were but one organism: comparisons are between its parts and not with other organisms.

Paleontologists interested in the History of Species, of Races, and of Groups of Organisms.—It is for the paleontologist to speak of the history of races and communities of organisms, that is, to look upon individual organisms as parts of some complex whole, to look at organisms as related to each other in the complex environment of the earth, the temporary world-surface, and in the consecutive time-relations which are recorded in the geological strata making up the surface of the globe. In the life-history of the individual, or Embryology, we have the body of the individual to bind together the various stages of development. For this history the hours of the clock or the days of the calendar are satisfactory time-divisions. The relations of the various organs or parts to each other are easily determined by noting the effect of artificial separation or excision; but we see no history of organisms until we compare those now living with others that lived unmeasured hours and days and even years ago. Comparison of living species with living species only, shows us differences which our classifications enumerate. While we might, theoretically, guess that the present living organisms came from others not like them, if we knew nothing of fossils this would be but a mere vague fancy, and could never find a place in true science. Paleontology, however, reveals to us a long series of organic forms, and when we speak of their history we assume that the series is connected genetically; the time-relations we read from the rocks, and in terms of subjacent strata. The relationship must be determined by comparison of entirely distinct forms; we must learn of organisms from their fossilized remains. These and many other facts must be presented before we have the data for defining the successive steps of the history.

Organisms and Environment.—Our subject, then, divides itself into two grand divisions, *organisms* on the one hand, and, to use a very comprehensive term, *environment* on the other hand—living things, and the conditions under which they have lived. The environment or conditions of life are strictly included in the science of Geology,—for geography is but the present final product of geological processes. When we treat of Biology geologically and study the history of organisms, we assume the truth of two propositions which are not required in the study of the characters and the development of the individual organism. The propositions are: first, that long periods of time have elapsed separating the periods of living of the several organisms under our investigation; and second, that there is genetic affinity between the organisms now living and those that have lived in the past. We assume that series of organisms genetically connected have lived during geological time.

Geological Formations.—It will be necessary to particularly consider the nature of geological formations, for in them are found the fossils, and from them is derived the evidence of the history which we are to read. We must consider how the formations were made, how the chronological scale is determined and what reliance may be placed in it. We must consider the manner of deposition, and under what condition fossils have been preserved; we must examine into the perfection or imperfection of the record, what has transpired to destroy the record, and hence how we can supplement the record we possess. Hence, geological classifications must be critically examined and analyzed. This will occupy the earlier chapters.

The Organism.—The second step will be to learn what the organism is and what it is not: what is meant by species and genera; what is the nature of systematic classification; the meaning of generation, race, modification, struggle for existence, geographical distribution, and many kindred terms.

Races and their History.—This will bring us to the third part of our subject, the specific study of races, their geological history, and the laws to be gathered from their study. The history of the organism may be viewed under two lights; as we consider the development of the individual as it passes from the germ to the fully organized adult, or as we consider one particular kind of organism as assuming the features which now characterize it from some other different kind of organism which preceded it. In the one case that which is continuous in the history is the individual life which develops, in the other case that which is continuous is the race which evolves.

The Chronological Scale.—In any discussion of history the first and essential element of fact to be established is a reliable chronological scale by which to mark off the relations of successive events or epochs of the history. In studying the history of the development of the individual organism, as artificial time-measures the clock or watch, or the regular periods of day and night, satisfy the demand. When longer periods are recorded, the seasons and years, with their artificial names, are sufficiently definitive. Human history deals with still longer periods, marked by great events in the nations: the rise or fall of a dynasty, the founding of a city, the discovery of a continent, the living of some man of powerful influence—these constitute landmarks by which to measure the order of lesser events. These, as chronological measures, are now easily applied, but in our studies in natural history we soon pass beyond the reach of even such records. A very few centuries back and human history ceases altogether; therefore the time-scale for the history of organisms must rest upon an entirely different kind of evidence. Another reason renders the ordinary units of time useless for the study of the history of organisms. The animals and plants associated with the earliest known traces of man present only the most insignificant amount of divergence from their living representatives. In most cases the differences are not greater than differences presented by the known descendants of common ancestors within the memory of a single generation of men. The period of human existence, however long or short that may be, is too brief to record any but the more minute details of those modifications of which paleontology teaches. It is unnecessary to state that the records we are to study are buried in the rocks. Everybody knows that the rocks must be of considerable antiquity; but when we pass beyond the age of man, as an inhabitant of the earth, our ideas of time-relations are necessarily vague; even for scientific men these time-relations, both their actual length, in terms of human standard, and also their relative periods, are not matters of simple arithmetical calculation.

Theories regarding the Length of Geological Time.—The theories underlying the interpretation of the rocks are far more important than at first would appear. The common notion, up to a very few centuries, and in some quarters a few decades ago, was that the antiquity of the inhabitants, and the world itself, did not exceed six thousand years. We now believe that the time that has transpired since the first organisms lived upon the earth is measured by millions rather than by centuries of years, "tens of millions and not millions nor hundreds of millions," as Mr. Walcott maintains.* Sufficient evidence appeared to have convinced the earlier geologists of this statement; but the evidence is not direct testimony to the fact of the great antiquity of the earth and its inhabitants. The fact that fossils are in the solid rocks and that they are different from the shells or hard parts of any organisms now living, were facts well known long before the notion of six thousand years was considered inadequate for the history of the earth. But the opinion that the fossils were the remains of organisms of no great antiquity, and that they had been buried by some great flood, some extraordinary cataclysm, was held to be sufficient to explain the brevity of the assumed time; and the differences between the fossils and the living forms were mysteries which were simply not explained at all until about the beginning of the present century. The general belief that cataclysms are possible, that antiquity is the great reservoir for the remarkable, the extravagant, the unscientific, or the unknown, has been, and is to some extent now, the common excuse for mistakes made in interpreting the laws of nature. In the study of rocks, we need to learn how to use them as measures of the time-relations of the fossil contents. An analysis of the classifications which have hitherto been made to express the chronological relations of rocks will show us what the facts are, how these facts have been interpreted, and how far these interpretations are at present satisfactory.

* Vice-Presidential Address, Section E, Am. Assoc. Adv. Sci., 1893.

CHAPTER II.

THE MAKING OF THE GEOLOGICAL TIME-SCALE.

The Heterogeneous Names now in Use.—A critical examination of the nomenclature applied to the several divisions of the geological scale reveals a strange mixture of names, the reason for which is not evident to modern students of the science. In the list of system-names we find Carboniferous and Cretaceous, indicative of mineral characters, associated with Tertiary and Quaternary, meaning rank in some undefined order of sequence. The presence of these terms is no less mysterious than the absence of grauwacke and old-red sandstone, and primary and secondary, which were originally included. Triassic is the name of another system and records the threefold division of the system of rocks to which it was applied; and Devonian, the name of another, reminds us of the county in England in which its rocks were first named. Observing these things, one is tempted to call in question the reliability of a systematic classification so heterogeneously compounded.

Importance of a Systematic Classification.—Although the older living geologists can remember back almost to the beginnings of the science, those who now are beginning their study of geology may find profit in examining the foundation principles, and the systems which have been devised and have led to the construction and belief in the present classification—a classification the adoption and unification of which has been thought worthy of the organization and continuance of an international Congress of Geologists. It is needless to call attention to the necessity of some systematic classification of geological formations, but as a foundation for the scientific study of the history of organisms there is need of a time-scale running back into the past, the degree of accuracy of which is known as well as the extent of its unreliability. In early attempts to classify rocks the chronological element of the scale was not considered, but by degrees the classification has passed from a classification of rocks to a classification of periods of time.

Ancient Notions of Geology.—The ancients in many respects were keen observers; they knew much about plants, animals, physical and chemical phenomena, and astronomy. But, with all their learning, there appears to have been no conception formed of an ancient history of the globe and its inhabitants prior to the earlier centuries of the Christian era. One of the first geological phenomena to become generalized into a theory was that of the formation of mountains by earthquakes, as cited by Avicenus in the tenth century. The gradual change of relative level of land and sea, as seen in the encroaching of the sea or the departure of sea from the shore, gave rise to speculations regarding the great length of time required for the lifting of the whole land by that means. In the sixteenth century, Lyell reminds us, attention was drawn to the meaning of fossils, and dispute arose as to their nature. Leonardo da Vinci doubted the then current belief that the stars were the cause of the fossil shells and pebbles on the mountain-sides, and advanced the idea "that the mud of rivers has covered and penetrated into the interior of fossil shells at the time when these were still at the bottom of the sea near the coast."*

Beginnings of a Scientific System of Classification.—By degrees, as Lyell has described in such fascinating manner, one after another the foundation principles were announced, discussed, controverted, and finally, by their intrinsic truth, became established. But it was not till nearly the beginning of the present century that enough was known of rocks for the formation of a general systematic classification of geological formations. The belief in a limit of six thousand years for the formation of the world was prevalent. Catastrophe was the universal resort for explanation of phenomena not then understood. And for geological purposes the Noachian deluge was an in-

* Lyell's Principles, p. 34.

dispensable agent for the scientific explanation of any extraordinary phenomena. For these reasons inquiry did not reach far into the antiquity of the geological ages. And the first attempts at classification took little or no account of actual time-factors in geology.

Lehmann's Classification according to Order of Formation.—Lehmann * is generally credited with having first proposed a classification of rocks on the basis of the order of their formation, as Primitive, Secondary, and a third class, the modern or superficial rocks made by the deluge or ordinary river action. Lehmann recognized also a direct relation of origin for the Secondary from the Primitive rocks, and thus arose the beginnings of the geological time-scale. Lehmann described three originally distinct kinds of rocks, or rock formations. The volcanic were separated from the others because having no particular connection with either in origin. The distinction, however, between Primitive and Secondary was fundamental. The Primitive was strictly the original, basal rock formed by crystallization from chemical solution before organisms lived; and the Secondary rocks were of secondary origin, made out of fragments of the older and always lying above them. In the original classification of Lehmann, Secondary included all the stratified rocks, as we now describe them, and in the classifications for some years following Lehmann the term Secondary was applied, though in a restricted sense.

Cuvier and Brongniart's and Réboul's Contributions.—Cuvier and Brongniart † proposed the name Tertiary for the rocks classified as Secondary by Lehmann, but lying above what is now known as the Cretaceous system; and Quaternary was used by Réboul ‡ in 1833 for the rocks of superficial position and of glacial or fluviatile origin. Thus the nomencla-

---

* J. G. Lehmann, "Versuch einer Geschichte von Floetzgebirgen, etc.," Berlin, 1766 (Kayser), 1756 (Poggendorf). French translation cited by Lyell "Essai d'un Hist. Nat. des Couches de la Terre," 1759. See Lyell, "Principles," vol. 1. p. 72, and Conybeare and Phillips, "Geology," p. vi and p. xlii. Johann Gottlob Lehmann died in St. Petersburg, 1767.

† Cuvier and Brongniart, "Descr. Géol. des Environs de Paris," ed. 2, 1822, p. 9.

‡ Réboul, "La Géologie de la Période Quaternaire," 8vo, 1833. Morlot, Bull. Soc. Vaudoise des Sc. Nat., IV. 41, 1854.

ture of Lehmann, which was proposed originally to indicate the derivation of the Secondary from the Primitive, was expanded on the basis of stratigraphic succession, and we observe the anomaly of a retention of two names (Tertiary and Quaternary), formed on the principle of Lehmann's terms, but his own terms, as well as his theory as a basis of classification, entirely discarded.

Werner's Perfection of the Lehmann Classification.—Werner (1750–1817) elaborated Lehmann's scheme and modified it. He was the great teacher of geology at Freiburg, Germany, in 1815, and left his impress upon the geologists of the time, though he wrote little in the way of systematic exposition of his theories of classification. He adopted Lehmann's *Primitiv Gebirge*, but of the Secondary rocks he made a lower class, which he called transition rocks (*Uebergangsgebirge*); they were stratified, contained none or but few fossils, and were more or less oblique in position; these characteristics were observed in northern Europe, where he studied them. The remainder of the original Secondary rocks he called *Floetzgebirge*, or flat-lying formations, and these were the equivalents of Lehmann's Secondary in the classification of the early part of the century. Later, the Wernerian school called the formations above the Cretaceous *neues Floetzgebirge*, to which, as they were studied in the Paris basin, Cuvier and Brongniart, in the latter decade of the last century, applied the name *Tertiary*, which still remains in the scheme. Werner called the looser, overlying, unconsolidated rocks *angeschwempt Gebirge*, or alluvial formations, which were afterwards, as above stated, called Quaternary by Réboul and Morlot.

The classification of Lehmann, as perfected by Werner, was then as follows:

German Names.	*English Equivalents.*
IV. Angeschwempt Gebirge.	Alluvial formations.
III. *b.* Neues Floetzgebirge.	Tertiary "
a. Floetzgebirge.	Secondary "
II. Uebergangsgebirge.	Transition "
I. Urgebirge.	Primitive "

These were the formations which made up the geological series as then recognized. Volcanic rocks were looked upon as local formations, and of small account in general classification. But they came to be more deeply studied by Werner, and his notion that trap was of aqueous origin led to much controversy, and gave chief prominence to his views (the Neptunian theory) and to that classification of rocks which will be next considered. The rocks of igneous origin, although sometimes interstratified with sedimentary rocks, do not enter into the present geological time-scale, and for the present purpose further consideration of their classification is unnecessary. There has always been a remnant of rocks at the base of the scale, the consideration of which may be discarded here, because it is chronologically known only as below those rocks of which distinct evidence of their relative age is apparent. The name Primitive has been changed to Primary, and finally to Archæan, a name which was proposed by Dana,* and is likely to be permanently retained for some of the basal part of the series.

This first comprehensive classification of rocks may be called the Lehmann classification. It was based upon a structural analysis of the rocks in the order of their actual positions. The nomenclature is applied on the theory of relative order of formation.

Richard Kirwan and Geology at the Close of the Last Century.—Richard Kirwan† claimed to be the first author to publish a general treatise on Geology in the English language. Although the book is written in a decidedly controversial spirit, the author appears to have had a thorough acquaintance with the various treatises in French, German, Latin, and English, in which were expressed contemporaneous opinions regarding geological science. He was a Fellow of the Royal Societies of London and Edinburgh, member of the Royal Irish Academy, and of Academies in Stockholm, Upsala, Berlin, Manchester, and Philadelphia, and Inspector General of his majesty's mines in the kingdom of Ireland. It is probable,

* Amer. Jour. Sci., VIII. 213, 1874.

† "Geological Essays," London, 1799.

therefore, that he presents a fair idea of the opinions which underlay the Lehmann classification. According to Kirwan's book the rocks were originally in a soft or liquid state, the centre of the earth was supposed to be hollow, or the whole earth was a solid exterior crust with immense empty caverns within. The materials of the earth were then in a state of fusion or solution, and by condensation, as time progressed, the solids were crystallized out and deposited from the chaotic fluid. The water contracted its surface and lowered upon it by sinking into the interior cavities. With the deposition of the primitive rocks from the chaotic fluid, the water became purer. Mountains were conceived of as the local points of original crystallization which drew to them, in the process, the minerals from the general fluid. As the waters gradually withdrew by evaporation and sinking into the interior caverns, they became clarified and capable of supporting organic life.

Kirwan says:* "The level of the ancient ocean being lowered to the height of 8500 or 9000 feet, then, and not before, it began to be peopled with fish." (Under the name fish he included shell-fish and all other petrifactions.) The plains were formed of depositions from the water of argillaceous, siliceous, and ferruginous particles, mingled with those derived by erosion from the already protruding mountains. All the rocks above the height mentioned, he observed, quoting from testimony of numerous travellers, "are lacking in fossils; even the limestones are crystalline or 'primitive' limestones and marbles." These observations were cited in refutation of Buffon's "error" in claiming that all limestones were derived from comminuted shells. According to some authorities, primitive mountains should include rocks of even less height than 8000 feet, and the occasional presence of fossils at a greater elevation was by them accounted for by their transference to that elevation by the deluge.

Geological Mountains (Gebirge) and Formations.—This account of Kirwan's will suggest the way by which the rock formation first came to be called "*Gebirge*," or mountains. Rocks were supposed to lie as they were originally formed, and thus in

---

* "Geological Essays," p. 26.

classifying rocks the larger aggregates were naturally mountain masses. As the conception of movements in the earth's crust with folding and displacement came into the science, the idea of classification and grouping of rocks was retained, but that their grouping was based upon present massing above the surface as mountains ceased to be accepted as truth. In the German language the term "*Gebirge*" was retained, and apparently with restricted meaning. Kirwan apparently translated the term directly into English as mountains. *Formation*, however, took the place of *mountain*, as applied to rock classification, in the early part of the century.

The Formation of Sedimentary Rocks according to Werner and his School.—In the following cut is illustrated the conception of the Wernerian school of the mode of formation of the rocks and the reason for the relative positions each kind occupies. In the figure *a a' a* is the supposed fundamental basin of primitive rocks crystallized out from the chaotic fluid as described above by Lehmann, and these rocks were hence named *Urgebirge*, or Primitive rocks. When the ocean

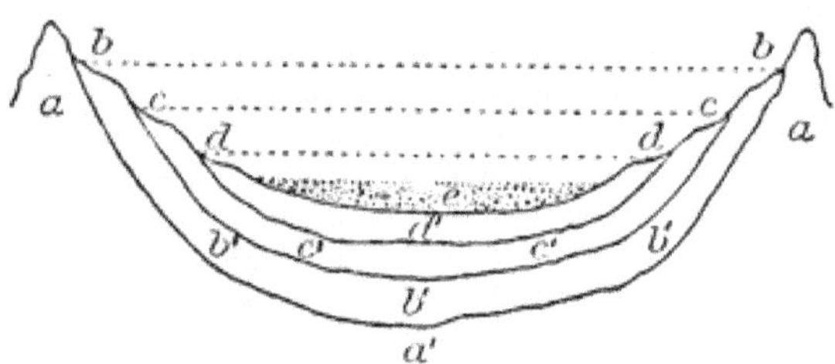

FIG. 1.—Diagram expressing the supposed mode of formation of the several formations (*Gebirge*) according to the Wernerians. (After Conybeare & Phillips.)

level had sunk to *b b*, deposition began and went on till the rocks *b b' b' b' b* were formed, the *Uebergangsgebirge* or transition rocks of Werner, whose position is oblique because of conformity to the sides of the original mountains as they stood in the original seas. As the surface of the ocean continued to sink, the deposits were accumulated lower and lower down on the mountain-sides, and more and more nearly horizontal, *c c'c'c* and *d d'd*, which represent the *Floetzgebirge* or flat-lying rocks; finally above the *neues*

Floetzgebirge (*d d' d*) were deposited the loose-lying gravels and soils of the valleys, *e*, of the rivers (alluvial) and of their flood-plains (diluvial).*

Lehmann's classification, in so far as it goes, expressed established facts of nature. There are Primitive, Secondary, Tertiary, and Quaternary formations, but the theory that they may be defined and determined by physical structure and present relative position is only approximately true. All crystalline rocks are not primitive, all the secondary rocks are not merely consolidated fragments of primitive rocks. Some of them are fully metamorphosed. All Tertiary rocks are not unconsolidated, as the Tertiaries of California illustrate, and we now know that altitude above the sea, or relative position of the various formations, is by no means uniform and forms no criterion for their determination.

Werner's Classification of Rocks by their Mineral Characters.—The next important advance in the classification of rocks was started by Werner and his pupils. It was a classification based upon the mineral constitution of the rocks. As the study of geology advanced Lehmann's classification was found difficult to apply with precision, and it was found to be unnatural in that rocks of apparently similar kind were dissociated, while rocks of unlike character were brought into the same class. And the mineral character and composition of rocks was found to be an accurate means of defining them. As the mineral characters became clearly understood, the rock masses received their names from the chief minerals in them, and finally the mineral nomenclature entirely superseded the nomenclature of Lehmann, and a second classification arose in which the theory of the original order of formation of the rocks gave place to the actual sequence of mineral aggregates, one after another, in examined sections of the earth's crust. In this study of minerals Werner was a conspicuous leader, and the classifications at the beginning of the present century were mainly his or adaptations of them.

---

* W. D. Conybeare and William Phillips, "Outline of the Geology of England and Wales, with an introductory compendium of the general principles of that science, and comparative views of the structure of foreign countries," Part I. p. xix.

Conybeare and Phillips's Perfection of the Wernerian System.—The form which the geological scale assumed in English geological systems is seen typically in Conybeare and Phillips's Geology of England and Wales (1822). Arranged in order from above downwards, it is as follows:

I. *Superior order.* (Neues Floetzgebirge of Werner.)

II. *Supermedial order.* (Floetzegebirge of Werner.)

(1) Chalk formation.

(2) Ferruginous sands.

(3) Oölitic system or series.

(4) { Red marle or New Red sandstone.
Newer Magnesian or conglomerate limestone.

III. *Medial, or Carboniferous order.*

(1) Coal-measures.

(2) Millstone, grit and shales.

(3) Mountain limestone.

(4) Old Red sandstone.

De la Beche.—De la Beche* carried out the system more completely, calling the first, or superior order, *Supercretaceous group*, and applying the terms *Cretaceous*, *Oölitic*, and *Red sandstone* to three groups into which he divided the second order, and giving the third the name *Carboniferous group*. Below these he recognized Werner's Grauwacke group, for what was the lower part of the original *Uebergangsgebirge* of his earlier classification, and below this were the *inferior stratified or non-fossiliferous rocks, and the unstratified rocks.* All of the names, it will be observed, are names indicative of mineral characters.

Maclure's Application of the System to American Rocks.—If we turn back to the year 1817 we find the same Wernerian system applied to the classification of North American rocks by William Maclure.† The author writes: "Necessity dictates the adoption of some system so far as respects the classification and arrangement of names. The Wernerian seems to be the most suitable, first, because it is the most perfect and extensive in its general outlines; and secondly, the

* "A Geological Manual," 3d edition, 1833.

† "Observations on the Geology of the United States of America," Philadelphia, 1817.

nature and relative situation of the minerals in the United States, whilst they are certainly the most extensive of any field yet examined, may perhaps be found the most correct elucidation of the general accuracy of that theory, so far as respects the relative position of the different series of rocks." * The classification there set forth is as follows (in the order from below upwards):

Class I. *Primitive rocks.*

Class II. *Transition rocks*—including (1) transition limestone, (2) transition trap, (3) greywacke, (4) transition flinty slate, (5) transition gypsum.

Class III. *Floetz or secondary rocks*—including (1) old red sandstone, (2) 1st floetz limestone, (3) 1st floetz gypsum, (4) 2d variegated sandstone, (5) 2d floetz gypsum, (6) 2d floetz limestone, (7) third floetz sandstone, (8) rock-salt formation, (9) chalk formation, (10) floetz-trap formation, (11) independent coal formation, (12) newest floetz-trap formation.

Class IV. *Alluvial rocks*—including (1) peat, (2) sand and gravel, (3) loam, (4) bog iron ore, (5) nagel fluh, (6) calc tuff, (7) calc sinter.

Notice that in this classification the "coal formation" is placed near the top of the secondary rocks, the "rock-salt formation" near its middle, and the "old red sandstone" at its base. Later investigations did not confirm Maclure's opinion of the accuracy of Werner's system as applied to American rocks.

Amos Eaton's Classification of the New York Rocks.†—Amos Eaton's classification of the New York rocks is an elaboration of the same system.

Principles involved in the Wernerian System of Classification.—In each of these classifications, except in a few cases of the retention of distinctions based upon the structural analysis, the whole nomenclature and classification is based upon mineralogical composition of the rocks. In the succeeding progress of the science a great part of the nomenclature has been replaced by other names composed on a different prin-

* "Observations, etc.," p. 28.

† As exhibited in his "Geological and Agricultural Survey of the district adjoining the Erie Canal in the State of New York," Albany, 1824.

ciple, but many of the divisions here recorded are still retained. This latter fact we may interpret to mean that distinctions based upon mineral or lithological characters are of some real and permanent value in geological classification. The history of development of this system from the first, or Lehmann's system, shows that the linear order of the series of formations in the list is based on the conception of a time-scale and a natural order of succession of the several formations. The Wernerian classification in this respect was a correct one for the rocks in Northern Germany for which it was constructed. The English scale expressed the facts of sequence, so far as known, for the English rocks, but the attempt to fit either of them to the facts in North America emphasized their imperfection. The fundamental error in the Wernerian system was the assumption that the scale of Northern Germany was a universal scale, or, expressed in general terms, that the mineralogical constitution of a rock bears some necessary relation to its place in the stratigraphical series.

Fossils substituted for Minerals in classifying Stratified Rocks.—The next step of progress in making the geological time-scale arose from the study of fossils. Fossils had been observed and recognized as organic remains for centuries before Lehmann and Cuvier. Lehmann, and he not the first, observed that Primitive rocks did not contain fossils, while Secondary rocks contained some, and what are now called Tertiary rocks contained them abundantly. But it was not until fossils were closely studied, their characters examined, and the species compared and classified that their importance was realized.

Cuvier and Brongniart.—Cuvier and Brongniart are generally credited with being the first to establish the scientific importance of fossils.* In 1796 Cuvier had called attention to the fact that elephant bones discovered by him in the Paris basin were different from the bones of living species. In thus drawing a distinction between living and extinct animals, as implying present and past groups of living beings, the foundation was laid, not only of Palæontology, but of the

* "On the Mineral Geography and Organic Remains of the Neighborhood of Paris," 1808.

whole field of investigation into the history and evolution of organisms. Cuvier and Brongniart, applying their methods of analysis to the rocks of the Paris basin, succeeded in classifying them into strata, and in defining the separate stratigraphical divisions in terms of the contained fossils. The Paris basin rocks, being found to lie above the Cretaceous rocks of France and England which represent the top member of the secondary formation of the Lehmann classification, were named Tertiary to indicate their geological importance and their relative position in the geological scale. These naturalists did not, however, perfect the geological classification which their biological studies suggested.

William Smith and Lyell.—William Smith in England * emphasized the value of fossils as means of identifying strata in different regions, and others had some part in the elaboration of the principle involved, but Lyell, more than any one else, perfected the scheme of classification of geological formations on the basis of their fossil contents.

Lyell's Classification of the Tertiary into Eocene, Miocene, and Pliocene.—The first attempt to use fossils as the fundamental basis of a classification of geological formations was made by Lyell in the classification of the Tertiaries of England. In the second edition of his "Elements of Geology," published in 1841, we find him saying: "When engaged, in 1828, in preparing my work on the Principles of Geology, I conceived the idea of classing the whole series of Tertiary strata in four groups, and endeavoring to find characters for each, expressive of their different degrees of affinity to the living fauna." † A mathematical comparison was made between the proportionate numbers of recent and of extinct species in the several divisions of the Tertiary rocks of England. The result is given in the following table: ‡

Period.	Locality.	Per cent of Recent Species.	Number of Fossils compared.
Post-Pliocene,	Freshwater Thames Valley,	90–100	40
Newer Pliocene,	Marine Strata near Glasgow,	85–90	160
Older Pliocene,	Norwich Crag,	60–70	111
Miocene,	Suffolk, red and coralline Crag,	20–30	450
Eocene,	London and Hampshire,	1–2	400

* "Tabular View," 1790, and in unpublished maps and sections of the first and second decades of this century.

† p. 280. ‡ Copied from his "Elements," 2d ed., vol. 1, p. 2[illegible]4.

In the nomenclature here proposed Eocene is derived from the Greek *ἠώς*, dawn, and *καινός*, recent; Miocene from *μεῖον καινός*, less recent; Pliocene from *πλεῖον καίνός*, more recent; and the definite meaning of the nomenclature and the classification is to signify that the strata called Eocene contain the first traces of the fauna now living, the Miocene strata a small proportion of the living species, the Pliocene and Post-Pliocene more and still more of the living types, and that the whole of the Tertiary is distinguished from the Secondary and all older beds by containing some representatives of the faunas now living.

In this earliest attempt to estimate time-relations by biological data, Lyell, like his contemporaries, considered species to be sharply defined natural groups, and therefore it was that the relations between a fossil fauna and its recent representatives could be expressed in mathematical terms, indicating the number of identical species. The principle underlying the classification, however, was of a deeper nature, and concerned the orderly succession of faunas and floras in time.

Extension of the Lyellian System by Forbes, Sedgwick, and Murchison.—From the application of this method of time-analysis to the Tertiary beds, it was extended to an analysis of the whole series of geological formations on the basis of their organic remains, and the Lyellian classification took the place of the older Lehmann classification as follows:

In place	of	Tertiary	we have	Cainozoic.
"	"	" Secondary	"	Mesozoic.
"	"	" Transition	"	Palæozoic.
"	"	" Primitive	"	Azoic.

This latter classification and nomenclature was gradually built up, and mainly by English geologists, as the Lehmann and Wernerian classification was largely elaborated by German and French geologists.

Edward Forbes proposed to divide the known faunas and floras into two great groups, Neozoic (modern) and Palæozoic (ancient). The two terms Palæozoic and Protozoic were proposed about the same time. Palæozoic by Sedgwick, for the formations known to be fossiliferous, extending from his lower Cambrian upwards to include Murchison's Silurian system, and Protozoic was a provisional name proposed for pre-Cambrian rocks which might be found to contain fossils.*

In his "Silurian System," Murchison proposed Protozoic in the following words: "For this purpose I venture to suggest the term 'Protozoic rocks,' thereby to imply the first or lowest formations in which animals or vegetables appear." †

Without entering into the delicate question of apportioning the honors due to each of these great English geologists,‡ it may be said that in this early usage of the terms, the distinction between Protozoic and Palæozoic was ideal—and in later developments Paleozoic has been retained for that lower great division of the scale containing distinct remains of organisms, with the Cambrian system at the bottom. To show the connection with the older nomenclature, it may be noted that Paleozoic is equivalent to Primary fossiliferous, and in the "Silurian System" Azoic was applied to the Primitive rocks of the Lehmann system.

Phillips's Scheme.—John Phillips, in 1841, proposed to extend this method of classification to the whole geological series; and as his scheme was apparently the first complete classification constructed on this basis, it is offered as it appeared in "Palæozoic Fossils of Devon and Cornwall." §

Proposed Titles depending on the Series of Organic Affinities.		Ordinary Title.
Cainozoic strata:	Upper =	Pliocene Tertiaries.
	Middle =	Miocene Tertiaries.
	Lower =	Eocene Tertiaries.
Mesozoic strata:	Upper =	Cretaceous system.
	Middle =	Oölitic system.
	Lower =	New Red formation.
Palæozoic strata:	Upper?	Magnesian limestone formation.
		Carboniferous system.
	Middle?	Eifel and South Devon.
	Lower =	Transition strata.
		Primary strata.

* Sedgwick, Proc. Geol. Soc., vol. II. p. 675, London, 1838.

† Murchison, "Silurian System," p. 11.

‡ See *American Journal of Science*, vol. XXXIX. p. 167, 1890.

§ London, 1841, p. 160. See also Penny Cyclopædia, articles "Geology," "Palæozoic Rocks," "Saliferous System," etc.

‖ The terms are founded on the verb ζάω or ζώω, to live; combined with καινός, recent; μέσος, medial or middle; and παλαιός, ancient.

Joseph Le Conte proposed Psychozoic, on the same principle, for the latest geological period in which man has appeared.*

Chronological Succession included in Lyell's System.—Lyell proposed to make, on this basis, a geological time-scale, and he applied the term Period to each of the several divisions of the scale. Thus we find in his Geology,† second edition, published in 1841, a recognition of the time element in classification, without, however, the adoption of the biological nomenclature. He gives a table "showing the order of superposition, or chronological succession, of the principal European groups of fossiliferous rocks." Under the heading "Periods and Groups" we find the following:

Periods	Groups
I. Post-pliocene Period:	A. Recent. B. Post-pliocene.
II. Tertiary Period:	C. Newer Pliocene. D. Older Pliocene. E. Miocene. F. Eocene.
III. Secondary Period:	G. Cretaceous group. H. Wealden group. I. Oölite, or Jura Limestone group. K. Lias group. L. Trias, or New Red Sandstone group. M. Magnesian Limestone group. N. Carboniferous group. O. Old Red Sandstone, or Devonian group.
IV. Primary Fossiliferous Period:	P. Silurian group. Q. Cambrian group.

Later Lyell adopted the biological nomenclature, and was prominent among geologists in developing and elaborating the idea of the successive appearance of new types of organisms coördinate with the progress of geological time.

Dana's Elaboration of a Geological Time-scale.—Dana was the first to classify and teach the facts of geology from a purely

* See Le Conte, "Elements of Geology," first edition, New York, 1878.

† Lyell, "Elements of Geology," second edition, London, 1841, vol. II. p. 178.

historical point of view. In 1856* he wrote: "Geology is not simply the science of rocks, for rocks are but incidents in the earth's history, and may or may not have been the same in distant places. It has a more exalted end—even the study of the progress of life from its earliest dawn to the appearance of man; and instead of saying that fossils are of use to determine rocks, we should rather say that the rocks are of use for the display of the succession of fossils. . . . From the progress of life geological time derives its division into ages, as has been so beautifully exhibited by Agassiz."

Referring to the nomenclature he used in the classification of American geological history he speaks of having adopted for the subdivisions of the Paleozoic the names given by the New York geologists; but, he adds, "I have varied from the ordinary use of the terms only in applying them to the periods and epochs when the rocks were formed, so as to recognize thereby the historical bearing of geological facts." The nomenclature proposed by Dana in 1856 is given in the following table:

I. Silurian Age.

1. Lower Silurian.

1. Potsdam Period.	1st epoch.	Potsdam sandstone.
	2d "	Calciferous sand-rock.
2. Trenton Period.	1st epoch.	Chazy limestone.
	2d "	Birdseye.
	3d "	Black River.
	4th. "	Trenton.
3. Hudson Period.	1st epoch.	Utica Shale.
	2d "	Hudson River Shale (Hudson River shale and Blue limestone of Ohio in parts of the West).

2. Upper Silurian.

1. Niagara Period.	1st epoch.	Oneida conglomerate, etc.
2. Onondaga Period..	1st epoch.	Galt limestone, etc.
3. Lower Helderberg Period, etc.		

II. Devonian Age.

1. Oriskany Period . . .	1st epoch.	Oriskany sandstone, etc.

* *American Journal of Science*, vol. XXII. pp. 305 and 335.

2. Upper Helderberg Period. } 1st epoch. Schoharie grit, etc.

3. Hamilton Period.... 1st epoch. Marcellus shales, etc.

4. Chemung Period.... 1st epoch. Portage, etc.

5. Catskill Period.... { Catskill red sandstone and shales, etc.

III. Carboniferous Age.

1. Subcarboniferous Period. } 1st epoch. Conglomerates, etc.

2. Carboniferous Period. } 1st epoch. Millstone grit, etc.

3. Permian Period, etc.

This classification was further elaborated in his manual, the first edition of which appeared in 1863,* and it has become the standard classification for American geology. Here we find the larger divisions, called *times*: I, Archean; II, Palæozoic; III, Mesozoic; and IV, Cenozoic times. The Palæozoic time is classified into *ages*, viz.: The age of Invertebrates, the Cambrian and Silurian; the age of Fishes, the Devonian; the age of Coal Plants, the Carboniferous. The Mesozoic is called the age of Reptiles. The Cenozoic time includes the age of mammals and the age of man.†

Each of the ages is subdivided into *periods* and *epochs*, in which the stratigraphical groups and formations form the basis, and the particular faunas and floras of each constitute the data of determination for the time-divisions.

The following chart shows the modifications in the nomenclature through which the classification now in use has grown out of the classifications of earlier authors:

---

* James D. Dana, "Manual of Geology; treating of the principles of the science, with special reference to American Geological History," 1st edition, 1862; 2d edition, 1874; 3d edition, 1880; 4th edition, 1895.

† In the article of 1856 the following periods were named (i.e., Triassic, Jurassic, Cretaceous, Tertiary, and Post-tertiary), but divisions into epochs were in this paper proposed only for the latter. The divisions of the Post-tertiary were the Glacial Epoch, the Laurentian Epoch, and the Terrace Epoch. Quaternary has been substituted, in the manual for Post-tertiary, and Champlain epoch for Laurentian.

In the last edition (1895) Era has taken the place of Age in the former editions, a Cambrian Era has been recognized in addition to Lower Silurian, and Carbonic Era has been substituted for Carboniferous Age; the name Carboniferous being applied to the formations included under the terms Coal-measures and Millstone grit of the early classifications.

CHART SHOWING THE HISTORICAL DEVELOPMENT OF THE SYSTEMS OF CLASSIFICATION OF GEOLOGICAL FORMATIONS.

Lehmann, 1756.	Werner, 1800.	English Wernerians.	French Authors.	Conybeare & Phillips, 1822.	Lyell, 1828.	Lyell Perfected Classification, 1841.	Systems, 1895.
Volcanic							
Alluvial	Ange-schwempt Gebirge Neues Floetz	Alluvial Tertiary	Quaternary Réboul (1833) { Alluvial / Diluvial } Tertiary (Cuvier & Brongniart) 1810–1822	Superior order	Cenozoic { Post-Pliocene Newer Pliocene Older Pliocene Miocene Eocene }	A. Recent, B. Post-Pliocene } Quaternary — Pleistocene C. Newer Pliocene, D. Older Pliocene, E. Miocene, F. Eocene } Tertiary { Pliocene, Miocene, Eocene } Neocene	Quaternary Tertiary
Secondary	Floetzgebirge	Secondary		Supermedial order { Chalk formation Ferruginous sands Oolitic series Red Marle or New Red sandstone Magnesian limestone }	Mesozoic	G. Cretaceous group } Cretaceous H. Wealden, J. Oolite, K. Lias } Jurassic L. Trias, or New Red sandstone, M. Magnesian } Triassic	Cretaceous Jurassic Triassic
	Uebergangs-gebirge	Transition		Medial or Carboniferous { Coal measures Millstone grit Mountain limestone Old Red sandstone } Grauwacke Inferior stratified or Non-fossiliferous rocks	Palæozoic	N. Carboniferous O. { Old Red Sandstone, or Devonian } P. Silurian, Q. Cambrian } Primary Fossiliferous Period — Upper Silurian, or / Lower Silurian, or	Carboniferous Devonian Silurian Ordovician Cambrian
Primitive Gebirge	Urgebirge	Primitive		Unstratified rocks (De la Beche) or Primitive	Azoic	Primitive / Archæan (Dana 1874) — Huronian (Logan 1858), Laurentian (Logan 1854).	Algonkian (U.S.G.S.) Archæan

The distinctions upon which the above divisions are based are primarily stratigraphical, and we have still to seek a time-classification on a purely biological basis for the whole geological series.

Biological Classification of Oppel.—One of the earliest attempts at systematic classification upon a purely biological basis was made by Dr. Oppel in classifying the Jurassic formations on the basis of the successive Ammonites characterizing the beds.* Oppel divided the lower part of the Jurassic system (the Lias) into 14 zones or beds, characterized successively from below upwards by their dominant fossil forms, chiefly ammonites.

Thus the successive zones were those of: 1, *Ammonites planorbis*; 2, *A. angulatus*; 3, *A. Bucklandi*; 4, *Pentacrinus tuberculatus*; 5, *A. obtusus*; 6, *A. oxynotus*; 7, *A. raricostatus*; 8, *A. armatus*; 9, *A. Jamesoni*; 10, *A. ibex*; 11, *A. Davœi*; 12, *A. margaritatus*; 13, *A. spinatus*; 14, *Posidonomya Bronnii*. Later classifications, elaborations or revisions of Oppel's system, have been made by Wright, in 1860; Judd, 1875; Tate and Blake, 1876, etc. This method of classification recognized the principle of temporary continuance of species and of associated faunas · and it has been applied with greater or less success all through the geological scale of formations for the definition of the lesser divisions.

As early as 1838 the importance of the biological evidence in determining the time-scale was clearly enunciated by Murchison, who wrote in the introduction to the Silurian System, "that the *zoological* contents of rocks, when coupled with their order of superposition, are the only safe criteria of their *age*." †

Geological Terranes and Time-periods Contrasted.—The making of the geological time-scale has now been traced far enough to clearly demonstrate the fact that the ordinary classification of geological formations, as found in our text-books, includes two distinct series of facts: (1) *geological terranes*, arranged stratigraphically and classified by their positions relative to

* A. Oppel, "Die Juraformation, Englands, Frankreichs und des südwestlichen Deutschlands" (1856–1858).

† "The Silurian System," p. 9.

each other and by their lithological characters; and (2) *chronological time-periods*, which may be locally marked by the stratigraphical division-planes, but which depend, fundamentally, upon biological evidence for their interpretation and classification. Gilbert * has concisely expressed the important fact of the purely local nature of the division-planes separating the formations stratigraphically into stages, series, systems, or groups in the words: "There does not exist a world-wide system nor a world-wide group, but every system and every group is local." "The classification developed in one place is perfectly applicable only there. At a short distance away some of its beds disappear and others are introduced; farther on its stages cannot be recognized; then its series fail, and finally its systems and its groups."

If we accept the correctness of this statement, it is evident that geological terranes and the stratigraphical division-planes by which they are marked, although indicative of time succession, present nothing in themselves to indicate the particular place they occupy in a time-scale. Even were the age of a particular stratum in one section accurately determined by other means, there is no stratigraphical or lithological mark upon the rock stratum by which the corresponding age can be recognized in another section. This is not meant to imply that it is impossible to trace a stratum or formation from one section to another in the same general geological province, for in such case it is a process of tracing with slight interruption the continuity of the one terrane. But when we pass from one basin to another the physical continuity is broken, and the stratigraphy and lithology were made on a separate basis. Hence we reach the conclusion that the perfecting of the geological time-scale must be wrought by the means, primarily, of organic remains. Chronological time-periods in geology are not only recognized by means of the fossil remains preserved in the strata, but it is to them chiefly that we must look for the determination and classification of the rocks on a time basis.

---

* G. K. Gilbert, "The Work of the International Congress of Geologists." Proc Am. Assoc. Adv. Sci., August, 1887, vol. XXXVI. p. 191.

United States Geological Survey Definitions of Formation and Period.—This principle is clearly enunciated in the rules adopted by the United States Geological Survey for the direction of the Survey.* "Among the clastic rocks there shall be recognized two classes or divisions, viz.: structural divisions and time-divisions." "The structural divisions shall be the units of cartography, and shall be designated *formations*. Their discriminations shall be based upon the local sequence of rocks, lines of separation being drawn at points in the stratigraphic column where lithologic characters change. . . . The time-divisions shall be defined primarily by palæontology and secondarily by structure, and they shall be called *periods*" (p. 65). We have thus reached the stage in the making of the geological time-scale at which the ideas of the *geological formation* and the *geological period* have become thoroughly differentiated. The geological period as a time-unit is primarily defined by the characters of the fossil remains in the rock, so that the elaborating further and making more precise the geological time-scale must come from a direct study of the life-history of organisms as recorded in the stratigraphical formations.

The classification of time-divisions made on this principle by the United States Geological Survey is expressed in the Tenth Annual Report as follows:

Period.	Letter Symbol.	Color used in Mapping.
Neocene	N	Orange
Eocene	E	Yellow
Cretaceous	K	Yellow, Green
Jura-Trias	J	Blue, Green
Carboniferous	C	Blue
Devonian	D	Violet
Silurian	S	Purple
Cambrian	C	Pink
Algonkian	A	Red

English Usage.—The English geologists maintain the dominance of the systems as the basis of classification, and deal with the geological formations as prime factors, considering the periods as secondary and as dependent upon the forma-

* Report of the Director in the Tenth Annual Report, 1890, pp. 63-65.

tions. In geological text-books and in other geological literature of America, when Silurian, Jurassic, or Cretaceous is used alone, the Silurian, Jurassic, or Cretaceous Period is meant. In English literature, however, system is generally understood when not otherwise specified.

Geological Systems the Standard Units of the Time-scale.—The final result of these attempts to arrange chronologically the geological formations is found in the standard classification of the systems. The systems were originally groups of successive rock-formations; their limitation was therefore determined, in the first place, by the extent of the rocks in the particular region where they were first defined. Hence the series of formations constituting an original system is in each case a standard of reference, and its general application is accomplished by determining its equivalent formations in other regions.

The time-periods are the periods represented by these systems; hence the periods of time-duration receive the names of the systems which were formed during the periods. The expression, the Cambrian Period, means the period of time during which the Cambrian system of rocks was forming, or the period in which the Cambrian faunas and floras lived. It is all-important to know what formations make up these standard systems; for only as other rocks contain the same faunas or floras can they be identified as of equivalent age, and therefore as belonging to the same system. The real time-indicators are, therefore, the fossils, although the rock-formations which held the fossils give us the names for the chief divisions of the time-scale.

THE GEOLOGICAL SYSTEMS.

Cambrian System.—The CAMBRIAN SYSTEM was defined by Sedgwick, and the name was applied to formations studied in North Wales. In the original definition of the system (1835), in a paper by Sedgwick and Murchison,* the extension of the

* "On the Silurian and Cambrian Systems, exhibiting the order in which the older sedimentary strata succeed each other in England and Wales." British Assoc., August, 1835.

system was too low, including rocks later recognized to be older than the Cambrian system (the "Lower Cambrian group" of 1835), and too high, in that the "Upper Cambrian group" of this first paper was claimed also by Murchison in his original Silurian system; and in fact the Upper Cambrian of Sedgwick is the same stratigraphically with the Lower Silurian of Murchison, as at present used. The Cambrian system includes the "Middle Cambrian" of the 1835 paper, which is composed of the following formation, viz.: Longmynd, Harlech, Menevian, Lingula flags, and Tremadoc. The rocks at first were believed to contain no fossils; later, fossils were found, and were more fully elaborated in Bohemia by Barrande, and defined by him as the "first fauna." In later correlations, and in other countries this first or primordial fauna of Barrande has been the distinguishing evidence of the Cambrian period of time.

The Cambrian system in North America includes three divisions: the earliest, or lowest, the (1) Georgian group, typically represented in the shales and limestones of that name in Western Vermont, and containing a fauna characterized by the presence of the *Olenellus*, a genus of Trilobites. The second or middle division is the Acadian group, typically seen in the form of shales and slates in Eastern Massachusetts, in New Brunswick, and in Newfoundland, and containing the Paradoxides fauna, or the fauna with the genus *Paradoxides*. The third division is the Potsdam group, and is typically represented in sandstones about the base of the Adirondack mountains, and contains the genus *Dicellocephalus*.

Ordovician System.—The ORDOVICIAN SYSTEM is a name proposed by Lapworth, in 1879, as a substitute and compromise for the Upper Cambrian of Sedgwick and the Lower Silurian of Murchison, both of which covered the same interval, and the original usage of which in current geological literature the geologists of the two schools have, since the death of the authors, strenuously maintained. The standard series of rocks are in Wales and Western England, and are the Arenig, Llandeilo flags, and Bala or Caradoc. The fauna is the "second fauna" of Barrande, and the standard system in North America includes the Calciferous group, typically represented around the borders of the Adirondacks, the Chazy, also at the eastern part of the Adirondacks; and the Trenton, expressed typically at Trenton Falls and on the western slopes of the Adirondacks, and extending southwestward.

Silurian System.—The SILURIAN SYSTEM, as now restricted, is the Upper Silurian of Murchison as defined in 1835 to 1838 and later. It includes typically the Mayhill, Wenlock, and Ludlow formations, as defined in the Silurian system of Murchison, of Western England. The fauna is characterized as the "third fauna" of Barrande. It is typically represented in North America by the Niagara, the Salina, and the Lower Helderberg groups of New York State.

Devonian System.—The DEVONIAN SYSTEM is a name, also, first proposed and defined by Sedgwick and Murchison in 1838. The typical rocks were found in North and South Devonshire, England. The limits of the system, stratigraphically, were not so definitely fixed as in the previous cases, the system having been founded originally on the distinction of the fossils, which by Lonsdale were determined as constituting a group intermediate to the Carboniferous and the Silurian faunas. The fossils from which the original determination was made were from the limestones of Plymouth and Torbay, South Devonshire. Later investigations have shown them to be of Mesodevonian age. The Devonian system was originally intended to include the rock series from the top of the Silurian to the base of the Carboniferous. The lowest member of the system in South Devonshire is the Foreland sandstone, and the highest are the Pilton beds, near Barnstaple, North Devonshire. There are what are known as Lower, Middle, and Upper Devonian faunas, and recent investigations have led certain European geologists* to set the lower limit of the Devonian system low enough to include part of what in North America are called Lower Helderberg group faunas. In America the standard Devonian system comprises the Oriskany sandstone, the Cornifer-

---

* See Kayser, "Die Fauna der ältesten Devon-Ablagerungen des Harzes" (Berlin 1878), and papers on the Hercynian question.

ous, the Hamilton, and the Chemung, including the Catskill group, all typically represented in New York State.

Carboniferous System.—The CARBONIFEROUS SYSTEM, as now limited, was first defined by Conybeare in 1822.* In his original grouping he included with the Coal-measures the Millstone grit, the Carboniferous or Mountain limestone, and the Old Red sandstone of England, typically represented in north England in the Pennine range, and not fully represented in any other one section in England. In England the Permian was regarded as a distinct system by Murchison, and as lying unconformably upon the lower strata; but the Permian fauna and flora both have closer affinity with those of the Coal-measures below than with the later Mesozoic types, and on paleontological grounds the Permian is now classified as the upper group of the Carboniferous system. In North America the standard rocks of the Carboniferous system are the Mississippian series, formerly called Lower or Subcarboniferous, of the Mississippi valley, having for its lowest member the Kinderhook or Chouteau formation, and for its upper member the Kaskaskia or Chester limestones and shales. The middle member of the Carboniferous system is the Coal-measures and underlying conglomerates, typically represented in Pennsylvania; but in the western part of the continent it is not coal-bearing, but consists of massive marine limestone. The upper member is typically seen overlying the Coal-measures in Kansas and Nebraska and farther westward and southward; it contains a marine Permian fauna, and is represented in Pennsylvania and Virginia by a plant-bearing series terminating the Coal-measures.

The Post-paleozoic or Appalachian Revolution.—The chronological division-line between the Carboniferous system and the Triassic is a very important one, both geologically and palæontologically. In America the point is indicated by the Appalachian revolution. It constitutes the division between the terranes of the Palæozoic and the Mesozoic times in the history of organisms. After the close of the Carboniferous

* Conybeare and Phillips, "Outlines of the Geology of England and Wales," London, 1822.

in North America the great part of the eastern half of the United States was raised permanently above water, and is therefore, except at its margins, devoid of records of later marine life. A border of a few hundred miles on the east and south contains Mesozoic and Cenozoic deposits and their characteristic fossils; but the larger part of the areas covered by Triassic, Jurassic, and Cretaceous deposits are found west of the 97th meridian, or of a line running from the western border of Minnesota to the western shore of the Gulf of Mexico. The disturbances recorded in this elevation of the quarter of a continent were felt in other parts of the world, and occasioned great shifting of marine conditions of environment, causing migration or extinction of great numbers of organisms, and opening up new regions with new conditions to which the organisms of the Mesozoic were rapidly adjusted.

Triassic System.—The TRIASSIC SYSTEM was first defined by Alberti in 1834.* The rocks which were grouped together to constitute the Trias system are the Bunter sandstone, overlaid by a middle calcareous member, the Muschelkalk, followed by sandy shales, and the Keuper; and they are well represented in central and southern Germany.

This system is poorly represented in eastern America—so poorly that the United States Geological Survey proposes to join the Trias and Jurassic of America into a common group, calling it the Jura-Trias system, distinguished by a common continuous fauna and flora. In the Rocky Mountain region there are thick deposits, mainly sandstones, with few fossils, which are intermediate between the Permian, or closing formation of the Paleozoic age, and the Cretaceous formations; but it is difficult to determine in particular cases whether the rocks should be classed with the European Triassic or Jurassic systems. In California Diller† has recently described fossiliferous Triassic terranes containing typical Triassic marine faunas.

* "Beitrag zu einer Monographie des Bunter sandsteines, Muschelkalkes, und Keupers," Stuttgart u. Tübingen, 1834.

† "Geology of the Taylorville Region of California." Bull. Geol. Soc. Am., vol. III. pp. 369–394, July, 1892.

Jurassic System.—The JURASSIC SYSTEM was originally applied by Brongniart, in 1829,* to the Jura limestone of the Jura Mountains and the associated formations—the Lias, thin, regular-bedded argillaceous limestones, known in many places in Europe and England, and the Oolitic rocks, or Oolite, so named on account of its resemblance to the roe of fish (*oon*, egg, and *lithos*, stone; roe-stone).

This Jurassic system is rich in its Ammonite faunas in Europe. In America the system is not characteristically represented, but in Texas, in the Rocky Mountain area and in California are seen typical exhibitions of the Triassic and Jurassic systems of the American type.

Cretaceous System.—The CRETACEOUS SYSTEM is an expansion of the "Chalk formation" to include the system of rocks associated with it; the Chalk of the shores of the British Channel was described in literature under that name before it became established as the name of a geological division. The Dutch geologist J. J. d'Omalius d'Halloy described as *terrain crétacé* the third division in his geological classification of the secondary strata of northern Europe, in an essay on the geological map of Holland, etc., in 1822.†

His classification of the secondary rocks was as follows: 1, *terrains peneens* (todte-liegende of the Germans); 2, *terrains ammoneens* (the Jurassic); 3, *terrains crétacé*; 4, *mastozootique* (the Tertiary of others); and 5, *pyroide* (for the rocks having igneous origin). Fitton, in 1824,‡ grouped together into a continuous series the rocks which were afterward recognized as constituting the typical Cretaceous system, but he did not name them at the time. The typical Cretaceous rocks of England and Europe were the Wealden, the lower Greensand, the Gault, the upper Greensand, terminating with the Chalk. In North America our standard has been determined by comparison of contained fossils, and the typical Cretaceous

---

* "Tableau des Terrains qui composent l'écorce du globe," p. 221.

† "Observations sur un essai de carte géologique des Pays-Bas, de la France, et de quelques contrées voisinés": Memoires, etc., Namur, 1828, p. 23.

‡ "Inquiries respecting the geological relations of the beds between the Chalk and the Purbeck limestone in the southeast of England": Ann. of Phil. His., vol. VII p. 365, 1824.

system is found along the Atlantic and Gulf borders in beds of clay, sands, green sands, and chalky limestone containing typical Cretaceous fossils.

Tertiary System.—The TERTIARY SYSTEM was first defined and the name specifically applied by Cuvier and Brongniart to rocks of the Paris Basin.* The system so named included the fossiliferous rocks lying superior to the Cretaceous system, the upper member of the Secondary. As now understood, it includes the three divisions, defined by their fossils, and named by Lyell, Eocene, Miocene, and Pliocene, typically exhibited in France and England in several places. The three divisions of the typical Tertiary of eastern North America are the Alabama, Yorktown, and Sumpter formations. For the West the series consists of Laramie, Wahsatch, Green River, Bridger, Uinta, White River, and Niobrara beds, some of which are fresh-water deposits.†

Quaternary System.—QUATERNARY SYSTEM was applied by Morlot, in 1854, to the rocks or geological material lying above the typical Tertiary deposits.‡ But the term "Quarternaire" was used by Réboul as early as 1833.§ The system is divided into the Pleistocene and Recent, and is distinguished by the presence of traces of man. In North America the deposits so named are classified as *Glacial*, or drift, from the evidence of glacial agency in their arrangement, *Champlain*, or Diluvial and Alluvial, or materials distributed by waters of melting glaciers and by river action, and the *River Terrace*, or Recent Period.

Fossils the Means by which the Age of a System is Determined.—These systems, although actually arbitrary groupings of the stratified rocks of particular regions, have come into common use as the primary divisions of the rocks whenever chronological sequence is considered. In describing any newly discovered fossiliferous strata in any part of the earth, the first step to be taken, in giving them a scientific definition, is to

---

* "Descriptions géologiques des environs de Paris," 2d ed., 1822, p. 9.

† See for details and nomenclature of the subdivisions of the systems, in this and other cases, Dana's "Manual of Geology," 4th ed., 1895.

‡ Bull. Soc. Vaudoise de Sci. Nat., IV. p. 41.

§ "La Géologie de la Période Quarternaire," 1833.

assign them to one or other of these systems upon evidence of the fossils found in them. The character of the rocks themselves, their composition, or their mineral contents have nothing to do with settling the question as to the particular system to which the new rocks belong. The fossils alone are the means of correlation. It thus happens that each geological system, which is a local aggregation of strata, of particular composition, structure, and thickness, becomes a standard of chronological period and duration by virtue of the fossils which it contains. The fossils are characteristic of some particular period in the history of organisms, and the strata containing them were deposited during that period.

CHAPTER III.

THE DIVISIONS OF THE GEOLOGICAL TIME-SCALE AND THEIR TIME-VALUES.

The Systems and Geological Revolutions. — The systems, although they are arbitrarily limited and classified, represent certain grand events in the history of the earth. Without explaining how the series of stratified rocks came to be divided into these particular ten systems, it may be said that their retention as the great units of geological classification and nomenclature is mainly due to the relatively sharp boundaries which each system exhibits in its typical locality. The systems thus serve as known and definite standards of comparison in the construction of the time-scale, as the dominance of nations, or the dominance of dynasties, in each case serves as a time-standard for the discussion of ancient human history. As the period of each dynasty in ancient history is marked by continuity in the successive steps of progress of the country, of the acts of the people and of the forms of government, and the change of dynasties is marked by a breaking of that continuity, by revolutions and readjustment of affairs, so in geological history the grand systems represent periods of continuity of deposition for the regions in which they were formed, separated from one another by grand *revolutions* which interrupted the regularity of deposition, and disturbed, by folding, faulting, and sometimes by metamorphosing them, the older strata upon which the succeeding strata rest unconformably and constitute the beginnings of a new system.

Geological Revolutions Local, Not Universal.—Geological revolutions were not universal for the whole earth; from which it results that these typical systems and their classification are not equally applicable to the geological formations of all lands. It is important also to note that the geological revolution was not a sudden catastrophe, but the culmination of slowly progressing disturbances bringing the surface of the region concerned ultimately above the level of the ocean, the ocean-level being a pivotal point in geological rock formation. The area whose surface is below the sea-level may be accumulating deposits and making rocks, but so soon as the region is lifted above the surface it becomes a region of erosion, destruction, and degradation. Whenever, therefore, in the oscillations of level, any particular part of a continental mass of the earth's crust passes permanently or for a long geological period of time above the sea-level, a great event in geological history has culminated. In case the elevation is only temporary the event is marked by unconformity, or a break in the continuity of the formations; when it is permanent, the geological record for that region ceases, except so far as fresh-water deposits in lakes may continue independent records. Hence it is that these periods of revolution are of such importance in the history of the continents, and constitute the most satisfactory marks for the primary classification of geological history.

Revolution Expressed by Unconformity and Disturbance of Strata.—The natural geological system is theoretically a continuous series of conformable strata. A geological revolution is expressed by unconformity and more or less disturbance and displacement of the strata from their original position. The grander revolutions are also recorded in the permanent elevation of mountain masses or extensive continental areas above the level of the sea, and thus out of the reach of later strata accumulation.

Appalachian Revolution.—The most widely recognized revolution in geological time, since the close of the Archæan, separates the Carboniferous from the Triassic system. In American classification, following Dana's usage, it may be called the Appalachian revolution. It terminated the series of formations which, with only minor interruptions, had been continuously accumulating in the Appalachian basin from the early Cambrian period onward. It left above the sea-level not only all the Appalachian region, but the great part of the eastern half of the continent, extending westward beyond the Mississippi River to a line running irregularly from Texas to western Minnesota. This revolution produced the Allegheny Mountains and those flexings and faultings which are still recognized in the line of lesser ridges extending from Pennsylvania to Georgia. In England, northern Europe, and northern Asia like disturbances took place at the same general period of time. In Australia, southern Africa, and South America the indications are that the revolution was not so extensive, if it took place at all at the same time. The probabilities are that while it was almost universal for the northern hemisphere, it was mainly confined to this half of the earth. The Appalachian revolution was not limited to a brief geological period, but, beginning near the close of the coal measures of the east, it did not become effective in the region of Kansas and Nebraska till the close of the Permian. The wide extent of the disturbance of strata and, consequently, of records at this point in the time-scale has led to making here a primary dividing-point of the scale, marking off Paleozoic from the following Mesozoic time. Several lesser, more or less local, revolutions have left their permanent marks in the grander structure of the rocks or in conspicuous geographical features of the restricted region of the continental area.

Although revolutions of the same kind, and perhaps producing greater effects upon the final condition of the crust, may have occurred previous to the deposition of the Cambrian system, as time-marks only those revolutions which occurred after fossils appeared in the rocks, and in stratified rocks, are here noticed; and their names and the particular events recorded are those affecting the history of the North American continent.

Taconic Revolution.—The first of these was the Taconic revolution, which separated the (Lower Silurian) Ordovician from the (Upper Silurian) Silurian, in the eastern part of North America. The elevation, disturbance, and metamorphism of the rocks of the Taconic mountain range along western New England, and extending from Quebec on the north to New Jersey, stand forth as monuments of this event. The Cincinnati uplift, extending from the western part of

Ontario, Canada, into Tennessee, marks a contemporaneous disturbance. Evidence of the same revolution is seen in unconformability between Ordovician and Silurian rocks in Nova Scotia and New Brunswick. The revolution is not sharply distinguishable in the rocks of the more southern or western regions.

Acadian Revolution.—The second of these lesser revolutions is expressed most sharply in elevation and unconformity terminating the Devonian formations of Maine, New Brunswick, and Nova Scotia, and may therefore be called the Acadian revolution. In the continental interior it may be indicated by the remarkable thinning out of the Devonian rocks toward the southwestward. In Tennessee, Alabama, and Arkansas they are represented by a thin sheet of black shale, a few feet thick, or by but little more than a line of separation between the rocks of the Silurian below and the Carboniferous beds resting scarcely unconformably upon them. This seems to indicate an elevation of the region still further south, toward the close of the Devonian, sufficient to produce extensive erosion, uncovering the lower Silurian rocks, which were again depressed to receive the marine deposits of the early Carboniferous period upon their eroded surfaces.

Appalachian Revolution.—The Appalachian revolution closed the Paleozoic time and left the great part of the eastern half of the continent above sea-level. It forms the natural interval between the Carboniferous and the overlying system, whatever that may be. Its characteristics have already been described (p. 40).

Palisade Revolution.—A revolution which affected the rocks along the eastern border of the continent during or closing the period in which the Triassic sandstones were being deposited may be called the Palisade revolution. It is expressed by the trap ridges in the Connecticut valley, the Palisades and other similar tracts distributed inside the coast from Nova Scotia to North Carolina, and by the uptilting and in some cases faulting of the underlying red sandstone and shale, and the resulting unconformity with the succeeding formations. The evidences of the revolution are not widely extended, nor is the time-relation of the termination of the revolution sharply defined, but it is sufficiently so to form a natural boundary-line separating the Triassic-Jurassic from the Cretaceous. After this point of time there occurred nothing in the eastern half of the continent which deserves the name or rank of a geological revolution, except the glacial revolution which is defined further on. The western part of the continent is conspicuous for the late occurrence of its geological construction, which was chiefly after the Triassic; along the western coast the Sierra Nevada revolution marked the same general interval of time recorded by the Palisade revolution of the East. These events on the opposite borders of the continent are alike at least in preceding the Cretaceous and in terminating the formations which are of Jura-Triassic age.

Rocky Mountain Revolution.—The Rocky Mountain revolution, which resulted in the elevation and disturbance of all the rocks in the region of the Rocky Mountains, and extended from them to the border ranges, is distributed along the time from the close of the Cretaceous to the Miocene, or possibly later. It is altogether probable that the actual length of time taken in elevating, tilting, and disturbing the strata, after the last marine deposits of the pre-Laramie formations, which resulted in the permanent adding to the continent of its western third, was not longer than that consumed in the various events terminating the Paleozoic and making into permanent land the great mass of the eastern half of the continent.*

This Rocky Mountain revolution resembles the Appalachian revolution in extending over and affecting a large area of the continent, in its general upward-lifting of that area, which process extended over a long period of time, and in the great accumulation of coal or lignite which was associated with the gradual emergence of the continental mass above the sea-level. Another feature in which the two revolutions resemble each other is the wide extent of the disturbances recorded. The elevation of the mountain ranges, from the Pyrenees eastward to the Himalayas and to the islands

* See further regarding this revolution Dana, "Manual of Geology," 4th ed., 1895, p. 875, etc., paragraph on "Post-Mesozoic Revolution: Mountain-making and its results," also pp. 932–939.

beyond, took place chronologically at the same general period, and that this series of disturbances may have affected the whole of the northern hemisphere is further suggested by the occurrence of gigantic erratic blocks of granite in the midst of Eocene strata in the neighborhood of Vienna and other places: Vezien* has suggested that an ice age is indicated by them.

The Division-line between the Cretaceous and the Tertiary.—This Rocky Mountain revolution marks the period of the second great break in the life of the geological ages. The Mesozoic time began with the close of the Appalachian revolution, and closed with the elevation of the marine Cretaceous beds above ocean-level. In our classification the division-line between the Cretaceous and the Tertiary was arbitrarily placed at the top of the chalk formations conspicuously developed on both sides of the British Channel. The difficulty, which American geologists have found in drawing the precise line to separate the Mesozoic from the Cenozoic, has resulted from the change in the character of life of the beds in the western interior from marine to brackish, fresh-water, and land types. This change was incident to the Rocky Mountain revolution, which had already begun and was slowly lifting the whole region while the fresh-water sediments were being laid down. Several stages may be marked in this grand revolution, but the facts connected with them are not so well developed as to serve for general purposes of classification of the time-scale. The amount of elevation produced by these epeirogenic movements after the deposit of the marine Cretaceous, in the western half of our continent, is estimated to have been not less than 32,000 or 35,000 feet.†

Columbia River Lava Outflow.—At the close of the Miocene a great outflow of lava in the northwestern part of the United States took place, and continued with interruptions through the Tertiary into the Quaternary time. About the Columbia River, where it cuts through the Cascade range, the basalt is over three thousand feet thick, and the outflows cover a vast extent of territory, estimated at 150,000 square miles. This

* Rev. Sci., vol. XI. p. 171, 1877.

† G. M. Dawson, Am. Jour. Sci., vol. XLIX. p. 463.

was incident to the vast earth disturbance which raised, to the amount of at least five thousand feet, a large part of the western half of the continent. The long line of volcanoes along the western coast of the two Americas had their origin in the same general period of time.

Glacial Revolution.—There was, still later, a revolution which has left little record in the way of disturbance or discordance of strata, but was of particular importance in life-history, as it introduced the recent period or the age of man. It constituted the combination of events marking the glacial epoch. In general, it consisted geologically of oscillations of the northern lands. for the northern hemisphere, and was associated with the accumulation of ice upon the surface and its continuance as a great ice-sheet for a long period of time.

Erosion of River Cañons as Gauges of Time Duration.—Some of the more definite estimates of the length of geological time are based upon the rate of erosion or gorge-cutting of rivers, and the period so measured dates back to the last uncovering of the river channels, coincident with the northward withdrawal of the ice-sheet. Standard examples of such estimates of the length of geological time are those made regarding the cutting of the Niagara River gorge, the retreat of the falls of St. Anthony from Fort Snelling to their present position, and the cutting of the cañons of the Yellowstone and Colorado rivers. But the unsatisfactory nature of these estimates is shown by the fact that different authors reach such divergent results from the same data. The time taken in the cutting of Niagara gorge is estimated by Upham to be 6000 to 10,000 years, by Spencer (1894) to be at least 32,000 years.

Continental Value of Revolutions as Time-Breaks in the History of North America.—The above revolutions are selected, not as the only revolutions interrupting the regular course of sedimentary formation of stratified rocks, but as chief examples of such interruptions in the North American column of deposits. All along the course of geological time there are evidences to show that there were constant oscillations of the relations between land- and ocean-level, and at some localities these oscillations were passing across the datum-plane of the ocean surface.

Wherever this happened, on one side rocks were forming, and on the other erosion and degradation were obliterating them as time-records. The Appalachian and the Rocky Mountain revolutions constitute the two grander revolutions. The first closed the Paleozoic life period, the fossils being chiefly marine until the Devonian, and being associated with marine forms up to the close of the Carboniferous. The deposits are distributed across the continent, with local interruptions. After the Appalachian revolution the eastern half of the continent, except its Atlantic and Gulf borders, became permanently above the sea-level. The period between the Appalachian and Rocky Mountain revolutions is the period of the Mesozoic life. In the faunas and floras of this period, land and fresh-water species take a prominent part. The marine life is distributed over the western half of the continent and along a narrow line of formations on the Atlantic and Gulf borders. After the beginning of the Rocky Mountain revolution, the deposits of marine origin and their faunas were distributed on the marine borders of the continent as it now is, and fresh-water and land deposits were accumulated over the plains and plateaus of the western half (with few exceptions) of the continent.

Time-scale and the Geological Revolutions of the American Continent.—Thus the grander revolutions recorded in the development of the American continent break up the geological time-scale, as expressed in the systems of stratified rocks, into a few natural sub-divisions, as may be illustrated by the diagram on the opposite page:

Revolutions made Interruptions in the Record.—In the use of the time-scale for the study of the history of organisms, the places marked by the revolutions are those in which are found the grander interruptions to the continuity of the record. They may represent periods of great relative magnitude. They do represent periods of marked change in the faunas and floras over extensive regions. Between the grander intervals of revolution the records of life-history are relatively continuous. There were series of successive faunas or even sub-faunas in which were expressed the general features of the evolution of life on the globe. The species preserved and known present but a very imperfect representation of the species that were living; but of those preserved in one formation there are generally found in the succeeding formations representatives of the same or closely allied genera; so that, for the kinds of organisms whose remains are best preserved, the record is fairly continuous for the grander rock-systems in terms of the generic, and in some cases of the specific characters.

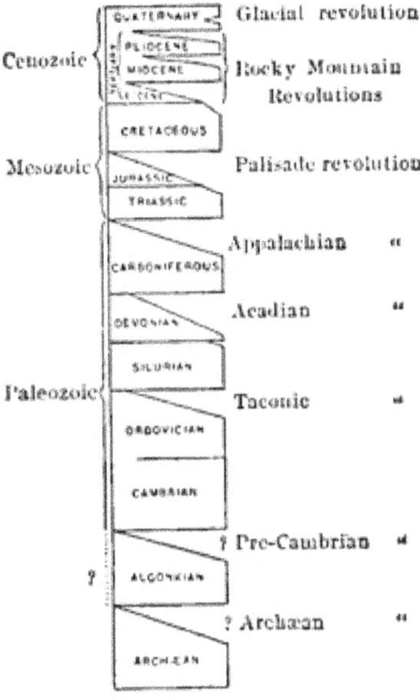

FIG. 2.—Diagram representing the order of succession from below upward of the formation of the geological systems in North America and the approximate time at which the grander revolutions eroded and disturbed the already made deposits.

Time-ratios, or the Relative Time-value of the Several Systems.—While the conditions of deposition for a particular region remained relatively constant and uniform, the strata were accumulated in successive beds one upon another; and thus the thickness of the deposits of the same kind, with proportionate thickness for deposits of different kinds, constitutes a scale of definite time-value: a foot of deposit representing a period of time, and the relative time-separation of two faunas is represented by the thickness of the strata between them. It was on this principle that the time-ratios of Dana were estimated. The maximum thickness of the known strata of each geological system was taken as a means of deducing the relative duration of their formation, as was first done by S. Houghton. The limestones were assumed to represent five times the time-value that is represented by the other sedimentary deposits per foot; or, in other words, every foot of limestone was estimated as equivalent to five feet of other sedimentary deposits in making up the time-ratios. Dana* estimated the time-ratio for the several geological periods to be as follows:

Period	Time-ratio	Era
Quaternary	$\frac{1}{4}$	Cenozoic 1.
Tertiary	$\frac{3}{4}$	
Cretaceous	1	Mesozoic $3\frac{1}{4}$.
Jurassic	$1\frac{1}{4}$	
Triassic	1	
Carboniferous	2	Paleozoic $12\frac{1}{4}$.
Devonian	2	
Silurian (Upper)	$1\frac{1}{4}$	
Ordovician (Lower Silurian)	6	
Potsdam	1	

Ward's Estimate.—Lester Ward, in the fifth annual report of the United States Geological Survey, has proposed to adjust these proportions as follows:

Group	Division	Time-ratio
3.	Quaternary-Recent	1
	Miocene-Pliocene	1
	Eocene	1
2.	Cretaceous	1
	Jura-Trias	1
4.	Permo-Carboniferous	1
	Devonian	1
	Silurian	1
	Cambrian	1

thus forming nine divisions of equal length.

* Dana's "Manual of Geology," 3d edition, 1874. In the latest edition, 1895, these estimates are revised and the following remark is made: "There is great doubt over conclusions based on this criterion [i.e., maximum thickness], because thickness is dependent so generally on a progressing subsidence—*no* subsidence giving little thickness, however many the millions of years that may pass. But as it is the only available method, it is still used," p. 716; also see beyond on p. 49.

Corrections and Elements of Uncertainty in these Estimates. —Since Dana's estimate was published additions have been made to the known thickness of the Cambrian rocks of North America, which may lengthen the Cambrian ratio to 5 in the above table, and duplications of thickness due to confusion in regard to the Quebec group may reduce the Ordovician (Lower Silurian) to 5, and the Cretaceous ratio may be somewhat enlarged. The Tertiary estimate in Dana's ratios assumes the thickness to be of less (½) time-value because of the increased rate of deposition due to transportation of rivers. This and many other factors enter in to complicate the time-value of thickness of strata: and it must be granted that the thickness of the sediments is the prime factor in determining these time-values of the geological scale.

In the last edition, 1895, of the "Manual," Dana expresses the following opinion: "The evidence at present obtained appears to favor the conclusion that the relative duration of the Cambrian and Silurian, the Devonian and the Carboniferous eras, corresponds to the ratio 4½ : 1 : 1, or perhaps 4 : 1 : 1, the ratio hitherto adopted; and for the Paleozoic, Mesozoic, and Cenozoic, 12 : 3 : 1."* However, the conditions of deposition, the fineness or coarseness of the clastic fragments, the abundance or rarity of supply of materials, and other variable conditions must be taken into consideration in an accurate reduction of thickness of strata into length of time. Errors, also, whose value is almost impossible of estimation, arise from the intervals between strata, particularly those where unconformity exists. However, after all these uncertainties are weighed the time-ratios formed on this general basis are of great importance in studying the history of organisms, and the value of accuracy in the time-scale is a sufficient reason for calling attention to the points in which greater accuracy may be attained by further investigation.

Estimates of Actual Length of Time Highly Hypothetical.—It is doubtful if it is possible with our present knowledge to reach an estimate, in years or centuries, of the actual length of geological time which is within 100 or perhaps 200 per cent of the truth. We may accept Dana's estimate of at

* p. 1023.

least 48,000,000 of years, or Geikie's of from 100,000,000 to 680,000,000. We find at one extreme the ancient theory of 6000 years, and at the other McGee's possible maximum of 7,000,000,000 years. The rate of accumulation of sediment over the bottom of the sea may vary between the limits of one foot in 730 years and one foot in 6800 years, as pointed out by Geikie, the figures being based upon the estimated proportion between the annual discharge of sediment in cubic feet and the area of river basins in square miles, in the case of the rivers Po and Danube. The estimate of 680,000,000 of years, quoted above, is dependent upon the assumption that the total thickness (maximum) for the sedimentary deposits is not less than 100,000 feet, and that the average rate of accumulation was not more rapid than that now going on at the mouth of the Danube, based upon Bischof's determination of the amount of sediment and matter in solution in the Danube at Vienna. It may be a query worth considering whether the estimates based upon the examination of the amount of suspended and dissolved matter in river water are not likely to err in the direction of too small amount of matter by reason of the abnormal precipitation along the course of the river incident to the presence of salts and acids put into the river by man. If the rate of the river Po were taken, the length of time would be 73,000,000 of years instead of 680,000,000.

The actual length of time in years, however, is of less importance to the geologist than the relative length of time for each of the eras, and these latter, the time-ratios of Dana, are deducible from the physical thickness, and size of constituent particles, of sedimentary rocks. Relative thickness is certainly one of the elements in the determination of the time-values of the geological formations, and the fields for investigation, along which greater accuracy is to be reached, include the problems of the rate of accumulation of muds, sands, and pebble beds, and of the formation of limestones, in relation to each other and under varying conditions, and the detection of the marks in the strata recording the conditions incident to the varying rates of accumulation. In making estimates of time, as represented by thickness of deposits, there should also be considered the effects of elevation or depression of the interior of land masses upon the amount of detritus carried down to the sea borders, there to be made into sediments. With all the errors of estimation there is, however, a real value to the time-ratios of Dana, and to legitimate corrections deduced from study of the same facts, which cannot be denied; the principle of time-ratios may be used as a "working hypothesis" until something better is devised.

Systems the Standard Units of Geological Chronology.—In the preparation of a universal time-scale for the history of organisms, systems are the actual facts in nature which are accepted everywhere as standard units of chronology.

Geological Eras and Times and their Names.—Whatever may have been the actual length of time occupied in the making of any one of them, or however much the estimates of the relative value of each may differ, it is certain that each system in its particular region represents that particular part of the geological time-scale during which the fauna and flora whose remains it contains lived.

This portion of time may be called the life era of the organisms which make up the fossil fauna and flora of each system.

The names of these systems may then be applied directly to the eras, and we thus have in the time-scale ten eras, viz., the *Cambrian era*, the *Ordovician era*, the *Silurian era*, the *Devonian era*, the *Carboniferous era*, the *Triassic era*, the *Jurassic era*, the *Cretaceous era*, the *Tertiary era* and the *Quaternary era*, including the present time. In a time-scale we know of the eras before the Cambrian only as *precambrian*, i.e., those that were earlier than the Cambrian.

The first five of these eras are classed together as *Paleozoic time*, the next three eras are called *Mesozoic time*, and the Tertiary and Quaternary constitute *Cenozoic time*.

Division of the Eras into Periods.—Fossils have been collected and studied with different degrees of precision for the several eras and in different parts of the world, but, taking the present stage of knowledge of fossil organisms, the paleontologist is able to distinguish about twenty different successive fauna-floras, or sets of organic species, which can be recognized wherever on the face of the globe they are found.

The time-duration of each of these fauna-floras may be called a *period*, and the successive periods thus distinguished constitute the divisions of the eras which at present are recognizable in each of the continents with greater or less fulness.

Locally greater precision in classification has been attained, but differences arising from adjustment of the organisms to conditions of environment, and in living species expressed in geographical distribution, make it doubtful if we are able to correlate fossil faunas or floras, the world around, with greater precision than to recognize the marks of the same period in each district.

Period a Recognized Division of an Era.—For the present, also, it seems more likely to conduce to real progress of knowledge to consider the periods to be divisions of the standard era, rather than absolute units of time-duration, dependent on their own criteria alone for definition.

In naming them, therefore, the subdivision of the era into early, middle, and later divisions is preferable to the adoption of separate distinctive names, and each continent or geological province will then be free to adopt its own interpretation of the local limits and marks of the period in its series of strata.

Standard Periods and their Names.—There are already defined such divisions in the several eras, as follows: in the Cambrian era, an early or *Eocambrian period*, a middle or *Mesocambrian period*, and a later or *Neocambrian period*. Walcott has called the faunas of these periods the *Olenellus*, the *Paradoxides*, and the *Dicellocephalus* faunas.

In the same way the Ordovician era is made up of the *Eoördovician*, or period of the Calciferous formation of American geology, and a *Neoördovician* period, or the period of the Trenton group of North America.

The Silurian era is composed of the *Eosilurian* period (Oneida, Medina, Clinton, Niagara), and a *Neosilurian* period (Salina and Lower Helderberg).

In the Devonian era are the *Eodevonian* period (Oriskany, Corniferous), *Mesodevonian* period (Hamilton) and *Neodevonian* (Chemung). In the Carboniferous era are the *Eocarbon-*

iferous period (Mississippian or Subcarboniferous), *Mesocarboniferous* period (coal measures), and *Eocarboniferous* (Permian).

In the *Triassic* and *Jurassic* eras no divisions have been defined which can be recognized in other continents than where described; hence the periods are equivalent to the eras, one period for each.

The division of the Cretaceous into *Eocretaceous* and *Neocretaceous* periods is fairly well recognized in several continents.

In the Tertiary era *Eocene* is the first period, and the *Neocene* period includes the Miocene and the Pliocene. And finally the Recent period may be regarded as geologically the time of the living of the fauna associated with man.

Use of the Term Epoch in the Time-Scale.—The term Epoch may be appropriately applied as an expression for the time-duration of each local formation: thus we may speak of the epoch of the Iberg limestone of the Hartz; of the Psammites of Condroz in France; of the Marwood beds of England; of the Dominik slates of Russia; of the Chemung of eastern North America; of the Lime Creek beds of Iowa. These are each of them well-defined formations in separate regions, each having a distinct geological structure, thickness, and relative stratigraphic position, and the period of each is *neodevonian;* but the faunas, although distinctive and constituting the means of determining the geological age, are not alike; and in time-values it is not possible to say that one is or is not the exact equivalent of the other. An epoch, upon this basis, would be a definite division of a period, distinguishable in the history of the organisms of a restricted region, but not of universal application.

With the present means of correlation it is impossible to attain a greater degree of precision, in comparing the fossil fauna-floras of widely separate regions, than to distinguish the periodsby their characteristic species.

A Comparative Time-scale for the Study of the History of Organisms.—The tabulation of these facts and nomenclatures produces a standard geological time-scale for use in discussing the history of organisms.

A GEOLOGICAL TIME-SCALE, PREPARED FOR THE COMPARATIVE STUDY OF THE LIFE-HISTORY OF ORGANISMS.

Times.	Eras.	Periods.	Percentage of the Whole Scale.	Time-ratios according to—			
				Dana (1874).	Ward.	Walcott.	Dana (1895).
CENOZOIC. 15%	QUATERNARY.... 5%	Recent .. { Modern, Pleistocene }	5%	¼	1	2	1
	TERTIARY... ...10%	Neocene .. { Pliocene, Miocene }	5%	¾	1		
		Eocene	5%		1		
MESOZOIC. 20%	CRETACEOUS.....10%	Neocretaceous	5%		1	5	3
		Eocretaceous	5%	1			
	JURASSIC........ 5%	Jurassic	5%	1¼	1		
	TRIASSIC.... ... 5%	Triassic	5%	1			
PALÆOZOIC. 65%	CARBONIFEROUS.15%	Neocarboniferous	5%	2	1	12	12
		Mesocarboniferous	5%				
		Eocarboniferous	5%				
	DEVONIAN. 15%	Neodevonian	5%	2	1		
		Mesodevonian	5%				
		Eodevonian	5%				
	SILURIAN........10%	Neosilurian	5%	1¾	1		
		Eosilurian	5%				
	ORDOVICIAN.....10%	Neoördovician	5%	6			
		Eoördovician	5%				
	CAMBRIAN......15%	Neocambrian	5%	1	1		
		Mesocambrian	5%				
		Eocambrian	5%				
		Total	100%	16½	9	19	16
	PRECAMBRIAN			?	?	?	

The periods are taken as the smallest divisions of time which can be universally recognized, and hence it is assumed that they are units of equal length. This assumption probably exaggerates the length of the more recent periods.

Importance of a Standard Time-scale.—For the comparative study of the history of organisms this time-scale may be used irrespective of estimates of actual length of time represented by each period.

The division of the eras into twenty successive periods is

a scheme which is actually recognized in the classification of the geological formations throughout the world, where the criteria of classification are the fossils contained in them. Geologists dealing with distinct series of strata have named the individual members of the series differently for different regions of the earth. Therefore, as the systems are made up of formations presenting local features, of stratification, of petrographic composition, of structure, and of thickness, which are given local names, the fossil fauna-floras representing each one of the periods are found in formations which have different names in separate regions.

In using such a scale it becomes necessary to correlate the faunas of formations having different names. While the formation names may well be retained, in the discussion of the time-relations of organisms it is essential to use a uniform scale of time-divisions expressed in a single series of names: the scale and names above given supply us with such a standard time-scale.

Actual Length of Geological Time.—That geological time is immensely long, as compared with any human standards, all modern geologists admit; but as to how much time, in centuries or years, has elapsed since the beginning of the series of sedimentary rocks, opinions greatly differ. A few facts may be mentioned to illustrate what is meant by great length of time in terms of geological work accomplished:

(1) Since the close of the Cretaceous Period the greater part of the mountain elevation along the southern part of Europe and extending to the extreme southeastern part of Asia was accomplished; and the Himalayas were raised, so that at least 16,000 feet thickness of their mass is composed of marine strata of Tertiary or earlier era.

(2) The large part of the Rocky Mountain region was under marine water in the Cretaceous time. Since the close of the Eocene, or beginning of the Middle Tertiary, as Captain Dutton estimates, the region of the Colorado cañons has been elevated approximately 10,000 or 11,000 feet, and 10,000 feet of erosion has taken place. G. M. Dawson estimates the total amount of elevation which has taken place since Cretaceous time, in British Columbia, to have been

32,000 to 35,000 feet. There are now cañons from 5000 to 6000 feet deep, excavated entirely since the Eocene period.

(3) It is believed that all the lava outflows in the Northwest, which cover 150,000 square miles along the Columbia River and the neighboring states, and through which the Columbia has cut a channel, in some cases, from 3000 to 4000 feet deep, were erupted and laid down since Miocene Tertiary time.

(4) Niagara River gorge, from the falls down to the whirlpool, and thence to the cliffs of the lake at Lewiston, it is estimated, was cut out since the retreat of the glacial ice from the surface of the northern part of the continent, and this is believed by many geologists to represent closely the length of time since man first appeared upon the earth. The gorge is 7 miles long, one fourth of a mile wide below, narrower above the whirlpool, and varies from 200 to 500 feet in depth.* The length of time required for its excavation is estimated to have been from 10,000 to 32,000 years. Taking Dana's general estimate of relative length of time, it is seen that the time since the Cretaceous is not over one sixteenth of the time from the beginning of the Cambrian, and that the length of Quaternary time is not over one third that of the Tertiary. Whatever be the actual length of time taken for these and similar geological processes, it is evident that the same forces working at the same rate would require but the extension of time to include the whole history of the earth.

Data upon which Time-estimates are Made.—Although we cannot go into full particulars respecting the theories proposed to determine the time-limits and extent of the geological ages, a few of the prominent attempts may be cited. The principal data upon which the theories have been based are as follows:

(1) *Physical and Astronomical.*—Estimates from the earth's heat, its rate of cooling, and the radiation of heat into space. (Kelvin.)

Estimates from influence of tidal friction, and thence to the length of time since the moon was separated off from the earth. (Darwin, G. H.)

*See J. W. Spencer, "The Duration of Niagara Falls:" Am. Jour. Sci., vol. XLVIII, p. 455. December, 1894

From the rate of the sun's loss of its stores of heat. (Tait.)
From other physical data. (Croll and others.)

(2) *Geological.*—(*a*) Calculations based upon the estimated thickness of the geological deposits of the total series of stratified rocks and the estimated rate of accumulation of deposits along the shores of continents at the present time. (Houghton, Dana, Croll, Wallace, Lyell, Humphreys and Abbott, etc.)

(*b*) Calculations based upon rate of erosion since the retreat of the glacial cover at the close of the Tertiary era. (Dana, Lyell, Hall, Gilbert, Winchell, etc.); and general estimates and sundry hypotheses as to the time since the glacial age. (Geikie, McGee, Croll, Prestwich, Wright, LeConte, and others.)

Method of Computing Time from Thickness of Rocks.—The elaborate report of Humphreys and Abbott on the "Physics and Hydraulics of the Mississippi River" furnishes the kind of evidence required for making the kind of calculations mentioned under (2*a*) above—that based upon the rate of deposition, or formation of deposits, at the mouth of rivers. The amount of silt borne down and deposited by the Mississippi River annually is estimated by Humphreys and Abbott to be equal to a mass with 1 square mile base and...... 241 feet deep,
the earthy matter pushed along.. .... 27 " "

or a total of sediment 1 mile square by....... 268 " "

But upon Humphreys and Abbott's estimate, and distributing the sedimentary deposit along the coast for a distance of 500 miles, and giving the strip 100 miles width (or spread it out for 1000 miles, and make it 50 miles wide), assuming the area of distribution of the product of erosion of the whole river to be 50,000 square miles,—on such assumptions the deposit in 6000 years would reach a depth of approximately 32 feet, or 53 feet in 10,000 years; or, if we put it in round numbers, 50 feet in 10,000 years. The thickness of sediments for the Devonian era is, according to Dana, 14,300 feet of clastic sediments and 100 feet of limestone; estimating the 100 feet of limestone to be equivalent in time-ratio to 500 feet of ordinary fragmental sediment, we thus obtain in terms of fragmental sediments a total of 14,800 feet. Reducing this 14,800 feet of thickness of sedimentary deposits into time-equivalent, on the basis of the above rate of formation of sediments, we have 2,960,000 years for the duration of the Devonian era. If now we assume the Devonian to be approximately 10% of the whole time-duration from the base of the Cambrian to the present, the total time-duration would be 29,600,000, which is a little over one half the estimate suggested by Dana, viz., 48,000,000 years since the beginning of the Paleozoic time—Paleozoic 36,000,000, Mesozoic 9,000,000, and Cenozoic 3,000,000.*

Forshay's estimate makes the amount of annual deposit 964 instead of 268 feet on a base 1 mile square in 1 year's time, which is about four times as rapid accumulation as the estimate of Humphreys and Abbott, and the effect upon time-duration expressed by rock-thickness would be to reduce the time one fourth, making the Devonian 740,000 instead of 2,960,000 years long. This would bring the age of the earth, as a solid globe, nearer to the estimate of Clarence King (24,000,000 years), to which Lord Kelvin gave approval as lately as March, 1895.†

Errors arising from Estimated Values in the Computations.—According to this estimate we notice that there are several important data which are assumed, and not observed or known.

(1) The thickness of the deposits themselves. Formations, as may be noticed, vary greatly in thickness for even the few localities or regions of America in which they have been studied. We find that the maximum thickness of the North American Paleozoic series is given as 55,000 feet, the general thickness of these deposits in the Appalachian region is 40,000, and in the interior of the continent it varies from 6000 to 3500. Since this estimate was made, Walcott has claimed for the Cambrian 7000 feet of fragmental rocks and 200 of limestones; in the Rocky Mountain province 10,000 feet of fragmental and 6000 feet of limestones, which, reduced to time-ratios ($\frac{5}{1}$ for limestone), gives, instead of (7000 + 1000 =) 8000, (10,000 + 30,000 =) 40,000, or five times the

---

* See "Manual of Geology," 3d edition, p. 591.

† See *Nature*, vol. LI. pp. 438–450.

time-duration expressed. The maximum thickness of the whole series is estimated to be about 100,000 feet, or 20 miles.*

Samuel Houghton estimated that the time represented by the intervals between the strata, when deposition was not going on at the locality where the strata are examined, was as great as that recorded by them. This will fully make up for the error from overrating the maximum thickness. Measuring the greatest thickness recorded on the earth for each of the various formations, Houghton estimated the aggregate to be 177,200 feet.

Upham proposes to increase this figure to account for undiscovered strata, and places the total maximum thickness of stratified rocks at 50 miles, or 264,000 feet. Thus, regarding thickness, we have estimates ranging from 100,000 to 264,000 feet. It may here be stated that the average thickness of the total known strata of the world is somewhere near 80,000 feet.

(2) Another element entering into the question of rate of accumulation of deposits is the rate of removal of mineral substances carried from the continents into the ocean in solution (see Dana, Geikie, and others).

Mr. Reade estimates that the soluble minerals removed from England and Wales in this way, mainly Calcium and Magnesium Carbonates and Sulphates, would equal 1 foot removed from the whole surface of the area in 12,978 years. Prestwich estimated 1 foot in 13,000 years for the area of drainage of the Thames, and for the world an average of 100 tons per square mile annually, with an assumption that the amount removed mechanically is six times as great, or total (600 + 100 =) 700 tons per square mile.

Houghton,† adopting the estimate of the rates of denudation of river-basins required to lower the entire rain-basin a thickness of one foot to be as follows:

---

* See Walcott, "Geologic Time, as indicated by the Sedimentary Rocks of North America": Proc. A. A. A. S., vol. XLII. 1893, pp. 129–169.

† See *Nature*, vol. XVIII. 1878, pp. 266–268.

Ganges	2358	years.
Mississippi	6000	"
Hoang-Ho	1464	"
Yangtse-Kiang	2700	"
Rhone	1528	"
Danube	6846	"
Po	729	"
found the mean rate to be	3090	"

From this table he concluded that "atmospheric agencies are capable, at present, of lowering the land-surfaces at the rate of 1 foot per 3000 years; but since the sea bottoms are to the land surfaces in the proportion of 145 to 52, the rate at which (under present circumstances) the sea bottoms are silted up, that is to say, the present rate of formation of strata, is 1 foot in 8616 years. If we admit (which I am by no means willing to do) that the manufacture of strata in geological times proceeded at ten times this rate, or at the rate of 1 foot for every 861.6 years, we have for the whole duration of geological time, down to the Miocene Tertiary epoch, $861.6 \times 177,200 = 152,675,000$ years. The coefficient 177,200 is the total number of feet of maximum thickness of all the known stratified rocks."

In this same paper Houghton expresses, in concise terms, the following conclusion, viz.: "The proper relative measure of geological periods is the maximum thickness of the strata formed during these periods."

If this sediment be distributed over a strip 30 miles wide and 100,000 miles long—the estimated coast border of deposition, amounting to an area of 3,000,000 square miles, or $1\frac{1}{19}$ of the land area, on this area the accumulation will be nineteen times as fast as estimated for the whole area, or 1 foot in about 158 years. Assuming this to be a more correct estimate of the actual depositing-ground, Wallace, taking Houghton's estimate of 177,200 for the total maximum thickness of the stratified rocks, gets for the time-period of the deposition of their thickness, approximately, 28,000,000 of years.

(3) The proportion between fragmental sediments and

limestones is an uncertain quantity, and the rate of deposition of limestones is a matter of vague estimation.

Errors Affecting the Values of Actual, not Relative Time-lengths.—But allowing that the various data are quantities of only approximate values, in making the estimates the errors are of such a nature that they do not materially affect the time-ratios. These time-ratios, it must be remembered, are the reliable facts that we get from the computation; whether the total time be 48,000,000 or 480,000,000, the probability is that the proportions derived by this method of calculation are correct to the degree of accuracy of our knowledge of the facts themselves.

By comparing the three series of values, assigned upon this principle to the several divisions of the time-scale, by Dana, Walcott, and the author, as tabulated in the above scheme (p. 54), it will be seen, reducing them to percentages, that there is a general agreement in the results.

The percentages for the three grand divisions are, according to the three computers, as follows:

	Dana.	Walcott.	Williams.	Average of the three estimates.
Cenozoic.....	6.25	10.526	15	10 +
Mesozoic....	18.75	26.315	20	21 +
Paleozoic....	75.00	63.156	65	68 −
	100.00	99.997	100	99 +

Various Estimates of the Length of Geological Time.—Many estimates, varying greatly in amount, have been made as to the total length of time represented in the formation of the present stratified crust of the earth. The extremes are seen in McGee's estimate * that the demands of evolution and the facts of geology warrant the assumption that 7,000,000,000 years have passed since the earliest fossiliferous rocks were formed, and twice as long, 14,000,000,000, since the earth began its planetary form, and in the old conception, on the other hand, which was supposed to be interpreted from the Bible record, of 6000 years from the beginning of creation to the present time. Both of these are probably far outside the limits of fact.

* *Am. Anthropologist*, October, 1892, vol. v pp. 327-344.

Sir Archibald Geikie, the Director of the Geological Survey of Great Britain, has expressed the opinion that the formation of all stratified rocks of the earth's crust required between 73,000,000 and 680,000,000 of years.*

Sir Wm. Thomson (Lord Kelvin), on the basis of radiation of heat from the surface, and the present underground temperature of the earth, estimated that the time since the consolidation of the crust is between 20,000,000 and 400,000,000, and that all geological history showing continuity of life must be limited within some such period of past time as 100,000,000 years.†

A more recent estimate made by Clarence King gave approximately 24,000,000 for the same period; this estimate has recently been approved by Lord Kelvin, after the debate arising from Prof. Perry's criticism of the validity of Kelvin's primary assumptions.‡ Geo. H. Darwin estimated, from the rate of retardation of the earth's rotation by tidal friction, that not over 57,000,000 years have elapsed since the moon separated off from the mass of the earth; and Prof. Tait, from these and other physical grounds, estimates not over 10,000,000 years for all the geological work on the surface of the earth. Houghton's estimate from erosion gives 28,000,000 for the deposition of the rock strata; Wallace accepts approximately the latter estimate.

Dana's estimate, as we have seen, is 48,000,000 years. Upham's § estimate, based upon glacial phenomena, finds Glacial and Postglacial time to be 30,000 to 40,000 years, Quarternary 100,000, and thence, by estimating the relative length of the faunal life periods, Tertiary 50 or 100 times longer than the ice age, or 2,000,000 to 4,000,000 years; this brings the mean approximately to the same figures given by Dana.

* Address before British Association, in 1892. See *Nature*, August 4, 1892, vol. XLVI. pp. 317–323.

† Address before the Geol. Soc. of Glasgow, February 27, 1868. See "Popular Lectures and Addresses of Sir Wm. Thomson (Baron Kelvin)," vol. II. p. 64.

‡ *Nature*, vol. LI. pp. 224, 341, and 582. See also Lord Kelvin's reply, pp. 227 and 438.

§ *Am. Journal of Science*, vol. XLV. pp. 209–220.

To these may be added Prestwich's* estimate of the division of the 30,000 or 40,000 years of the Glacial and Postglacial period into 15,000 to 25,000 years for the former, and 8,000 to 10,000 for the latter. This estimate approaches the amount derived from the rate of erosion of the Niagara River gorge, and the retreat of the falls of St. Anthony.†

Mr. C. D. Walcott‡ thinks the Mesozoic and Cenozoic are in relation to the Paleozoic proportionately longer periods than as estimated by Dana (that is, 1, 3, 12 for the Cenozoic, Mesozoic, and Paleozoic times respectively).

Walcott suggests the following as probably nearer the truth: Paleozoic 12, Mesozoic 5, Cenozoic (including the Pleistocene) 2. He places the estimated duration of these geologic divisions of time as 17,500,000 years for the Paleozoic, 7,240,000 years for the Mesozoic, and 2,900,000 years for the Cenozoic, or 27,650,000 years for the time since the beginning of the Cambrian. He further estimates that the Algonkian was not over 17,500,000 years, and the Archæan not over 10,000,000 years more.

Average of the Estimates of only Hypothetical Value.—Examining the estimates from all these various sources, of the length of time required to account for the deposition of all the stratified rocks in which the geological record of the history of organisms is preserved, we reach the conclusion that an average of opinions lies somewhere between 25,000,000 and 75,000,000 of years from the beginning of the Cambrian to the present time. Although it should be held as an extremely hypothetical belief, the probabilities are considerable that the time represented is within these limits rather than outside them either way.

Provisional Units of the Time-Scale Assumed to be of Equal Value.—But so long as the estimated value of the time-lengths in geology must be considered highly hypothetical, it may be

* "Geology," vol. II. p. 534.

† See, further, papers by Gilbert and by Spencer on the length of time represented by the erosion of Niagara Falls; and, regarding the St. Anthony Falls estimate, see Winchell, vol. II., "Final Report of Geology of Minnesota;" and a *résumé* of the subject in Dana's "Manual." 4th edition, pp. 1023, etc.

‡ "Geologic Time, as Indicated by the Sedimentary Rocks of North America": Proc. Am. Ass. Adv. Sci., vol. XLII. 1893, pp. 129–169.

as satisfactory in dealing with the time-scale to discard them altogether, and to consider the divisions as units which, added together, make up the total duration of time from the foot of the scale to the top, or to present time.

Adopting this plan, each of the periods in the time-scale on page 54 may be considered as a unit of time of uncertain length, but of definite position in the scale; and the several periods may, until evidence is found for a closer estimate, be considered to be of equal value. This makes the time-ratios to approach nearly the estimate made by Walcott, dividing the whole scale from the base of the Cambrian into 20 such units and assigning 13 of them to the Paleozoic, 4 to the Mesozoic, 3 to the Cenozoic time. Walcott's values were 19 units, and 12, 5, and 2 for the Paleozoic, Mesozoic, and Cenozoic times respectively. In the scale here adopted there is one probable exaggerated error, i.e., the more recent units were probably relatively shorter than the more ancient units which are represented of equal length.

The time-scale as provisionally adopted is as follows: Dividing the total time represented by the faunas and floras from the earliest Cambrian to the present time into one hundred units, there are found to be twenty distinguishable and pretty universally recognized biological life-periods, which for convenience may be assumed to represent equal periods of time, each period representing one twentieth or five per cent of the whole. There are three of these periods in the Cambrian era, two in the Ordovician, etc.; therefore the eras represent in percentages: the Cambrian, 15%; Ordovician, 10%; Silurian, 10%; Devonian, 15%; Carboniferous, 15%; Triassic, 5%; Jurassic, 5%; Cretaceous, 10%; Tertiary, 10%; Quaternary and Recent, 5%. Paleozoic time is thus 65%, Mesozoic 20%, Cenozoic 15% of the whole.

These estimates, for the purpose of measuring the relative duration of organic forms and thus the progress of the history of organisms, have a rough approximation to the truth according to the cumulative evidence from all sides at present before us, but they must be accepted as provisional estimates to be perfected by evidence which will come with the progress of knowledge.

CHAPTER IV.

STRATIFIED ROCKS—THEIR NATURE, NOMENCLATURE, AND FOSSIL CONTENTS.

The Common Usage in Classifying Stratified Rocks.—As defined on a previous page, geological systems are the primary units of the time-scale; they are also the grand divisions made in classifying stratified rocks. When terms indicating lapse of time are applied to these divisions, the meaning is lapse of time during which the system was forming. There is a Carboniferous period only as it is the unknown lapse of time during which certain strata included in a Carboniferous system were forming. The limits of that time are determined only by the unknown points of time when the first and the last strata of the system were laid down. The thickness and kind of rock, or other phenomena, may give us a clue to the possible duration measured between the two points, but it is a mistake to imagine that we know anything of the particular geological time, period, era, or epoch at which a particular stratum was made, except as indicated by the fossils which record the age. The laying down of a particular sandstone at a particular place marked a definite point in time, though we may not know in terms of years, or centuries, or millions of years, how long ago it was, and it is the stratum, and not the period, that is definite.

Fossils of Higher Value than Strata for Determining Time-relations.—According to general usage the fossils are not supposed to be the time-indicators, but the stratum is supposed to be the indicator of the age of the fossil. This common usage is defective, in that fossils, when considered as the remains of races of organisms regularly succeeding one another, record the steps of progress made in their evolution and may thus become independent sources of information regarding time-succession. From this point of view we find fossils to be the marks of the stages of progress in life-histories upon the earth, and the strata then serve, as the sand in the hour-glass, to measure the length of the time-intervals spanned by the life of particular species, genera, or families, or of faunas or floras.

The Necessity of Two Scales; Strata Furnishing the Data for the Formation-scale and Fossils Forming the Basis of the Time-scale.—This new point of view will lead to the separation of the time-scale from the formation-scale, and the making of a dual nomenclature and classification. The fossils, independent of the thickness or succession of the strata holding them, have a definite time-value, as indicated by the classification of the scale into Paleozoic, Mesozoic, and Cenozoic times, and the Eocene, Miocene, and Pliocene divisions of the Tertiary, proposed by Lyell.

The extension of this method of dividing the time-scale results in the formation of a pure time-scale, based upon the stages in the life-history of the fossil faunas, quite independent of the formations of any particular section, although adopting the systems, arbitrarily, as well-known and recognized units of universal distribution.*

Use of the Terms Period and Formation.—In treating of historical geology we speak of the age of invertebrates, the age of fishes, the age of coal plants, etc., but the application of time-designations to the rocks or formations is always perplexing and often leads to confusion of ideas. The terms Silurian, Cretaceous, Permian, Trenton, or Miocene were names of rock formations before they could be applied to the periods of time in which the formations were made. This double usage was introduced as early, at least, as 1828, when Lyell proposed to divide the Tertiary formation into "four groups or *periods* to which they belonged," calling them Eocene, Miocene, older Pliocene, and newer Pliocene. Although the science demands two classes of designations, a time-scale and a formation-scale, it certainly will tend to

---

* See p. 52 also "Dual Nomenclature in Geological Classification," *Journal of Geology*, vol. II., February–March, 1894, pp. 145–160.

clearness, and definiteness of thought and language, to retain the nomenclature now in use for the classification of rock formations and to apply names derived from the fossils to the time-divisions, since the fossils are the means by which the time-divisions are recognized.

A geological formation, made up of clastic fragments of other rocks, has in itself nothing by which to determine its time-relations; it is only its position, geographical and stratigraphical, in relation to underlying or superimposed strata that indicates its relative time-relation; when considered abstractly, or irrespective of its position, it loses its time-indicating characters.

Strata Parts of a Geological Formation, Fossils the Marks of a Geological Period.—It is not scientific, therefore, to speak of a rock or stratum as belonging to a particular period, the rock belongs to a formation. The fossil imbedded in it, however, does belong to the period, is characteristic of the period, and thus, in nomenclature, it is actually taken as the mark of the time-division. Just as we speak of the Chemung group, as the name for the upper Devonian rocks of New York State, so with like propriety we may say the *disjuncta epoch*, or the epoch of the *Spirifera disjuncta* and the fossils associated with it; and for the same reason. The application of Chemung to the group is appropriate, because one of the most typical outcrops of the rocks so named is along the valley of the Chemung River, at Chemung Narrows, in southern New York. Not that it is not exhibited elsewhere, and not that it is all exhibited at Chemung Narrows, but the group of rocks of which the cliffs at Chemung are a good example is appropriately and distinctly defined by the name. So to call the epoch the disjuncta epoch is appropriate, because the *Spirifera disjuncta* is a characteristic shell in the fauna of the epoch, and the designation *disjuncta* as a specific name is permanently applied to those characteristics of the genus which are peculiar to the closing part of the Devonian age, in all regions from which the fossil has been obtained; and although not the only fossil, and this one not always present, still it may be used whenever found as indicative of the time-epoch which is so named.

The "Hemera" of Buckman.*—Buckman has recently proposed the term *hemera* (ἡμέρα, a day) to indicate a time-division of this nature. He writes: "The term 'hemera' is intended to mark the acme of development of one or more species. It is designed as a chronological division, and will not therefore replace the term 'zone' or be a subdivision of it, for that term is strictly a stratigraphical one. . . . Successive 'hemeræ' should mark the smallest consecutive divisions which the sequence of different species enables us to separate in the maximum development of strata. In attenuated strata the deposits belonging to successive hemeræ may not be absolutely distinguishable, yet the presence of successive hemeræ may be recognized by their index species, or some known contemporary; and reference to the maximum developments of strata will explain that the hemeræ were not contemporaneous, but consecutive."

Again he writes: "Our present 'zones' give the false impression that all the species of a zone are necessarily contemporaneous; but the work of Munier-Chalmos in Normandy, and my own labors in other fields, show that this is an incorrect assumption. The term 'hemera' will therefore enable us to record our facts correctly; and its chief use will be in what I may call 'palæo-biology.'" †

The Terms Age of Reptiles, Planorbis Zone, etc.—The nomenclature at present in use in geological classification, it will be seen, is a nomenclature for the classification of formations, and is applied to the time-classification for want of a better. We have in use names for a few of the grander divisions of time properly chosen, as the ages of man, of mammals, of reptiles, etc., and in a few cases subdivisions of the finer kind have received names after the same plan, as the *planorbis zone* and the *angulatus zone*, before referred to in the classification of the Ammonite beds of the Jurassic. The selection of time-designations by this method can only come through careful study of the characteristic fossils on the

---

* S. S. Buckman, "The Bajocian of the Sherborne District: Its Relation to Subjacent and Superjacent Strata": Q. J. G. S., vol. XLIX. p. 481, November, 1893.

† *L. c.*, p. 482.

basis of their succession in chronological sequence. Although the relative position of the strata is the only infallible mark of time-sequence, it is the fossils in the strata that are the only infallible marks of time-periods.

Nomenclature of the International Congress of Geologists.—In general usage the time-designations have been applied directly to the formations, as in the nomenclature proposed by the International Geological Congress, where the formation-names *stage, series, system, group*, have their corresponding time-names *age, epoch, period, era*. In a similar way various other terms, which apply to the strata of formations, have their corresponding terms for the fossils of such formations.

Fauna and Flora.—A particular bed, stratum, or formation is said to have its fauna or flora, in the same way as a particular geographical region or province has its fauna or flora. A particular rock stratum marks a particular faunal horizon, as the Tully limestone may be called the horizon of the Cuboides fauna. We find an admirable definition of fauna in the Century Dictionary: "*Fauna*, the total of the animal life of a given region or period; the sum of the animals living in a given area or time." *Flora* is used similarly for the plants of a region or period.

Horizon.—We find under the word horizon an equally apt definition of that term. A geological *horizon* is defined as "A *stratum*, or *group of strata*, characterized by the presence of a particular fossil, or a peculiar assemblage of fossils, not found in the underlying or overlying beds."

Zone and Stratum.—The term *zone* is applied in geology to the stratum or the strata in which a particular fauna or flora is distributed. In some cases authors speak of the zone of a particular species; but whether a single species, or that one and other associated species, be taken as the distinguishing marks of a geological zone, the difference between a *zone* and a *stratum* is found in the distinction that the zone is characterized by continuity of the same life and the stratum by continuity of the kind of stratified deposit.

Facies.—The term *facies* is used in a particular sense in geology to apply to the particular composition or condition of a formation in a given region; for instance, the Hamilton formation in western New York is calcareous and finely argillaceous; in eastern New York the same formation is arenaceous and flaggy; although representing the same formation, one may be called the argillaceous or calcareous facies, and the other the arenaceous facies, of the Hamilton formation.

Area, Province, Region.—Again, the terms *area, province, region*, when applied geologically, refer to the geographical districts in which there was greater or less uniformity in the kind and succession of sedimentation for a given geological period. Thus, the Appalachian province or the Mississippian province may be spoken of. These same terms when used in zoology or botany refer to the districts which, separated by more or less sharp physical boundaries, are characterized by distinct faunas or floras.

Geological Range and Geographical Distribution.—A convenient distinction may be drawn in the usage of the terms *range* and *distribution*, which are now almost synonymous. In speaking of the separation of species, or genera, or faunas, or floras, when separated in space, *distribution* will be used; when separated in time, *range*. Thus, according to Ulrich, the Vitulina fauna of the Middle Devonian may be said to have a *distribution* limited to South and North America and Africa; its *range* is Lower and Middle Devonian.

Variations and Mutations.—Waagen has proposed to distinguish the changes of form observed on comparing the same species from different places. When the specimens compared belong to the same geological horizon, but come from the same or different geographical areas, the differences of form are called *variations;* when the specimens come from different geological horizons, thus representing time-range, the differences of form are called *mutations.*

Development and Evolution.—Another analogous distinction, which is explained more fully elsewhere, is observed in the restriction of the term *development* to the processes of expansion of characters of the individual in ontogenetic growth, and the term *evolution* to the changes expressed in the individuals succeeding each other in phylogenetic succession.

Initiation and Origin.—Another distinction, in the way of greater precision, is in the use of the term *initiation* in place of *origin*, when speaking of the first appearance of a new type of structure in the geological formations. It is difficult not to associate some idea of causation with the terms *origin* and *originate*, but the term *initiation* refers simply to an incoming or a beginning to appear, leaving other questions open for discussion.

System.—This is the name for one of the larger geological divisions, but there is no uniform rule for its application. Originally, as proposed by Murchison, system was applied to a series of rocks continuously exposed in some geographical region. Thus, Silurian system was the series of rocks exposed in Wales and western England at one time inhabited by the Silures. The Devonian system was the series of rocks exposed in south and north Devonshire; Permian system, certain fossiliferous rocks first thoroughly studied in Perm, Russia; etc. The term system was afterwards adopted as a name for a large and prominent series of stratified rocks, as Carboniferous system, Tertiary system, etc.

Systems have been arbitrarily determined, and the list as given, including those in which fossils have heretofore been found, is as follows: Cambrian, Ordovician, Silurian, Devonian, Carboniferous, Triassic, Jurassic, Cretaceous, Tertiary, and Quaternary or Recent, or including Recent. These, as has been said, are arbitrarily fixed, and there is in each case a typical system expressed in the rocks of a particular region.

These systems are applied with an approximate degree of uniformity in all countries, although arbitrarily: and era is the time-designation which is applied to indicate the lapse of time during the formation of the rocks of a single system.

Geographical Conditions Determining the Local Characters of Stratified Rocks.—There are a few particulars, regarding the way in which these rocks were formed and their present condition and order, which help to explain the conditions under which the organisms lived in the past, and may explain why we have full records in some cases, very little record in others, and in many cases very sparse and greatly broken records of the life-histories we are seeking to read.

The stratified rocks, as already stated, are the result of water-action; First, erosion from already formed rocks; second, transportation of the fragments by water; and, in the transportation, third, separation of fine from coarse and further rounding of the individual grains; fourth, sedimentation under water in layers or strata. The materials for each stratum have gone through these various processes of water-action. The result is that the present characters of the strata have been determined by (*a*) the nature of the source of materials, (*b*) the rate, direction, and force of the activity of the water, and (*c*) the relations of the bottom of the ocean to the surface, or the depth of the water. Each of these three conditions is variable and generally is the same for only a limited area. To illustrate: We know from observing the phenomena of an ocean beach that the beach material where the shores are low and composed of soil is made up of the wash of the shore. If a large river empties in the vicinity, the shore is made up of fine silt and mud; if, on the other hand, the shores are hard rocks, the beach is composed of coarse pebbles and gritty sand, the result of the disintegration of the rocks themselves. If we examine the shore material of Florida, where calcareous rocks alone are exhibited, we find the sand composed of broken shells and corals. This, when filled by deposited calcite carried into the interstices in solution and hardened, becomes a calcareous rock, called coquina, and finally a compact limestone.

Again, if we examine the materials lying on the beach at high tide and those on the bottom out to a depth of a hundred fathoms, we find that the coarse pebbles and boulders are distributed along the line of most violent wave-action near shore, then gravel, and further out only fine sand, and finally only the finest silt appears. This sorting is entirely co-ordinate with the change in violence and rapidity of normal motion of the water in waves and currents. The more rapid and violent the motion of the water, the larger the particles moved and transported by it, and, hence, the farther out from its source the material is borne, the finer and less in amount will be the resulting deposit.

For all fragmental material the land surface, where it comes in contact with water in motion, may be regarded, in a general sense, as the source, and, in a general way, distance from such source determines the relative size of the particles making up the sediment. The source may be far up in the interior of the continent where river erosion or lake erosion is eating away the land, or it may be on the ocean-shore, but in general it is true that local geographical conditions are fundamental in determining the lithological character of geological formations.

Varying Conditions of Environment in Relation to Time-estimates.—The conclusion from these observations is that all sedimentary rocks may be supposed to have been formed within about a hundred miles of the shore from which the sediments were derived. This theory is supported by the deep-sea soundings, which show very small amount of material accumulated on the bottom of the present ocean at great distances from land. From these considerations we turn to our classification of formations, and see why it is that we cannot expect to find uniformity of details in either the structural or stratigraphical order, or in the lithological composition of the formations. (1) At the same time there may be in process of formation a limestone, a sandstone, a conglomerate, and a mud-shale, and all may be forming within a relatively short extent of coast. (2) In the same period of time the thickness of material accumulated may greatly vary; while an inch of limestone may be deposited in one place, a hundred feet of sandstone may be formed in another. Thus the limestone of one locality may be represented by a sandstone in another, and a thousand feet of strata in one place may be represented by a hundred or less in another not far distant.

Relative Order of Deposits in Relation to Depression and Elevation—Another series of facts may be considered in this place. The shore-lines do not remain constantly the same for the accumulation of sediments. The simple fact that there are marine fossils in rocks above the level of the ocean is evidence that there has been a change in the relative level of land and ocean surfaces; there has been an elevation of the land surface. Since the conditions of sedimentation vary with the distance from shore-line, a particular series of these conditions extending from shore-line out into deep water will be bodily shifted seaward by elevation, and landward by depression of the continental border.

Order of Deposits with a Sinking Land.—Other conditions remaining the same, for instance, on a shore with land to the westward and ocean to the eastward, a gradual continuous depression of the land would result as follows: The shoreline would gradually retreat westward; at each spot the water would gradually become deeper and further off shore; and, considering only this one law of sedimentation, the deposits forming at each spot would gradually become finer and finer with the progress of time: so that finally it would happen that the deposit forming directly over the place where the shore-line was at the outset would be the very fine silt peculiar to deep water far out from shore, the same which at the beginning of the period was being deposited only at a distance off shore.

To compare the sections taken at three localities we would get the following results, seen in Fig. 3:

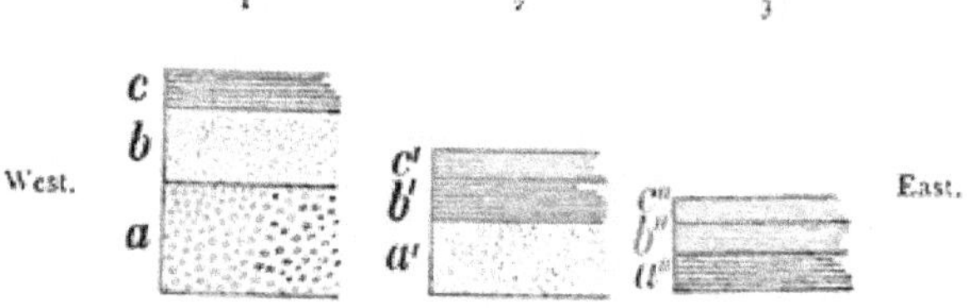

FIG. 3.—Three different sets of deposits formed during the same periods of time at three points, 1, 2, and 3, separate from each other, with a sinking of the land as the sediments are accumulated.

In which the section at 1 would exhibit a series of deposits, one overlying the other (*a b c*), presenting the same differences of sedimentation that would be exhibited on comparing the first deposits in the several sections (*a a' a''*). It is likely, too, that the general character of the fossils would correspond, but as a matter of age the deposits of like character in the three sections (*a'' b' c*) would represent consecutive periods, instead of the same period of time.

Order of Deposits with Elevation of the Land.—If we suppose a gradual elevation to take place, instead of depression, then the shore-line would advance gradually seaward,—eastward in the supposed case,—and the first locality (1) would cease to receive deposits, and would be eroded away by the action of the waves and partly redistributed over the other deposits, while the one farthest out (3) would receive first the finer deposits (*a''*), then still coarser (*b''*), and finally the shore conditions would prevail and their appropriate sediments would be deposited (*c''*). The following would result:

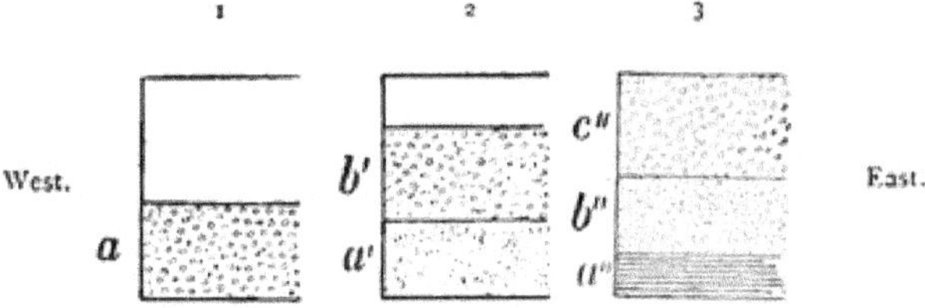

Fig. 4.—Three sets of deposits formed under the same conditions as those of Fig. 3, except that the land was gradually rising during the accumulation of sediments. In both figures the coarser sediments are represented by open dots, the sands by fine dots, the coarse muds by heavy horizontal lines, the finer muds by similar finer lines.

There would be a reversal in the order of the sediments; also a change in the relative thickness of the three sections; and number 3 would be the thicker section. Although gravel might appear at the top of each section, it would represent a later period in section 3 than in section 1, and all the period represented by *b''* and *c''* of the third section would be represented in the first section by an hiatus or line of erosion. It is essential to assume that such oscillations, upward or downward, were taking place constantly during the accumulation of the sedimentary deposits now called stratified rocks, and the above analysis exhibits the nature of the perplexities which must arise in a precise study of the relations of the formations of different regions to each other.

Characteristic Fossils.—In a general way fossils are characteristic of the age of the systems, but actually the systems represent great lengths of even geological time; and in many cases this time is long enough to include the beginning, the luxuriant abundance, and the extinction of a whole genus or a family of organisms. Such generic groups have had their stage of beginning, have spread about the earth, and during their distribution and adaptation to the various conditions of environment have become specifically modified, so that each of the systems is marked by the presence of certain genera which are characteristic of the fauna and flora for a long period, and thus serve as arbitrary marks of these great periods. The individual continuing beyond a certain specific zone in one section does not interfere with the general law that there are grand divisions of time which are characterized by peculiar types of organisms.

Although we cannot, in the present state of knowledge, draw sharp lines which shall be universal, between the formations, or between the several species represented in them, it is convenient to recognize these systems, and in each country the lines can be arbitrarily fixed, and the sub-divisions locally recognized.

SUMMARY.

Reference has been made to the difference between the history of the organism (Ontogeny) and the history of organisms (Phylogeny). It has been shown that there is a natural history of the development of the individual, and that there may be a history of organisms as a whole—a history in which all the species of the same kind are but as a unit in a great complex of organic life which has evolved with the geological ages. In this latter history time and the conditions of environment have played very important parts; but ordinary time-scales are practically useless, because they are not divided into long enough periods, and because they do not reach back far enough. A special time-scale was needed. This has been constructed by an analysis of the classification of rock formations. In this analysis we have seen that progress of science is as much a progress of ideas as it is an increase of known facts; that the accumulation of confirmatory facts has followed rather than preceded the formulation of speculative theories; that the theories about the earth have dominated in each proposed scientific classification of facts, and thus in the formulated science of each period.

The result of the analysis emphasizes a few laws, which may be stated as established, regarding the chronological aspects of the rocks.

First. There is *a natural succession in the original forma-*

tion of rocks. There are certain rocks that are relatively primitive; these are crystallized and compact, as granites and gneisses. There are other rocks that are of sedimentary origin; these are secondary in original formation; they are made of fragments of rocks, and are in stratified form, and lie upon primitive rocks whenever the two are in contact. There are still other geological formations that generally are not in compact form, but are composed of loose fragments, sand, and fine mud, or soil, and naturally lie above the others.

A second law established by experience is that (with exception explained by later disturbance) *for the sedimentary rocks natural order of superposition indicates relative chronological order of formation;* viz., in any given case of two stratified rocks the underlying rock is the more ancient.

A third law is that *the mineral character of any particular stratified rock bears no necessary relation to its age.* As, for instance, rocks of the same composition, structure, and color, but coming from separate geographical regions, may be of entirely different geological ages.

Fourth. It is an established law that *there is some definite relationship between the characters of the fossils and the relative geological age of the rocks in which they occur.* This law is formulated in the classification Paleozoic, Mesozoic, Cenozoic, applied to the respective geological formations in their chronological order.

In accordance with these laws a classification of formations has been formed (Cambrian, Ordovician, Silurian, Devonian, etc.) in which the relative antiquity of the systems is expressed. This constitutes the *formation-scale,* and it is based upon the series of strata, lying one upon another, composed of sedimentary materials of various kinds forming sandstone, limestone, shales, and conglomerites, etc., originally nearly horizontal in position, but now variously tilted and folded. In such rocks the fossils are found from which the *time-scale* proper is constructed. The recognizable units of this time-scale are the periods, characterized by fossil faunafloras, whose characteristic species may be distinguished the world over and thus form the marks of the standard timescale for the study of the history of organisms.

CHAPTER V.

FOSSILS—THEIR NATURE AND INTERPRETATION, AND THE GEOLOGICAL RANGE OF ORGANISMS.

Fossils of Vegetable and Animal Origin.—Having explained the nature of the series of geological formations, their classification into systems, the value of these as reservoirs of information regarding the history of organisms, we next inquire into the nature of the fossils, which are preserved in them and furnish the records of the separate lives whose history we would trace.

Fossils are any traces of any organisms which, having been buried in rock-forming muds, are preserved to tell of the life of the dead organisms. Vegetable fossils are remains of plants, leaves, stems, wood, fruit, nuts, or resin, gum, or carbonaceous matter, coal or bitumen, or oil or gas. Animal fossils are remains of animals, their foot-prints, tracks, trails, cases formed of particles of sand, as of the Caddis-worm, etc., skeletal or dermal hard parts, bones, teeth, spines, scales, shells, or corals, and secretions of various kinds, formed during life for protection or defense, or offensive weapons, or excretions, when of sufficient hardness to resist destruction, as the coprolites of fish and reptiles, preserving, in some cases, evidence of the shape of the intestinal canal (spiral) through which they passed.

Original Material of Fossils.—The original materials were as various as the hard parts now formed by living organisms. The great majority of the known fossils were originally composed of calcic carbonate, calcic phosphate, chiton, bone, silica, or, in the case of plants, bituminous matter. In some cases the whole animal may be preserved, as in the case of insects in amber, or the fossil elephants in the ice of northern Siberia, which have furnished abundant store of ivory to enterprising explorers; or in the case of minute organisms buried in the muds, the softer or destructible parts may decay and pass away as gases or in solution. Generally, however, fossils are but fragments or parts of the original structures united during the life of the organism. Again, the original substance of the fossil, when removed by solution after fossilization, may be replaced by other mineral substance brought in from without by infiltration. Or the mineral may be molecularly changed or replaced; an example is fossilized wood, in which the grain and structure of wood is preserved but silicified. This replacement may be by Silica, Calcite, Pyrite, Marcasite, Siderite, and rarely some other minerals.

Various Aspects of the Original Form represented.—In the fossil condition the form may differ from that we are accustomed to see in the corresponding part of a living organism. Thus a fossil snail-shell may be simply a fossil shell, that is, the shell itself buried in the rock. Or it may consist of the impression of the shell now removed, in which case it may be the reverse or cavity over the exterior of the shell, or, in case of flat shells, like clam-shells, similar impressions of the inner surface; or the cavity may be again filled with detrital matter, forming a cast of either the inner or outer form of the shell or object fossilized: in the former case it would be called a mould; in the latter, a cast.

Preservation of Fossils.—Fossils may have been covered under various conditions and at various places; and the fossils themselves are the best indication of the conditions. The fossils may consist of land species alone, or types of organisms adapted to live in air and not in water; but in order to be preserved it is almost universally necessary that the part fossilized be covered from the air: first, because atmospheric conditions are extremely destructive to any substances exposed to them, even quartz or glass suffering more or less by continuous exposure. The protection by soil will preserve the more insoluble matters, but here again decomposition and solution of any substance that can be decomposed or dissolved will take place with slower or faster rapidity. Entire exclusion from air and from circulation of acidulated and alkaline waters is the condition under which the more perfect fossils were preserved, and this condition is found only under muds, in marine conditions; in the bottom of lakes or in river bottoms fossilization may take place, but the fossils are then liable to some change of composition. Fossils preserved under the most favorable conditions, by long-continued pressure and the slight circulation of fluids in rocks, suffer change after their formation, particularly in the way of assuming a crystalline structure.

The Majority of Fossils are of Marine Organisms.—From the above remarks it is evident that the larger proportion of fossils must be those preserved under the surface of the ocean; next will be found those buried in land basins covered by fresh water; and only very rare will be the cases of fossils otherwise preserved. Hence marine organisms will naturally present in the rocks the fuller records of their history: fresh-water or brackish-water species will be recorded less perfectly; and the organisms normally living under land or air conditions will be recorded in fossils very imperfectly at the best. The great majority of even the hard parts of such organisms must be destroyed before reaching the position of a safe burial-place, and our studies will be directed by this law of preservation. Marine organisms, and largely marine invertebrates, will be selected as illustrations of the laws of the history of organisms, because the records regarding these are fuller than regarding any other kinds of organisms.

Various Kinds of Fossils enumerated. — To the question "What are fossils?" the concise answer is: Fossils are traces of organisms buried in the rocks. A full definition would be a descriptive treatise on Paleontology. As to their forms, fossils are as various as are organisms. A useful analysis, however, may be made of their composition. Fossils are composed (A) either of the original materials of the organism which made and left them; they are then strictly remains of dead organisms, or of parts of them. Or (B) fossils may be casts or moulds in the rocks where these structures were originally buried and afterwards removed. (C) The filling of the cavities thus formed constitutes other kinds of fossils. (1) The cavity may be filled by mineral matter carried in by infiltration and redeposited; (2) the cavity may be filled through molecular replacement of the mineral of the original structure by some other mineral, as calcite, silica, pyrite, etc.; (3) the cavity may be filled by detrital matter washed into the cavity from outside. (D) The original substance may be changed in molecular constitution, or even in chemical composition, losing a part of its elements, or gaining other elements; thus, a piece of wood may become coal, or a shell may become crystalline calcite, or aragonite. (E) Finally, the fossil may consist of traces left in the sediments while the animal was alive, as footprints or other marks of organic activity.

Fossils represent chiefly the Hard Parts of Organisms.—An important generalization may here be made regarding all fossils. Fossils represent organisms, but almost universally they represent the *hard parts* of living organisms; hence the most valuable lessons to be learned from fossils must be derived from the study of the hard parts of organisms. These hard parts are the parts which have attained definite and fixed form during the life development of the individual. Soft parts, or organs, are adjustable to changing exterior conditions, but its hard parts are already adjusted, and, therefore, they are an expression of the working adjustment of the species, to the conditions of its environment, at the particular time in which it lived.

Best and most perfectly adjusted Organisms of the Time left their Records.—The history of organisms, which we particularly trace in the study of fossils, is not the history of imperfect organisms struggling toward perfection, but it is the history, for each age and epoch, of the perfected adjustment of the organisms of the time to the particular conditions of environment in which they lived. They did not die before their time, overcome by the mythical fittest who are said to survive in the struggle. They were the fittest, and died natural deaths, having provided before they gave up the struggle for their progeny to succeed them. The hard parts record the history of adults which had endured the struggle, and thus represent the royal line of succession for the geological ages.

General Laws regarding the Occurrence of Fossils.—There are certain general laws, concerning the occurrence of fossils and the relations which their specific forms bear to the place they occupy in the geological scale, which indicate a definiteness in the order of their succession, quite independent of the evidence furnished by the stratigraphic succession of the rocks themselves; and it is this testimony of the fossils, pure and simple, as mere physical forms, upon or in the rocks, that confirms and helps to complete the chronological scale indicated by the successive geological systems.

Pictet* announced a number of propositions setting forth the more prominent of the laws of occurrence of fossils and their relations to time and place. In the "Handbuch der Palæontologie" Zittel † has condensed and culled them so carefully that we there have concisely formulated in a few sentences the chief facts regarding their occurrence.

They are as follows: (1) All stratified sedimentary rocks (with the exception of metamorphic rocks) enclose, more or less richly, fossils, and thus prove that the earth, for an immeasurable length of time before the appearance of man, was inhabited by organisms.

(2) The fossils of the oldest and deepest strata represent extinct species, and for the most part extinct genera; only in the more recent strata are found forms which are identical with those now living. The deeper down we penetrate in the series of strata the more divergent are the fossils from the forms now living; and, on the contrary, rising from the earliest to the more recent formations there is a continuously increasing resemblance to the present creation.

(3) The different fossil faunas and floras follow each other the world over in the same regular sequence; the formations stratigraphically nearer to each other contain the most similar fossils, and those most separated in age present the greater differences.

(4) Constant change characterizes the evolution of the organic creation. Species of one geological formation are either completely or partly replaced by other species in the next superimposed strata.

* Pictet, François Jules, "Traité de paleontologie: ou, hist. nat. des animaux fossiles, etc." 1853–57.

† Zittel, "Handbuch der Palæontologie," vol. I. pp. 17, 18.

(5) Each species, like the individual, has a certain shorter or longer life-period, after which it perishes, never to reappear.

(6) From these principles it arises that the approximate age of a stratum may be determined by the degree of similarity of its fossils to the forms of the present time. The fossils contained in the strata are the means of determining the equivalency (that is, likeness of age) of the strata themselves, and in general, identical fossils indicate contemporaneity of the enclosing strata.

Change of the Forms of Fossils with Passage of Time, and particular Form characteristic of Particular Periods of Time, undeniable Facts of Paleontology.—Thus it appears that whatever we make out of fossils, whether we consider them stones or organisms, however we account for their origin, whatever relation we conceive them to bear to each other, the fact is startlingly vivid to the paleontologist that the form of a fossil is intimately associated with the time in which it appeared on the earth: that the morphological characters assumed by fossils have been gradually and incessantly changing from the beginning of the world.

Inorganic Things, on the contrary, Unchangeable.—This is contrary to the law in respect of every inorganic thing. The chemical composition of things and the chemical properties are the same so far back as we can trace them and to the most distant star in space. Minerals in the Archaean ages, before any fossils had appeared, crystallized out into exactly the same forms which they assume to-day. We know of not the least fluctuation in the laws of physics for all time. Indeed, it is by dependence upon the absolute certainty and uniformity of these laws that the astronomer is able to calculate the position, the size, and the orbit of some unknown and unseen planet, and directing his telescope to the place where it should be, to discover it there.

Fossils characteristic of Particular Periods of Geologic Time.—The morphological combination of characters, which we call a fossil (as a Trilobite or an Ichthyosaurus), has its definite relationship to geological time, and each form is characteristic of a particular period of time. *A fossil becomes the unmistakable mark of the age of the rock in which it is enclosed:* the Trilobite is characteristic of the Paleozoic, and the Ichthyosaurus is characteristic of Mesozoic time, as truly as man is characteristic of recent time.

Stony Corals: the Zoantharia.—In order to emphasize and illustrate this law of the intimate connection between organic form and time, the statistics regarding the great order of the stony corals (the Zoantharia) may be chosen.

For the convenience of those who may have no special acquaintance with the scientific nomenclature of systematic Zoology, a few facts regarding the principles of classification and nomenclature are here offered. The classification of animals is based primarily upon differences in form, structure, and function. On this basis zoologists have classified animals under nine chief divisions, called, 1, Protozoa; 2, Cœlenterata; 3, Echinodermata; 4, Vermes; 5, Arthropoda; 6, Molluscoidea; 7, Mollusca; 8, Tunicata; 9, Vertebrata. (Claus.) Each of these divisions is called a Branch (Phylum or Subkingdom) of the Animal Kingdom, and each is characterized by a distinct type of organic structure.

Under each of these chief divisions the animals are associated by their greater degrees of likeness, and are separated by their lesser differences, into subdivisions, called respectively, from higher to subsidiary rank, Classes, Orders, Families, Genera, and Species. The Cœlenterata are thus at present known under four Classes, viz., Spongia, Anthozoa, Hydrozoa, Ctenophora.

The class Anthozoa (coral animals) is subdivided into two orders, Alcyonaria and Zoantharia. The order Zoantharia is subdivided into three suborders: The *Antipatharia*, the *Actinaria*, and the *Madreporaria*. The first two of these suborders develop no hard parts that have been recognized in a fossil state, and therefore we cannot speak of their historical relations. The *Madreporaria* are the polyps which secrete stony corals, and of their calcareous skeletons great numbers have been found in the rocks; many massive beds of limestone consisting mainly of them or their fragments.

The Madreporaria, or stony corals, have been classified in two groups of families, the most characteristic feature separating them being the arrangement of the septæ in one of them in fours or multiples of four, Tetracoralla, and in sixes or multiples of six in the other group, Hexacoralla.

Numbers of Genera of the Zoantharia recorded for each Era.—There are several thousand species of stony corals described, but for the present purpose it is sufficient to note that there are 448 genera of Zoantharia already described and recognized. (Zittel.) That is, there are 448 different combinations of form of the stony corals, which are sufficiently sharply defined and constant in their character to be classed under distinct genera. If we only note the numerical relation of these genera to the successive geological periods of time, the law above referred to becomes at once apparent. In the Lower Silurian 4 genera are reported by Zittel; 5 genera have been since reported in the Cambrian for America alone by Walcott; the Silurian has 54 genera, Devonian 39, Carboniferous 34, Triassic 17, Jurassic 84, Cretaceous 110, Tertiary 125, and Recent 132 genera.

Two Types of the Zoantharia indicated by the Two Maxima of Genera in Separate Eras in the Time-scale.—In this series of numbers of genera there are two maxima, one in the Silurian, one in Recent time. This is explained by another fact: the Order (Zoantharia) is divided, as above stated, into two biological groups, distinguished by a marked difference in the numerical arrangement of radiating divisions of the body. The first group, the Tetracoralla, has 81 genera; the second group, the Hexacoralla, has 367 genera. With the exception of a single genus, all the 81 genera of Tetracoralla are confined to the Paleozoic. The Hexacoralla are mainly later than the Paleozoic.

These statistics for the Madreporaria, arranged in tabular form, produce the following table (the figures in each column opposite each family expressing the number of genera of the family which made their first appearance in the geological system corresponding to the letter at the top of the column):

TABLE OF THE NUMBERS OF GENERA OF MADREPORARIA MAKING THEIR FIRST APPEARANCE IN EACH GEOLOGICAL SYSTEM, GROUPED IN FAMILIES.

		C.	O.	S.	D.	Cr.	T.	J.	K.	Ty.	Q.	R.
Tetracoralla	81 genera	5	4	42	10	19	0	0	0	0	1	
Hexacoralla	367 genera	0	4	17	11	7	22	72	75	81	78	
1. Favositidæ		0	3	14	8	3	0					
2. Poritidæ		0	1	2	0	2	1	2	5	8	2	
3. Madreporidæ		0					0			1	1	
4. Pocilloporidæ					2	1				1	1	
5. Eupsammidæ				1				2	2	6	7	
6. Fungidæ							6	13	9	9	10	
7. Astræidæ					1	1	13	46	44	33	28	
8. Stylophoridæ								2	0	1	1	
9. Oculinidæ								6	2	4	4	
10. Dasmidæ										1	0	
11. Turbinolidæ							2	1	13	17	14	
Total Madreporaria	448 genera	5	8	59	21	26	22	72	75	81	79	

Evolution Curve of a Group of Organisms.—These statistics may be so arranged as to express in graphic form the rate of generic differentiation.

For this purpose a table is constructed, composed of a series of ten perpendicular columns, each one, from left to right successively, representing the successive geological eras, Cambrian, Ordovician, Silurian, Devonian, etc. The name of the era is indicated at the top of each column by its initial letter. The length of time of each of these eras is represented roughly by the width of the spaces between the separating lines, according to the time-scale described in Chapter III. Thus, starting from the lower left-hand corner, the abscissa represent time-extension from the beginning of the Cambrian era.

Drawing an horizontal line from this point across the base of the several columns, the distance above this base-line or the ordinate expresses the degree of differentiation in terms of units of genera (or of species, as the case may be) appearing in each era.

Thus, by connecting together the points representing the amount of differentiation (the ordinate) for each geological era (the abscissa), we produce a curve representing the rate of generic differentiation for the particular order or class (as the case may be) under consideration. This curve may be called the *evolution curve*. In the following table are represented the evolution curves of the Madreporaria, and those of several divisions and families of the Madreporaria, based upon the statistics before us.

Construction of the Diagram.—This diagram was constructed as follows:

Extension laterally represents time-duration, beginning at the left-hand lower corner with the base of the Cambrian; the total length of geological time thence to the present is made to cover 100 spaces. The several geological system-eras are represented in their estimated proportionate lengths, thus: 15% is given to the Cambrian, 10% to the Ordovician, 10% to the Silurian, 15% to the Devonian, and 15% to the Carboniferous; to the Triassic 5%, Jurassic 5%, Cretaceous 10%; and the Tertiary and Quaternary together are given 15%, 10% to the former and 5% to the latter.

This, it will be seen, assigns to the Paleozoic, Mesozoic, and Cenozoic, respectively, 65%, 20%, and 15%, or 13, 4, 3 as the time-ratios, Dana's revised estimate (1895) being 12, 3, 1, and Walcott's estimate stands 12, 5, 2, as explained in the third chapter.

Vertical lines are drawn to separate off the time-scale into periods with these proportions. Vertical extension of the curved lines represents the number of new genera of each period. The curve *m* running highest is the curve of generic differentiation for the Order Madreporaria, and is compiled from the lists of genera in Zittel's "Handbuch," with some corrections based upon facts appearing since its publication, and the geological range there assigned to them, with a rearrangement of the genera of the first family of the Hexacoralla, Favositidæ, into Favositidæ and Poritidæ.

The differentiation curve is formed by making a vertical scale and placing the point representing the differentiation for each period above the base-line by the number of divisions corresponding to the number of new genera initiated during the period. In the same way separate differentiation-curves are formed for the genera of several of the families: thus *f* is the curve for the Favositidæ; *a*, the curve for Astræidæ; *b*, for the family Turbinolidæ.

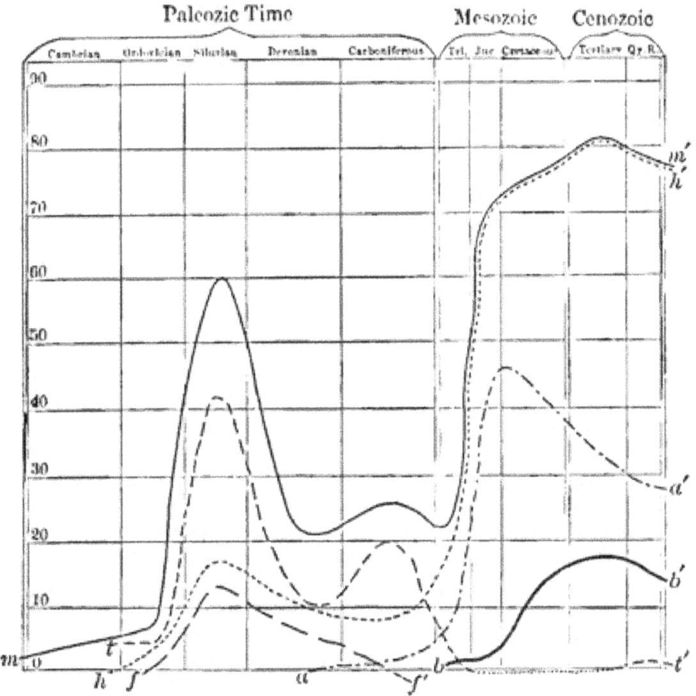

FIG. 5.—Evolution curves of the families of the Madreporaria. The vertical lines represent the points of time separating the several geological eras of which the names are at the top of the chart. The horizontal lines represent, by tens, the number of new genera first appearing in each era. The curved lines represent the rate of differentiation of each family type in number of genera first appearing in each successive era. *mm′* evolution curve for the whole Madreporaria, *tt′* for the Tetracoralla, *hh′* for the Hexacoralla, *ff′* for the Favositidæ, *aa′* for the Astræidæ, *bb′* for the Turbinolidæ.

Meaning of the:e Evolution Curves.—This diagram illustrates the following points: The curves express the rate and the degree of differentiation of generic form expressed in the suborder Madreporaria in geological time. This law for the whole group is expressed in curve *m*. The irregularities of the curve suggest at once that it is compounded of at least three independent curves, of which the nodes are at the close of the Silurian, Jurassic, and Tertiary, and this suggestion is verified by examination of the taxonomic classification. There we find that the systematists, studying the structure, have divided the genera into 13 families, grouped in two divisions, Tetracoralla (2), Hexacoralla (11), and the curves for each of these is separate. Thus we find that the curve of differentiation for the genera of the Favositidæ (curve *ff'* of the diagram), which begins with the Ordovician and ends with the Paleozoic, accounts for the main features of the Paleozoic part of the differentiation of the whole suborder. Although other families have their beginnings in the Paleozoic, it is only with a few genera.

If we examine the curve for the genera of the family Astræidæ (*aa'* on the diagram) it is evident that the chief differentiation for the early Mesozoic was within this family. This family and the Fungidæ will nearly fill out the total differentiation-curve. The third irregularity in the curve is again explained by the late culmination and differentiation of the family Turbinolidæ, *bb'*, which shows its first genus in the early Mesozoic (Lias), but presents 17 new genera as late as the Tertiary.

Chronological Value of Family Groups of Genera.—Thus it appears that groups of genera are not only families according to the taxonomist (that is, genera naturally grouped together because of the likeness of their general structure), but the genera composing them are naturally associated together by the time of their initiation among the organisms of the world; and the simple tabulation of the time-relations of the genera of an order reveals, by the irregularities of the curve of differentiation, that the order is made up of several families having separate evolution-curves or separate life-histories.

The Life-period of a Genus.—The numbers thus given do not refer to the same genera repeated, but in large measure to different genera for each system. Without going into details, this may be illustrated by the following statement: Of the genera above tabulated 182 are peculiar to a single geological system, 89 are found in only two contiguous systems, 40 have a range of three systems, and only 9 range through four systems; or, to express the fact in proportionate numbers, the life-period, or geological range, of $\frac{4}{5}$ of the known 448 genera of Stony-corals is not greater than that of a single geological system: $\frac{1}{5}$ of the genera have a life-period of one or two systems length; $\frac{1}{11}$ of them lived only through two periods and into a third; and only 9, or $\frac{1}{56}$, continued existence for more than the length of three geological systems.

Organisms express Evolution in their Geological History; a Fundamental Law.—These statistics are chosen only as a convenient illustration of a general law, which might be illustrated by any other group of which we have the facts. Without stopping to ascertain what the particular nature of the forms is, it is evident that *divergence of organic form is intimately associated with lapse of time.* We do not require to see every form that has lived on the earth to distinguish the working of this law; but the few imperfect evidences, as well as the fuller particulars we know respecting some of the better preserved organisms, emphasize the presence of the law whenever we examine the facts. Thus we are led to conclude that *morphological differentiation (evolution) is as characteristic of the history of organisms in geological time as organic growth (development) is characteristic of the history of the individual organism in its lifetime.*

The Meaning of Genus and Species.—We have been speaking of combinations of form which are defined as classes, orders,

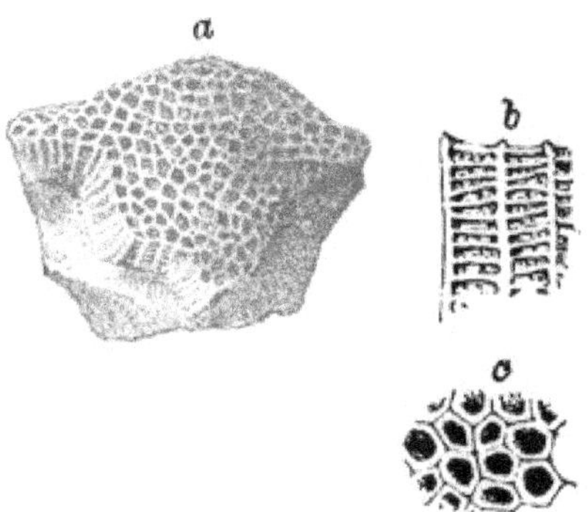

FIG. 6.—*Favosites niagarensis*, Hall. Original figures of the fossil coral from the limestone on Goat Island, in Niagara River: *a*, fragment of the coral showing the ends of the corallites; *b*, a magnified view of two corallites, showing the dissepiments and the perforations of the walls; *c*, end view of the corallites, showing the walls and perforations. (After Hall.)

genera, or species, and of genera as living at a particular time, and having a particular range, and differing one from another. In the study of fossils we do not actually see species and genera, or classes, or subkingdoms; but we see only certain shells, or impressions, or marks on the rocks: we say these fossils represent animals that have lived, and we give them particular generic and specific names. To take an example, we find a specimen in the Niagara limestone, illustrated in the accompanying figure. (Fig. 6.)

The Fossil Coral, Favosites niagarensis, as an Illustration.—It was named *Favosites niagarensis* by Hall, which means that its generic characters are those of the genus *Favosites*, its specific characters those of the species *F. niagarensis*, and that it was described by the paleontologist James Hall. It is a fossil coral (Actinozoan), of the order Zoantharia.

Analysis of the elements of form, which must be observed in classifying the specimen, will reveal somewhat more distinctly what is meant by saying that organic form and lapse of time are intimately associated. We notice, in the first place, that the fossil is made up of a large number of polygonal calcareous tubes attached together by their outer faces. This peculiar structure is the evidence for placing it in the order Zoantharia. Living corals (Zoantharia) secrete calcareous tubular bases, in and upon which each Zooid is supported, and in living corals these corallites are aggregated in the same manner as in the specimen before us. The radially symmetrical structure of the corallites is sufficient evidence that the specimen belongs to the subkingdom Cœlenterata, and we know of the existence of this subkingdom in the first or Cambrian period.

The continuous, hard, calcareous skeleton shows the fossil to be a Madreporarian, the structure of whose soft parts we assume to have been that of living Madreporarians, and therefore to be one of the class Anthozoa which is characterized as "polyps with œsophageal tube and mesenteric folds, with internal generative organs (no medusoid sexual generation)." The septa, which are rudimentary in the species before us (see Fig. 8), are twelve, and this character distinguishes the specimen from the subclass Tetracoralla, in which the septa are grouped in multiples of 4, and from the order Alcyonaria, which has 8 tentacles; and they show it to be an Hexactinia (or Hexacoralla), in which the septa are six or some multiple of six. True Hexacoralla have not been discovered below the Ordovician, or second geological period.

The diagram Fig. 7 illustrates the fundamental elements of a coral (Hexacoralla).

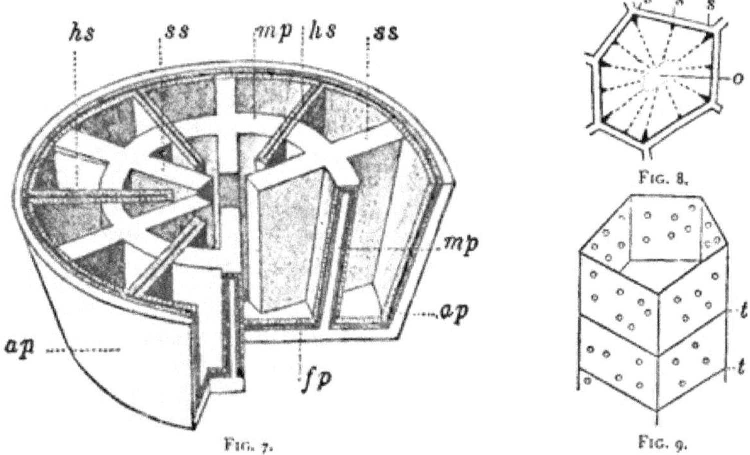

Fig. 7. Fig. 8. Fig. 9.

Fig. 7.—Diagram of the structure of a coral: *ap* = exotheca; *hs* = mesentery; *fp* = basal plate; *ss* = septa; parts in white = calcareous skeleton; shaded = ectoderm; black = mesoglœa; dotted = endoderm. (After McMurrich.)

Fig. 8.—Diagram of an end view of a single corallum of Favosites, showing the rudimentary septa *ss*, the dotted lines indicating the probable arrangement of the mesentery and the position of the mouth opening *o*.

Fig. 9.—Diagram of two chambers of a corallum of Favosites with the perforated walls, and the transverse dissepiment or tabulæ, *tt'*, separating the chambers.

The calcareous tube or support of each animal (polyp) is the *corallum*, the wall (*ap*) is the *theca*, the longitudinal partitions (*ss*) are the *septa*. The septa radiate toward the centre and are in multiples of 4 in the forms called Tetracoralla and in multiples of 6 in the forms called Hexacoralla.

The characteristics of the Hexacoralla cup are expressed in the specimen before us, the *Favosites niagarensis;* the septa are, however, in only rudimentary condition, appearing in the fossil forms only as faint ridges or rows of spinous projections on the inside of the tubes, as in the diagram Fig. 8. The transverse partitions (see Fig. 6, *b*) are basal plates, constructed as the corallum grows upward for the animal to rest upon, and are called tabulæ or dissepiments. The Favosites are characterized by the prominent development of tabulæ, from which character the corals of this type are called tabulata. The specimen presents another character (see Fig.

6); we notice the prismatic form of the corallites, their close crowding together to form a massive colony, like a honeycomb, and the septa are rudimentary, or reduced to mere striæ on the inside of the theca, and still further we observe that the theca are perforated by minute holes, and that the tubes are horizontally partitioned off by tabulæ, making each to consist of a series of superimposed chambers. These several morphological features are characteristic of the family Favositidæ, and we say, therefore, that the family to which the specimen belongs began in the Niagara, and 21 genera are assigned to this family, all restricted to the Paleozoic time. The specimen is also a particular kind of the Favositidæ; the coral is massive, the corallites are closely approximated and sharply polygonal, mostly six-sided, the pores are regular and of definite circular form, the tabulæ are regular, of nearly equal distance apart throughout the length of each corallum, and the septa are but rudimentary pseudosepta, and twelve in number. This is a more restricted combination of morphological characters and distinguishes the genus *Favosites*. The genus is limited in range to the Paleozoic, and in the genus there are 53 species found in American rocks. Each of these species has some special mode of growth or size of corallum, or other distinguishing morphological characters, and each species is confined mainly to a single geological epoch, or, at greatest, to a single period; to the Niagara in the case of *F. niagarensis*, or to the Hamilton, as *F. dumosus*, Winchell.

Geological Range and Taxonomic Ranks of the Characters.—Thus, we may say of *Favosites niagarensis*, Hall, that its specific characters (speaking only of morphological characters, or the arrangement of matter in a particular mathematical shape) are characteristic of the geological time when the Niagara series of rocks were forming, that is, the lower part of the Silurian system, or the Eosilurian period of time. Its generic characters—viz., the massive polygonal tabulate corallites, however, have a longer range; they began in the Silurian and range through the Devonian and Carboniferous eras. Again, its family characters—viz., the perforation of the walls, one of the characters of the Favositidæ—range from a little earlier and appeared in the Ordovician era, but ceased with the Paleozoic time, and its subordinal characters—viz., the six primary septa—date back as far as the Cambrian era, and are being repeated in the generation of species living at the present time. Thus, in the case of an individual specimen of *Favosites niagarensis*, we can point to one character and say, This character continued to reappear in other individuals until the close of the Niagara era, then it ceased; of another, This character continued to reappear until the close of the Paleozoic time; and of a third character, It is still appearing in individual organisms now living in the ocean. The facts in the case may be graphically expressed by the following table:

TABLE EXPRESSING THE GEOLOGICAL RANGE OF THE CHARACTERS OF THE FOSSIL *FAVOSITES NIAGARENSIS* (HALL), ARRANGED ACCORDING TO THEIR TAXONOMIC RANK.

	C	O	S	D	Cr	T	J	K	Ty	QR
Specific characters			—							
Generic " (Favosites)			—	—	—					
Family " (Favositidæ)		—	—	—	—					
Group " (Hexacoralla)		—	—	—	—	—	—	—	—	—
Subordinal " (Madreporaria)	—	—	—	—	—	—	—	—	—	—
Ordinal " (Zoantharia)	—	—	—	—	—	—	—	—	—	—
Class " (Anthozoa)	—	—	—	—	—	—	—	—	—	—
Branch " (Cœlenterata)	—	—	—	—	—	—	—	—	—	—

Time-values of the Characters of an Individual Differ according to their Taxonomic Rank.—We learn from this analysis that any particular fossil represents a particular living animal, whose time of living was identical with that of the formation of the rock in which it was buried; also that the fossil exhibits morphological characters of various taxonomic rank, and these characters have a time-range quite of the same order as their taxonomic rank. In any particular organism, fossil or living, the characters of highest rank in classification are historically the oldest, and the characters of lowest taxonomic rank, as the specific characters, are of most recent origin and their geological range is of the shortest duration. In studying fossils, therefore, and using them as time-indicators, or studying the history of the organisms represented by them, it is all-important to notice the taxonomic rank of the morphologic characters under consideration, since it is true that the less the taxonomic value of the character the sharper and more diagnostic is its time-value.

Although the successive eras are distinguished by change in the specific and in the generic types of organisms, and it may be supposed that some of them at each era are directly descended from those of different species of a previous era, it is not so clear that the succession should present any analogy to the succession of morphologic form exhibited by the individual in its various stages of growth, as will be seen by the following considerations.

Stages of Growth in Ontogenesis.—In the growth of the individual there are certain stages called (1) infantine, or larval, (2) adolescent, (3) adult, (4) senile, which may be sharply distinguished by morphological characters, and during the life of the individual by distinct physiological operations. These stages are found by Hyatt and others to be so characteristic of the period of time in the growth as to be precisely named; Bather* has called them terms of auxology. Hyatt, in a later article,† suggests the propriety of using the term bathmology, first proposed by Cope, for this classification of the stages of individual growth. The technical names proposed by Hyatt are slightly modified by Bather, and are as follows, viz., the infantine or larval stage or form is called *embryonic* and *brephic*, the adolescent stage is called *neanic*, the adult stage is *ephebic*, the old age or senile stage of development is called *gerontic*, with a declining, *catabatic*, and an *hypostrophic* or *atavic* substage. Bather proposed the application of these terms to the temporal stages in individual development by the addition of the prefix *morpho*—thus *morphephebic*—to denote the characteristics

* *Zool. Anzeiger*, Nov. 14 and 28, 1892, pp. 420, 424.

† Proc. Boston Soc. Nat. Hist., XXVI. p. 61, etc., 1893.

of the adult; and the prefix *phyl*—thus *phylephebic*—to denote the characteristics of adulthood in racial evolution, assuming, as these authors do, that races in evolution have their characteristic stages corresponding to the stages of development of the individual. There can be no doubt that in the growth of the organism there is this general law of progressive change of form and structure with its embryonic, adolescent, adult, and senile stages, more or less distinctly marked. To this process of progressive morphological change observed in the growth of the individual the term *ontogenesis* has been applied.

No Successive Stages of Functional Activity seen in Phylogenesis.—A comparison of living forms with fossils arranged in series in the order of their sequence in the rocks (i.e., chronologically) has led to a belief that races, like individuals, have their beginning, adolescence, maturity, and old age, and the term *phylogenesis* was suggested by Hæckel to express this idea. The fact must be emphasized, however, that in individual development there is a *change of function* associated with the several stages of ontogenesis; while it is difficult if not impossible to imagine any corresponding change of function in the successive representatives of a common race, and while there are many analogies between the stages of development of ontogenesis and the stages of evolution in the history of organisms (phylogenesis), great caution is necessary not to force this theory of correspondence between the ontogenetic stages of functional activity and the order of differentiation of new characters expressed in the phylogenetic history of organisms.

Contrast between the Developmental Stages of the Individual and the Succession of Species.—The two series of phenomena present this marked contrast, that in the one (ontogenesis) each particular phase of development is a repetition of phenomena which have been repeated in the same way from the beginning of organic life; in the other (phylogenesis) each change is a step in advance of anything that has occurred before; the series is a single progressive series, with modifications and increment, but with no cycles of repetition. Development begins anew with each individual organism.

Evolution was already progressing as far back as we can find fossils, and appears to be going on still. Organic development repeats itself over and over and over again, producing cycles of changes, each one of which constitutes the life-history of an individual organism, each cycle with almost imperceptible variations, the same from generation to generation in each series; but organic evolution, although it is by slow processes, constitutes a continuous series; there are no repetitions in the series. Looked at from the point of view of our knowledge, the series had a beginning, and the evolution has been continuous since the beginning, and is not stopped to-day.

But the evolution has been an evolution entirely of form and function, not of substance. The same substance, that is, matter, has been used over and over again: the materials have preserved the same chemical and physical properties, have been temporarily built up to form new combinations, have taken organic form, have performed their function, have died and gone back to their simple condition again. As far as can be ascertained, no change has taken place in the nature of matter; what it is to-day matter has been as far back in time as science can penetrate.

Evolution an Organic Process, and not Applicable to Inorganic Things.—Thus we reach the undeniable conclusion that organisms, which fossils represent, are something unique and distinct from other things in nature. The physical constitution of matter presents no evolution. What it is, it was back into the mists of eternity. Chemical properties of matter offer no law of evolution. We interpret the chemistry of the sun, or the most distant stars, by the same tests we use in our working laboratories upon the things about us. The crystalline properties of minerals offer no evolution: the angles of a quartz crystal and the system of its crystallization in the Archæan granite are precisely those exhibited by quartz crystallizing at the present day.

Fossils furnish the Direct Evidence of Evolution.— Fossils first exhibit to us true evolution; and this evolution, which we recognize as an orderly sequence or progress of events, becomes the fundamental characteristic of organisms, and is an essential peculiarity of organic activity. Fossils not only represent organisms, but fossils alone record for us and reveal to us the actual laws of organic evolution. But the paleontologist has ever to bear in mind that he has only the records, not the living organism, for study; and he has to look to the zoology and botany of living organisms for the interpretation of his records.

Living Organisms furnish Direct Evidence of Purposeful Development.—The zoologist finds the organism to be essentially a machine accomplishing a multitude of acts, which he calls functions, because every act of the organism appears to be purposeful, the end seems to be more essential than the means, and the organism grows to be a complex structure, with a variable number of parts, each constructed with adaptation to the function to be performed. This is what is found upon analysis of the living individual; the organism is already active, performing its functions, and building or constructing parts for the fuller performance of those functions, or for performance of other functions. As the individual organism is seen in activity, the changes it undergoes, or technically its development, is seen to be definitely purposeful.

When it is compared with other organisms it is looking forward to distinct functions to be performed in the future, and when we look backward along the course of its development we see it arising in the midst of a perfected individual like itself, and it imitates in its development the very steps taken by this earlier organism. Because of this imitation, because of a repetition of what was before, we assume this ancestral model to have determined the particular form, and function too, of the newly arising individual. In all this study we find the living organism to be incessantly changing. If we make histological examinations we find every particle changing, but relative integrity and solidarity of some of the parts which perform definite functions is preserved. These parts are called organs. These organs are the parts of the machinery with which the individual works. The active, living individual is thus between two forces. The ancestry behind it determines its development, but the conditions into which it comes determine it from before, and the product is the resultant of these two forces, ancestry and environment, working together.

The soft, active organs are the chief parts of study for the zoologist; they best express the stages of ontogenetic development, but the characters of the hard parts best record the phylogenetic evolution. So long as there is plasticity in the characters themselves there is possible adjustment, but when we find a rigid resisting body formed, it expresses a permanent step taken in the evolution and established.

Fossils and Geological Biology.—Geological biology treats of the organism as a unit, with its relations to its ancestors, to its race, to time, and to environment; zoological biology treats each organism as a complex bundle of organs with their numerous functions adjusted together, but ever distinguished by their specific histological and anatomical peculiarities. The zoologist studies organs and functions as they are combined in the individual organism; the paleontologist studies varieties and species as they are combined to make up faunas and races, and as adjusted to the varying conditions of time and place. In studying a fossil he asks, not only and not chiefly, what place has it in systematic classification? but how is it related to what has gone before, and what is its ancestry? and how is the organism related to what follows, or of what is it prophetic?

These questions lead us to seek such characters as will indicate, first, *genetic affinities* and, second, *effects of environment*.

Hard Parts express both Relation to Environment and Relation to Ancestry.—For these purposes the hard parts are of the greatest value, and why? The hard parts are such as teeth; organs of offence and defence, as horns, hoofs, spines, scales, shells; and skeletons, external and internal. They represent, not the active vital part of the animal, but some part built up between the living animal and nature; hence they have an outer and an inner surface, the outer suffers degradation with use; the inner expresses the form assumed by the animal in the natural function of animal growth. Fossils are the result of growth, and hence express the final morphological result of the living individual. As hard parts they express the effects of struggle with environment more accurately than do any others, for it is with the hard parts that the animal has met environment, struggled with and resisted it; hence, *fossils, so imperfect as evidence of the anatomical structure of the organisms, are the best of evidence of the effect of the interaction between the forces of ancestry, working through the laws of generation tending to repeat the ancestral characters, and the forces of the environment working through the laws of struggle for existence in modifying those characters by adjustment.*

Kinds of Hard Parts of the Animal Kingdom preserved as Fossils.—As we deal, then, with the hard parts only, a few words will be said regarding the kind of hard parts which are found in the several classes of the Animal Kingdom.

We glance over the Animal Kingdom and see that there are large groups of animals now living, which, if they were to die and every advantage were offered for their preservation in their natural habitat, would leave no trace of their existence a year after their death. It is important, therefore, to learn at the outset to what extent the paleontological record will be found silent because of impossibility of preservation of the evidence.

Protozoa.—Among the lowest group of animals, the subkingdom Protozoa, the Gregarinidæ, found mainly within other animals, would be absent because they form no hard parts nor framework which could be preserved.

Among the Rhizopoda, differing from the former class in

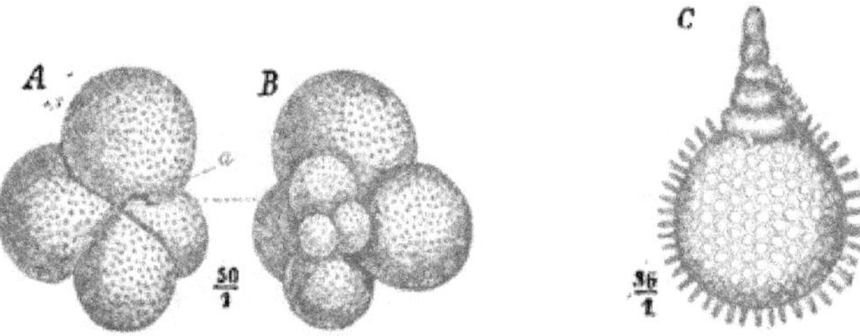

FIG. 10. FIG. 11.

FIG. 10.—Foraminifera. *Globigerina bulloides* d'Orb. Miocene. (S. and D.)
FIG. 11.—Radiolaria. *Stichocapsa Grothi* Rüst. Jurassic. (S. and D.)

the possession of pseudopoda, and leading a more active and independent life, the orders of *Monera* and *Amœba*, as far as known, do not develop any structure which would be likely to escape disintegration and resolution in the ordinary process of fossilization. But the other two orders of the class, *Foraminifera, Radiolaria*, develop hard skeletons of lime or silica, and great numbers of them are preserved in a fossil state. The Infusoria (a higher class than the others, in the possession of mouth and vibratile or contractile cilia) are not known to exist in a fossil state, though now abundant under proper conditions, and though most probably they lived in like conditions back to earliest geologic time. Figures 10, 11.

Cœlenterata.—Of the Cœlenterata the classes Spongia and

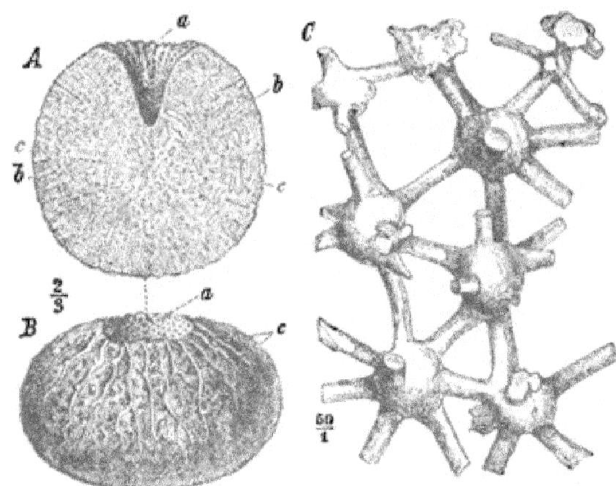

FIG. 12.—Spongia. *Astylospongia præmorsa* Gf. sp. Silurian. (S. and D.) *A*, vertical section; *B*, lateral view; *C*, silicious skeleton, greatly enlarged.

Anthozoa and the Hydroid Zoöphytes (Hydrozoa) are represented. All of the orders of the Anthozoa have families producing some hard parts, "corals," which are preserved in the rocks, but in each order there are some families not developing calcareous skeletons, hence not preserved; and in the Hydrozoa (class) several orders and a few whole subclasses (as the Lucernaridæ, Siphonophora, etc.) are of such a nature as to be wanting in any geologic record, and therefore in so far the history of the Cœlenterata is necessarily imperfect. However, Corals are among the most abundant fossils, and Graptolites (related probably to the Hydroid Polyps, or

Sertularidæ) are also abundant in a few zones in the Paleozoic rocks. Figures 12, 13, 14.

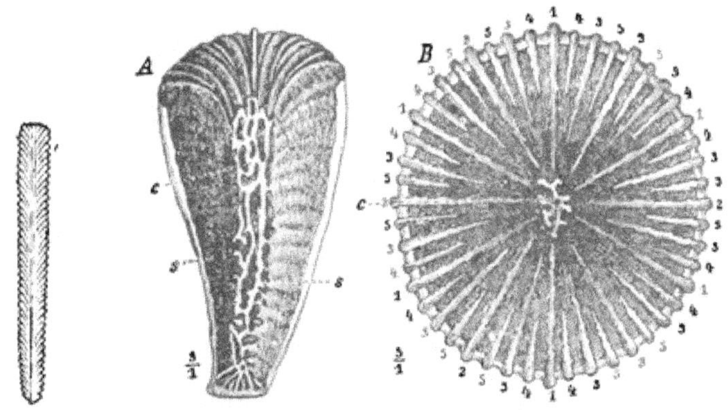

FIG. 13. FIG. 14.

FIG. 13.—Graptolite. *Diplograptus palmeus* Barr. Silurian. (S. and D.)
FIG. 14.—Coral. *Parasmilia centralis* Mant. sp. Cretaceous. *A*, corallite, longitudinally sectioned; *B*, the same seen from above; *s* and 1–5 = septa, *c* = columella.

Echinodermata were represented in fossil form, developing some hard parts in each order, viz.: Crinoidea, Blastoidea, Cystidea, Ophiuroidea, Asteroidea, Echinoidea, and even the Holothurioidea probably recognized in the spiculæ. The Solecida (parasitic worms, whether grouped with the Echinodermata, or with annelids under Vermes) are all soft, and do not come within the province of the paleontologist. Figures 15–19.

Vermes.—Among the Vermes (the leeches, earthworms, and sea-worms) there are some which produce earthy cases of mud, others have left their tracks where they bored through the tenacious mud; also teeth have been found, supposed to belong to this group. (See Serpula, Spirorbis, etc.) Still, these are rare fossils, and probably represent but very imperfectly the worms living in ancient seas. Figure 20.

Arthropoda.—Of the Arthropoda, including all those animals composed of definite segments arranged longitudinally, one behind the other, and the locomotor appendages of which are jointed or articulated to the body, we have four great classes: Crustacea, Arachnida, Myriapoda, Insecta. All of these produce a more or less enduring, horny or calcareous crust or case, within which the soft parts are contained, making the

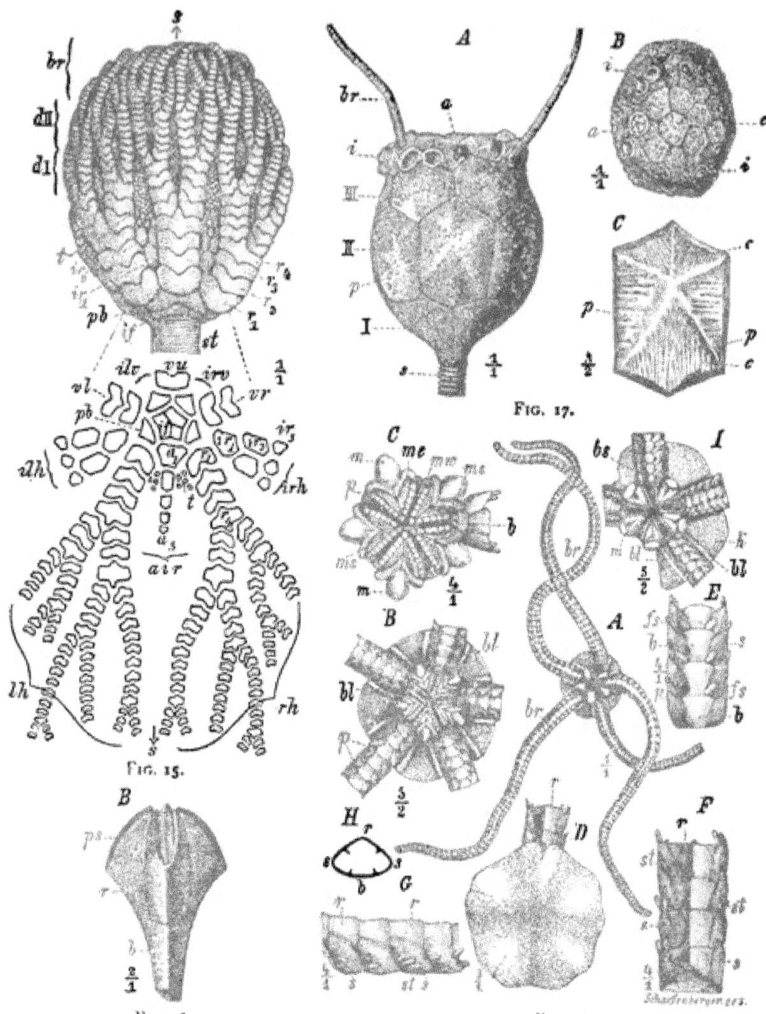

Fig. 15.—Echinodermata, Crinoid. *Taxocrinus multibrachiatus* Ly. and Cass. Carboniferous. Above: calyx with stem. Below: the plates of the calyx dissected. *st* = stem, *br* = free arms, *air* = anal interradial plates; *rh* = right posterior, *lh* = left posterior, *vr* = anterior right, *vl* = anterior left, *vu* = anterior medial radial plates; *irh* = right posterior, *ilh* = left posterior, *irv* = right anterior, *ilv* = left anterior interradial plates; *if*, infrabasalia; *pb* = parabasalia; r_1-r_4 = radialia; *d*I, *d*II = distichalia, first and second rank; *br* = brachialia; a_1-a_5 = anal plates; ir_1-ir_3 = larger interradial plates; *t* = smaller interradial plates; *ss*, plane of symmetry. (After S. and D.)

Fig. 16.—Blastoid. *Pentatremitidea Eifeliensis* F. Rö. sp. Devonian. *b* = basalia; *r* = radialia; *p* = ambulacra. (After S. and D.)

Fig. 17.—Cystoid. *Caryocrinus ornatus* Say. Silurian. *A*, calyx with stem *s*; *br* = arms; I, II, III, first, second, and third zones of plates of the dorsal capsule; *p* = porous plates; *i* = place of attachment of arms; *a* = anal opening. *B*, view of the ventral dome, *c* = central summit plates, *i* and *a* as above. *C*, inner surface of a plate of the second row, showing the pores of the Hydrospires (*p*) and their connecting canal (*c*). (After S. and D.)

Fig. 18.—Ophiuroid, A-H. *Ophioceramis ferruginea* Böhm. Jurassic. *A*, a complete specimen, from under side; *br* = arms. *B* = the disk from the inner side; *bl* = bursal shields; *p* = ambulacral pores. *C*, mouth-skeleton from below; *mw* = mouth-angle; *p* = papillæ; *me* = angle-plates; *ms* = side shields; *m* = oral plates; *b* = second ventral plate. *D*, disk from above; *r* = dorsal shield. *E*, a part of an arm from below; *b* = ventral shields; *s* = lateral shields; *p* = ambulacral pores. *F*, a part of an arm from above; *r* and *s* as above; *st* = spines. *G*, the same, lateral view. *H*, cross-section of an arm. *I*, *Geocoma planata* Qu. sp.; *bs* = bursal slit; *bl* = bursal shield. (Steinmann and Doederlein.)

possibility of fossil remains. But, except in the case of

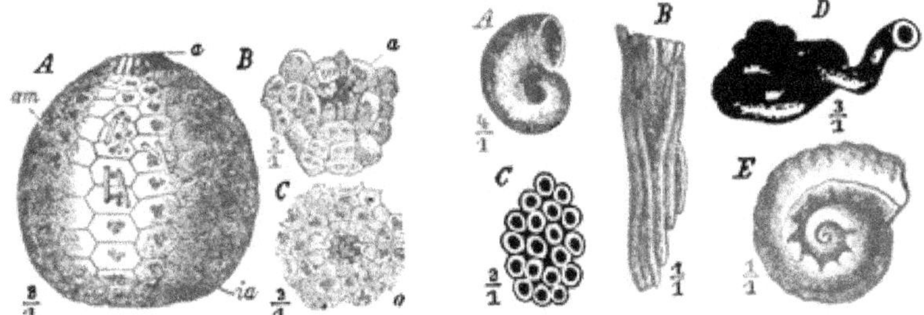

FIG. 19. FIG. 20.

FIG. 19.—Echinoid. *Bothriocidaris Pahleni* Schm. Ordovician. *A*, side view; *a* = anal opening; *am* = paired ambulacral plates with two double rows of pores and small spines; *ia* = single row of interambulacral plates. *B*, summit region with the anal opening (*a*). *C*, under side, with the mouth opening.

FIG. 20.—Vermes, Annelida. *Serpula*. *A*, *S.* (*Spirorbis*) *omphalodes* Gf. Devonian. *B*, *C*, *S.* (*Galeolaria*) *socialis* Gf. Jurassic. *C*, cross-section of the tubes. *D*, *Serpula gordialis* Schl. Cretaceous. *E*, *S.* (*Rotularia*) *spirulæa* Lam. Tertiary.

Crustacea, it will be observed that the animals belonging to these classes live mainly on land and in the air, and when we bear in mind that fossilization is a process usually requiring water for the preparation of the matrix (sand, mud, gravel, etc.), and for the covering of the body with the material when prepared, it is evident that all land and aërial animals, although possessing parts capable of fossilization, and living in abundance, run very small chance of being found in the deposits made under water, in which fossils are mainly preserved. Hence Crustacea, being water animals, are preserved as fossils in considerable numbers, while the other classes of Arthropoda, that is, insects, spiders, and Myriapods, although occasionally found, are rare, and prob-

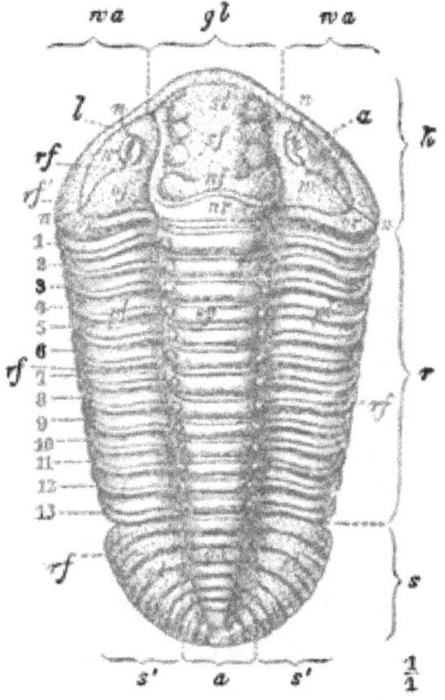

FIG. 21.—Arthropod, Crustacean. Trilobite, *Calymene Blumenbachi* Brgt. Silurian. *k* = cephalic shield; *r* = thorax; *s* = pygidum; *gl* = glabella; *wa* = cheeks; *wa'* = free part of the cheek; *n* = facial suture; *l* = border; *a* = eyes; *st* = frontal lobe; *sf* = lateral furrows; *nf* = neck furrow; *of* = occipital furrow; *nr* = neck-lobe; *or* = occipital ring; *rf* = dorsal furrow; *rf* = marginal furrow; *sp* = axis; *pl* = pleuræ; *a*, *ax* = axis; *s'*, *sl* = lateral lobes of the pygidum; 1–13 = the 13 thoracic segments.

ably represent in only the most meagre way the forms of these classes which lived in past ages. Figures 21, 22.

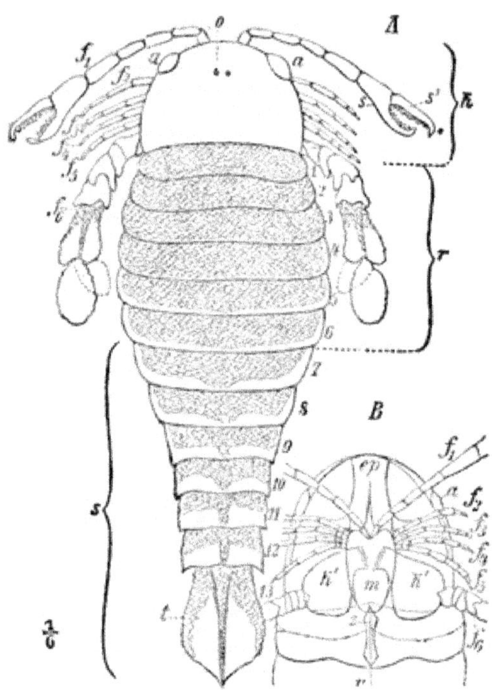

FIG. 22.—Arthropod. *A*, *Pterygotus anglicus* Ag. Devonian. Dorsal view. *B*, *Pt. osiliensis* Schm. Silurian. Under side of head. *k* = cephalic shield; *r* = thorax; *s* = abdomen; *a* = eyes; *o* = eyelets; f_1-f_6 = cephalic appendages; 1-6 = thoracic segments; 7-13 = abdominal segments; *t* = terminal segment or telson; *ep* = epistoma; *kl* = masticating plates of the sixth pair of appendages; *m* = metastoma; *z* = median plates; *n* = median suture.

Molluscoidea.—The Molluscoidea, including the Polyzoa and the Brachiopoda, is a group of much interest to the Paleontologist. The Brachiopods are well preserved, and are, perhaps, from the point of view of the scientific paleontologist, the most important group of animals he is able to study. Of their history, the record is more complete, the condition, as a whole, more perfectly preserved, the missing links fewer, than for any other group. They have been studied more thoroughly, are of greater value as marking geological horizons, probably, than any other. They develop a chitonous or calcareous bivalved shell, the external and internal form of which, and the intimate structure of the shell substance, are generally well preserved. Figures 23, 24.

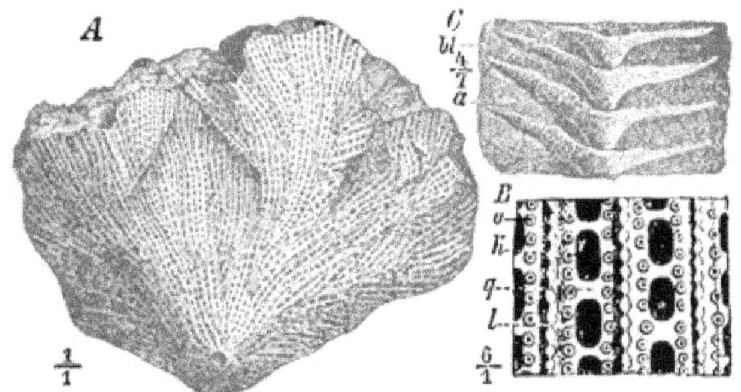

FIG. 23.—Molluscoida, Bryozoa. *A*, *B*, *Fenestella retiformis* Schl, Permian. *A*, a funnel-shaped stock from the outside. *B*, enlarged view showing the cell mouths (*o*) and the perforations (*l*) between the cell-rows of the cœnœcium. *C*, *Archimedes wortheni* Hall. Carboniferous. The stock consists of a broad cœnœcium (*bl*), wound spirally about a central axis (*a*). Fragments of the cœnœcium separate from the axis present a structure similar to that of Fenestella.

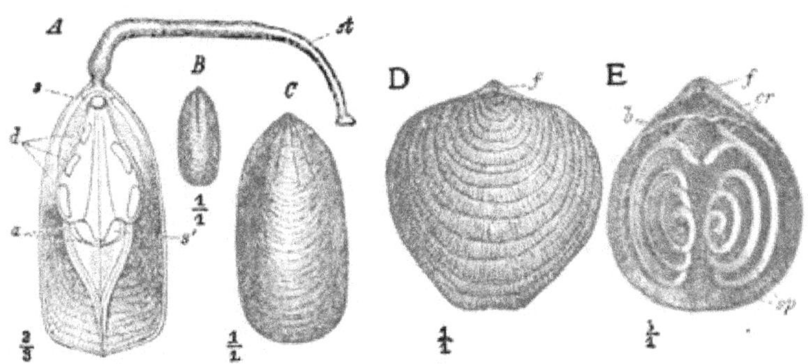

FIG. 24.—Molluscoida, Brachiopoda. *A*, *B*, *C*, Inarticulata. *Lingula*. *A*, *L*, *anatina* Brug., living, pedicle valve from within. *st* = pedicle; *s*, *d*, *a*, *s'*, muscular impressions. *B*, *L. tenuissima* Br. Triassic. *C*, *L. Beani* Phill. Jurassic. *D*, *E*, Brachiopoda articulata. *Atrypa reticularis* L. sp. Devonian. *D*, surface view of brachial valve. *E*, view of interior, the brachial valve being in great part removed; *f* = foramen for passage of the pedicle; *cr* = crura; *b* = jugal processes or jugum; *sp* = spires or spiral coils of the brachidium.

Mollusca.—Of the true Mollusks, all the four classes, Lamellibranchiata, Gastropoda, Pteropoda, Cephalopoda, construct, in most of their genera, calcareous or horny shells, external or internal, which are preserved, more or less perfectly, in a fossil state. Gastropods and Lamellibranchiates in the older rocks are very apt to be in the condition of impressions and moulds, the substance of the shell being dissolved and carried away; this is also the case with many families of the other two classes, so that very much is wanting to a complete record of these classes. Figures 25, 26, 27.

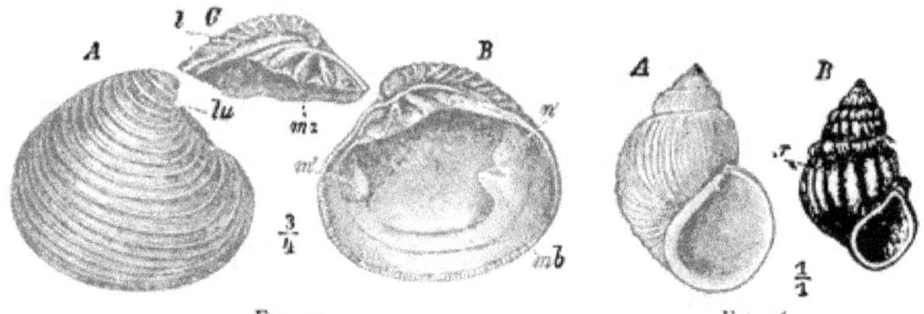

Fig. 25. Fig. 26.

Fig. 25.—Mollusca, Lamellibranch. *Venus multilamella* Lmk. Tertiary. *A*, a right valve, outer surface; *lu* = lunula. *B*, the same interior. *C*, hinge of the left valve. *m'* = anterior, *m''* = posterior, muscular impressions; *mb* = pallial sinus; *l* = ligamental pit; *mz* = cardinal teeth.

Fig. 26.—Gastropod. *A*, *Paludina pachystoma* Sdb. Tertiary Miocene. *B*, *P. avellana* Neum. Pliocene.

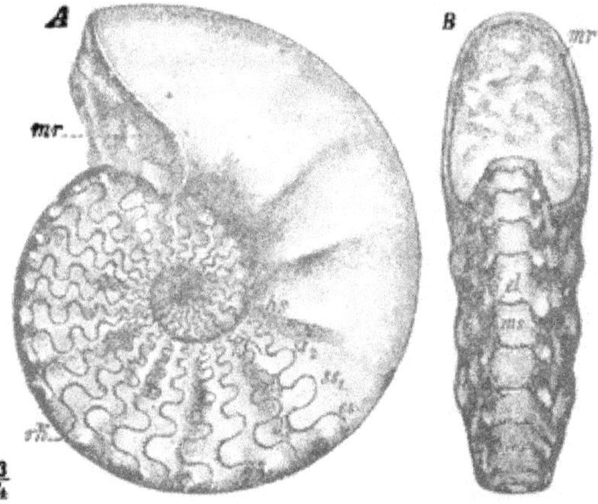

Fig. 27.—Cephalopod. *Ceratites nodosus* d. Haan. Triassic. *A*, complete shell from the side; *B*, front view of the same; *mr* = rim of the outer chamber; *ss*, ss_1, ss_2, *hs* = saddles of the sutures; *el*, sl_1, sl_2, *hl* = serrate lobes of the suture lines.

Vertebrata.—Of this branch there is scarcely an order that does not develop hard parts of some kind, which might be preserved in fossil condition under favorable circumstances. Among the lowest orders (Lancelot, Hag-fish, Lampreys) there is nothing likely to be preserved, except small teeth. In the cartilaginous fishes teeth are the main parts of sufficient hardness to resist decay and disintegration, while the bones and scales of other fishes are hard and enduring if well buried under water, but are easily destroyed if left exposed in contact with the atmosphere for a long time. So again, while many fish and reptiles and a few mammals are inhabitants of the ocean, birds and most mammals and many reptiles are inhabitants of land, and many fish and reptiles are fresh-water species only. Again, the remains of Vertebrates are subject to the destructive agency of lower animals and of themselves, so that it is not to be supposed that under the most favorable natural conditions

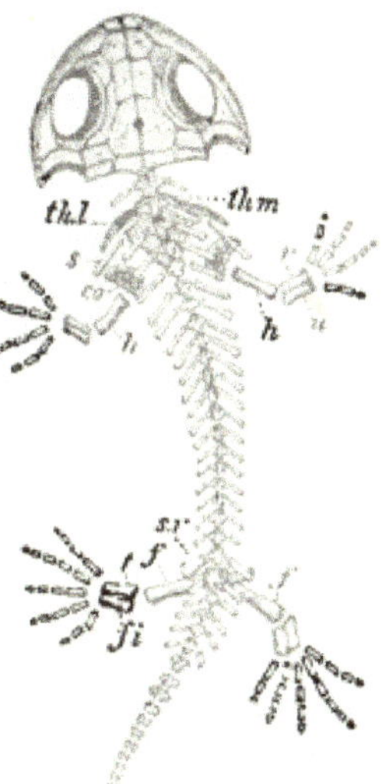

FIG. 29.—Vertebrate, Amphibian. *Brachiosaurus amblystomus* Credn. A young form (*B. gracilis* Credner). Triassic. *co* = coracoid; *f* = femur; *fi* = fibula; *h* = humerus; *r* = radius; *s* = scapula; *sr* = sacral rib; *t* = tibia; *th.l* = lateral and *th.m* = medial thoracic plates; *u* = ulna.

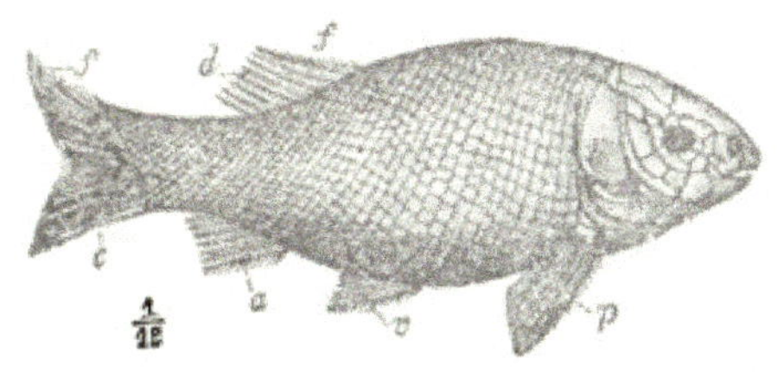

FIG. 28.—Vertebrate. Fish. *Lepidotus elvensis* Blv. Jurassic. *a* = anal fin; *c* = caudal fin, hemi-heterocercal; *d* = dorsal fin; *p* = pectoral fin; *v* = ventral fin; *f* = fulcra (on the front edge of all the fins); *k* = gill-covers.

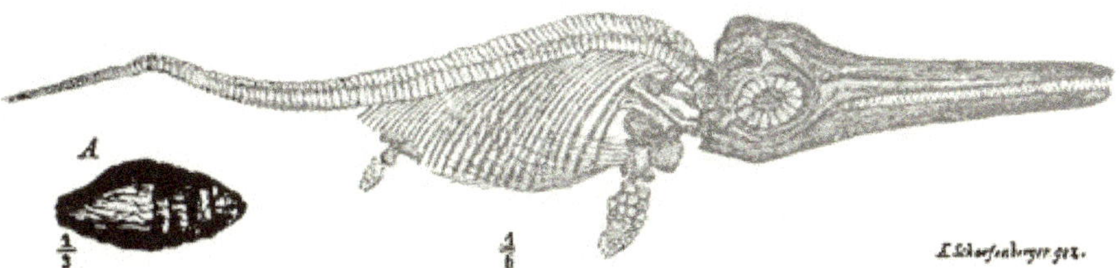

FIG. 30.—Vertebrate, Reptile. *Ichthyosaurus quadriscissus* Quenst. Jurassic. Skeleton of a young individual *A* = coprolite. (After Steinmann and Doederlein.)

anything more than the most meagre representation of the vertebrate life of the world would be preserved in fossil condition, and of those preserved, the more abundant would be reptiles, fishes, and larger mammals, with a few birds. (Figures 28–32.)

Looking over the Animal Kingdom, in this general way,

FIG. 31.—Vertebrate. Saurura. *Archæopteryx macrura* Owen. Jurassic. Restored in the position of the Berlin specimen. *c* = carpus; *cl* = clavicula?; *co* = coracoid; *h* = humerus; *r* = radius; *s* = scapula; *u* = ulna; I-IV = 1-4 fingers. (After Steinmann and Doederlein.)

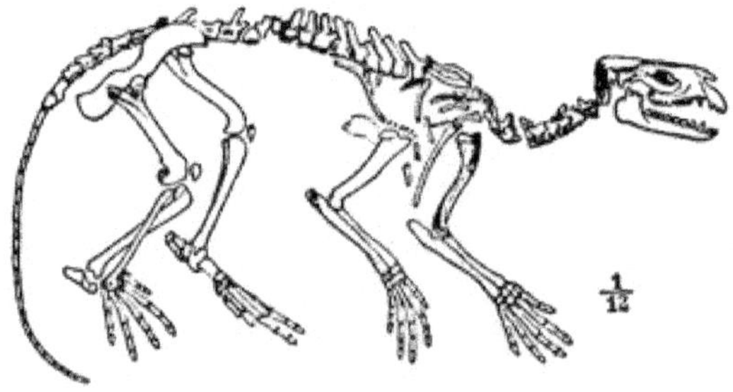

FIG. 32.—Vertebrate, Mammal. *Phenacodus Wortmanni* Cope. Eocene.

we find a few classes among the several subkingdoms, producing parts which could be preserved as fossils, but there are reasons why even these are not present in abundance except for a very few orders; the rest may be represented by here and there a specimen, but only rarely, and any conclusions drawn from their study will be conjectural to the extreme. In the study of the laws of organic history it becomes necessary, therefore, to make judicious selection of those classes of organisms whose records are sufficiently abundant and continuous to furnish the desired evidence.

Summary.—To summarize: When we study fossils in their simple physical aspect, as mathematical forms in the rocks, we find them presenting an orderly arrangement of sequence, one after the other, in strict chronological order. When classified by their likeness to each other into groups to form natural species and genera, and when separated from each other by their points of difference to form separate families, orders, and classes, we find that there is the closest relationship existing between the form they assume and the periods of time when they lived. Taking a single suborder of the Cœlenterata (the stony corals, or Madreporaria, with 448 known genera), of which fossil remains are found all the way along, from the earliest fossil-bearing rocks to the sea-shores of our modern ocean, we find all the genera relatively short-lived, rarely exceeding the period of two systems in length of duration, and the genera most nearly allied to each other in form are always found in the systems chronologically nearer to each other; and uniting the similar genera into families, the families presenting greater contrast are found farther separated chronologically from each other than from the families presenting less strong contrasts. When we carry our study further and interpret these fossils as the remains of organisms, and say that they represent living organisms, we come face to face with the fundamental law of organisms, that is, the law of change and variation. All organisms have a history. So unchangeable are the physical properties of matter, so invariable are the laws of crystallization of minerals, and so constant are the chemical properties of substances, that any irregularity in any of them at once suggests the influence of organisms. This fact is apparent to every one, and it is no new discovery in this age. Naturalists have for centuries known that animals and plants grew, and have been seeking for some mysterious living property by which to distinguish living form and living matter: they have been examining dead organisms, they have described hundreds of thousands of different organisms, looking always for, and in most cases only grasping, the dead products of life: they have examined the organic mechanism and observed its mode of action and the results attained: but it is only recently that, under the names *development* and *evolution* the fundamental characteristic of all vital phenomena has become an object of serious study and investigation. The morphological relations of organisms have been thoroughly studied, but their time-relations have only begun to be scientifically investigated.

CHAPTER VI.

GEOGRAPHICAL DISTRIBUTION—THE GENERAL RELATION OF ORGANISMS TO THE CONDITIONS OF ENVIRONMENT.

In the last chapter it was shown by an analysis of the characters of the genera of Madreporaria—a group of organisms well adapted to furnish this evidence (because of their living under the same conditions required for the making of the strata themselves, and producing hard parts, easily preserved from the earliest times onward)—that the form of an organism has an intimate relationship to the geological period during which it lived.

The natural conclusion from this observation is that the order of sequence in the appearance of organisms is the expression of a natural law of their succession in time, or that it is a law of nature for organisms to succeed each other in this observed geological order.

We observed that the classification of organisms by their morphological characters, as expressed in their arrangement in the classes, orders, families, and genera of the zoologist, shows that this relation of characters to time of appearance is expressed in every detail of structure, and the more minute our inspection the more distinctly is the truth of this principle brought to light.

A species or genus has not only a particular relationship to other species or genera, but every genus has a particular period in the time-scale when it lived, and a particular duration of geological time to which its living was limited, before which it did not exist, and after which it failed to reappear. This illustrates the general law that *the particular morphological characters assumed by an individual organism are immediately related to the ancestry which is behind it;* but if we turn our attention to the facts of geographical distribution, we shall find that organisms present as close a relationship to the conditions of the environment into which they are born.

The Importance of the Study of Geographical Distribution.—Geographical Distribution is a subject which no one has studied more thoroughly and with keener appreciation than Alfred R. Wallace, and a quotation will, in a few words, express the importance of the subject. He says: "So long as each species of organism was supposed to have had an independent origin, the place it occupied on the earth's surface, or the epoch when it first appeared, had little significance. It was, indeed, perceived that the organization and constitution of each animal or plant must be adapted to the physical conditions in which it was placed; but this consideration only accounted for a few of the broader features of distribution, while the great body of the facts, their countless anomalies and curious details, remained wholly inexplicable; but the theory of evolution and gradual development of organic forms by descent and variation (some form of which is now universally accepted by men of science) completely changes the aspect of the question, and invests the facts of distribution with special importance." "The *time* when a group or a species first appeared, the *place* of its origin, and the area it now occupies upon the earth become essential portions of the history of the universe. The course of study, initiated and so largely developed by Darwin, has now shown us the marvellous interdependence of every part of nature. Not only is each organism necessarily related to and affected by all things, living and dead, that surround it, but every detail of form and structure, of color, food, and habits, must, it is now held, have been developed in harmony with, and to a great extent as a result of, the organic and inorganic environments. Distribution becomes, therefore, as essential a part of the science of life as anatomy or physiology. It shows us, as it were, the form and structure of the life of the world considered as one vast organism, and it enables us to comprehend, however imperfectly, the processes of development and variation during past ages which have resulted in the actual state of things. It thus affords one of the best tests of the truth of our theories of development [evolution]; because the countless facts presented by the distribution of living things, in present and past time, must be explicable in accordance with any true theory, or, at least, never directly contradict it." *

In studying the geographical distribution of organisms the understanding of the nature of the conditions of environment can scarcely be overestimated.

The Natural Conditions of Environment—Nomenclature.—There are various conditions of environment which modify the growth and life of organisms. Among the chief of these are: The (1) *medium*—air or water; (2) *temperature*—or climate and limits of annual temperature; (3) *in water*—the depth, the purity, the salinity, the light, the motion; (4) *on land*—secondarily altitude as affecting climate and temperature; (5) *the other organisms*, because all animal life appears to require other animal or plant organisms for its own food, hence (5*a*) *struggle for existence;* also (5*b*) the amount of organic food determines the growth of higher organisms which require the food. *Medium* and *Habitat* are the names applied to the immediate conditions in which the organisms live. *Province* is the name of the region occupied by a group of organisms which are naturally adjusted to each other. *Zone* is the name of the tract of sea-bed between boundaries of depth, variously determined. *Flora* is the name applied to all the plants, naturally associated and adjusted to the conditions of environment, of a particular province or geographical area. *Fauna* is the name of the group of animals so associated and adjusted.

Natural-history Provinces.—The primary classification of the conditions of environment as affecting organisms is considered under the terms *Terrestrial* (land plants and animals), and *Marine* (those living in the ocean). It is found that the present life of the globe is divided into numerous floras and faunas, the boundaries of which are not absolutely fixed, either in species or in conditions; but the areas are distinct in some of their features, and the association of organisms is peculiar for each, although some of them may be common to neighboring areas. These provinces, both marine and terrestrial, differ in their outlines for different kinds of

---

* Article "Distribution," 9th ed. Encyclo. Brit., vol. VII. p. 267.

organisms. To distinguish them they are called *Natural-history Provinces.* We say, for instance, that the natural-history province marking the distribution of flowering plants, differs in its boundaries from the province marking the distribution of fresh-water Mollusca. The reason is apparent when we note that the limiting cause of the distribution is perhaps temperature and climate in one case, and community of fresh-water channels in the other. The boundary of the water-bed of a great river-system is the limiting cause of the distribution of the Mollusca, the conditions of temperature and rainfall that of the plants.

Normal Adaptation to Conditions of Environment.—We have spoken of distribution as applied to organisms. This term implies that each organism is normally adapted to a certain set of conditions, which is called by the general name *Environment.* Within limits the individual adjusts itself to slight change of the environment, but extreme change of the conditions of environment restricts the possible living of the particular organism, and for each particular organism the distribution is supposed to mark the particular extent of differing conditions in which it is normally adapted to live.

Specific Centre of Distribution and Varieties.—Theoretically, each organism is supposed to be qualified to live under a certain set of conditions, and to adapt itself to change of those conditions to a greater or less extent. While geologists do not find a species to be determined rigidly by any one criterion, general usage applies the name *Species* to those plants or animals which possess common morphological characters, and are confined in their distribution to one natural-history province (but taking this as a general definition, exceptions are recognized in the case of species distributed over two or many provinces). Practically, too, each species appears to have a *centre of distribution,* at which point (or specific centre) the combination of environing conditions are the more favorable; the species may be distributed from this centre, but it is not so abundant outside, and is often seen to present slight differences of form, size, color, or minor differences on the outskirts of the province of its distribution. These differences from the typical form at the centre constitute what are called *Varieties* of the species. The conditions of environment existing at a specific centre, or metropolis of the species, as Forbes called it, constitute the *normal habitat* for that species. The particular morphological and structural characters which the species express are called its *typical specific characters.* The modifications from these typical characters which are seen in representatives of the species on the borders of its specific distribution are its *varietal characters.*

The Distinctness of the Flora and Fauna of Distinct Provinces.—The species associated together in a natural-history province are the *flora* and *fauna* of that province, and as generally defined, not over one half of the species of two distinct provinces are identical; or, to put it in the converse form, about half the species of any province are distinct, or peculiar to that province. Such a rule is purely arbitrary, and will vary greatly as applied by different naturalists, but such a general rule is applied in the distinguishment of the provinces of marine species.

The Various Classifications of Natural-history Provinces.—In the classification of provinces in Woodward's "Manual of Mollusca" we find eighteen such marine provinces recognized, and the land regions are defined under twenty-seven names. Sclater (1857) defined six terrestrial regions, which were afterwards adopted by Wallace (1876)* and subdivided into twenty-four sub-regions. Fischer (1887) combined and extended the former classification, and defined thirty regions distributed in the following seven zones, viz.: Palearctic, African paleotropical, Eastern paleotropical, Australian, Neantarctic, Neotropical, Nearctic. Each of these regions is again subdivided into sub-regions with their special faunas; as, for example, the region Circamediterranean is the second of the Palearctic regions; this is subdivided into the sub-regions (*a*) occidental or Atlantic, (*b*) Meridional or Mediterranean (with the four Faunas, Hispano-Barbaresque, Egypto-Syrienne, Hellado-Anatolique, and Italo-Dalmate), (*c*) Centrale or Pontique, (*d*) Orientale or Caspique.†

* A. R. Wallace, "The Geographical Distribution of Animals," 1876.

† Paul Fischer, "Manual de Conchyliologie," Paris, 1887.

Marine Organisms Particularly Important to the Paleontologist.—Because of the fact that preservation of fossils is almost entirely dependent upon the covering by water of the remains preserved, the questions of distribution and environment of chief interest to the paleontologists are those of marine and fresh waters.

Haeckel's Classification of the Marine Conditions of Life.—Walther, in his "Bionomy of the Sea," presents a classification of organisms according to their bionomic character, as follows: The sum of the marine faunas and floras is called Halobios, corresponding to them the fresh-water life is called Limnobios, and the land organisms receive the name Geobios. The Halobios, or marine organisms, are further classified into (1) Benthos—those animals and plants living on the sea-bottom, distinguished further as (*a*) sessile, (*b*) vagile, (*c*) littoral, and (*d*) abyssal Benthos; (2) Nekton, or the life of open sea, with strong powers of active locomotion; and (3) Plankton, the more or less passive life of open seas. Haeckel (from whom Walther adopts the nomenclature) further subdivides the Plankton, or open-sea life, into the following five groups: The neritic Plankton includes the swimming flora and fauna of the coast regions of continents, archipelagos, and islands; the oceanic Plankton includes the swimming flora and fauna whose habitat is the open ocean; the pelagic Plankton inhabits the ocean surface and approximately 200 metres below; the bathybic Plankton inhabits the waters from the bottom for about 100 metres up, and between the latter two lives the zonaric Plankton.*

Walther's Further Analysis of Conditions of Environment.—Walther has amassed very interesting statistics to show the particular influence upon distribution of the various conditions of light, of temperature, of salinity, of tides and waves, of currents and ocean circulation, and has classified the floras and faunas of the seas in relation to these conditions.

The flora of the shores, littoral flora, are divided into that of the (1) dune and sand-plain zone, (2) flora of coast rocks, (3) of the mud zone, (4) the sand-plants flora. Four different

*Walther, "Einleitung in die Geologie als historische Wissenschaft, I. Theil, Bionomie des Meeres," 1893, pages 16–22; also Haeckel, "Planktonstudien," Jena, 1890, page 18, etc.

zones of coast vegetation are recognized in the tropics by Schimper: the Pescapræ formation, the Barringtonia formation, the Nipa formation, and the Mangrove formation.

The fauna of the coast is also determined in its composition by the conditions of the shore itself, and thus we find different kinds of animals associated with the rock beach, the bowlder beach, the pebble beach, the sand beach, and the mud beach.

Under the sea surface downward a number of zones have been distinguished, defined most easily by their depth, which present strong contrast in their faunas. We owe it to Edward Forbes that we have a nomenclature for these zones of depth. The divisions made by him are the littoral zone, the laminarian, coralline, and the deep-sea zones; the latter, as the result of deep-sea dredgings, has been divided into the zone of deep-sea corals, or brachiopod zone, and the abyssal zone.

In a Report of investigations made upon the faunas of the seas off the New England coast, Professors Verrill and Smith found it to be a fact "that there are in the waters of this region three quite distinct assemblages of animal life, which are dependent upon and limited by definite physical conditions of the waters which they inhabit."* These are described under the following divisions, viz.:

1. The fauna of bays and sounds;
2. The fauna of the estuaries and other brackish waters;
3. The fauna of the cold waters of the ocean shores and outer banks and channels.

This classification of environments is not bathymetric, but is chiefly on the basis of temperature and purity of the waters.

It is altogether probable that every kind of difference in the environment, which could be described as beneficial or otherwise to the vital functions of organisms, is also represented by greater or less adaptation of the organization, to profit by the favorable conditions or to avoid the evil effects of those which are unfavorable.

Relations of Organisms to Time and to Environment Equally Significant.—When we consider alone the historical relations

* United States Fish Commission, "Report upon the Invertebrate Animals of Vineyard Sound and Adjacent Waters," p. 5, etc.

of the organisms, as expressed in their geological sequence, the order of the phenomena appears like a mere unfolding of successive phases of organic life upon the globe, each phase preparing the way for the next; and had we no preconceptions, I think this evolution would seem to be the most natural thing in the world.

Gradual modification with each step of generation would be found in each case the sufficient explanation and cause of that which followed.

But when it is observed that each living organism is closely adjusted to a particular set of environmental conditions, and that specific organic form and specific conditions are closely co-ordinate factors, the question as to the influence exerted by environment upon the organism becomes a problem of equal importance.

An Explanation required for Succession of Species as well as for Adjustment of Species.—The study of the relations of organisms to geological time and to geographical space first brings out the simple fact that differentiation of organic form is actually related to both. There is an adjustment of the organism to each of the phenomena, time-succession and place-extension. If we turn from this simple statement of fact to seek for some reason why organisms differ in form, and why one organism has one form and another organism of another time and place differs from it, then there appears back of geological succession and of geographical distribution an element of causation. There are conditions in the succession and in the distribution which we may suppose have been the cause, or at least the occasion, of the changes of form exhibited by the organism.

Evolution and Adaptation both observed Facts.—We have already remarked that the examination of a series of forms in the rocks shows the modification and change in their form to be co-ordinate with progress of time, and on following them from the lowest rocks upward through the geological column to the present, each series ends in recognized living organisms: hence we conclude that it is a characteristic of organisms to pass through continuous change in time. This process of changing morphological characters, expressed in the history of organisms, is called *Evolution*.

Second, organisms now living are so distributed in relation to the conditions of environment that we are led to recognize this general law: that the morphological characters of organisms are in some way associated with or related to the physical environment in which they live.

Ancestry and Environment as Causes of Evolution.—Thus, by looking only at the superficial relation of organisms, i.e. those which may be expressed in number and ratio, we find a definite relationship existing between organic form (morphology) and both geological time and physical conditions on the earth's surface. We may express the relationship by the proposition, *that the morphological characters of any particular organism have come to be what they are through the operation of two sets of conditions: first, the organic conditions which were antecedent to the appearance of the given organism; and, secondly, the external physical conditions into which it was born.* The first set of conditions is expressed by the general term *Ancestry*, and the second by the term *Environment*.

Differences of Opinion respecting Interpretations not Facts.—So long as we confine our attention to the simple relationship existing between organic structure and the passage of time or the varying conditions of environment, we have touched only the fundamental facts of the real problem before us.

The series of correlated phenomena are as they are, whatever be our interpretation of them. The reason for first carefully spreading out the facts themselves is in order to show that they are not invented by any theory, that they exist independently of any preconceived view, and that the differences in opinions regarding them are not matters of observation, but are matters of philosophy.

Introduction of Causation into the Discussion.—And here we introduce a new element into the discussion. We assume that cause and effect are involved in their relationship. We assume that in the course of time the organisms which went before must bear the relation of determining cause to those that follow, and that in physical space or environment, the conditions of geographical locality are a determining cause in relation to the species adjusted to particular natural-history provinces.

Ancestry and Environment in Relation to the Beginning of Each Individual.—From this point of view we recognize two classes of phenomena which are all-important factors in determining the particular form and structure of every organism, and the fundamental difference between the two groups is found in the relation they bear to the beginning of the development of each individual. The one set of conditions have exerted their effect when the first germ of the new individual arises, and to them is applied the general name *Ancestry*. The other set begin to influence the individual only after development has begun, and to this set of conditions the general term *Environment* is applied. *Evolution* is the name given to the results, in structure and function of organisms, which are traced to Ancestry and Environment as determining causes.

It is from this philosophical point of view that the following definitions become appropriate:

Definition of the Terms "Ancestry" and "Conditions of Environment."—*Ancestry*, as defined in the Century Dictionary, is "the series of ancestors, or ancestral types, through which an organized being may have come to be what it is in the process of Evolution:" and in the same work the term *conditions of environment* is defined as "the sum of the agencies and influences which affect an organism from without; the totality of the extrinsic conditioning to which an organism is subjected, as opposed to its own intrinsic forces, and therefore as modifying its inherent tendencies, and as a factor in determining the final result of organization. It is an expression much used in connection with modern theories of evolution in explaining that at a given moment a given organism is the resultant of both intrinsic and extrinsic forces, the latter being its conditions of environment and the former its inherited conditions."* Ancestry and Environment are, in the abstract, names for these intrinsic and extrinsic factors of evolution.

If we examine only the paleontological series, we might conclude that the course of evolution was determined entirely by the first set of conditions, Ancestry; and, on the other

* Century Dictionary, vol. I. pp. 201 and 958.

hand, if we look alone to the relations of organisms to environment, this set of conditions appears sufficient to account for the course of evolution, because in both cases we find adjustment of morphological character to the conditions pre-existing at the beginning of each individual life.

Two Factors Producing the Effects of Evolution.—Assuming these definitions to be formulations of the truth in the case to such a degree of accuracy that they may be adopted as working hypotheses, the next step in our analysis is to ascertain what part each of these factors plays in bringing about differentiation of organic form and structure.

Three Views Possible.—There is practically but one of three opinions to take in the matter: either (1) the differences observed among organisms are accounted for entirely by ancestry—that is, the potency of all organic differentiation and evolution is found in the ancestry at any particular moment of the process; or (2) environment is the efficient factor in bringing about all modification of organic structure; or (3) the actual course of evolution as it takes place is the resultant of the co-operation and antagonistic action of both factors.

The extreme old school (of Cuvier, for instance) adopted the first opinion, the extreme natural selection or Darwinian school holds substantially the second view. It is believed by the author that the truth will be found in the third position.

First Cause of some sort Essential to any complete Theory of Evolution.—The discussion of evolution has for the past fifty years chiefly centred about the theory of the origination of species. Ancestry, in the general sense here used, includes all the antecedent intrinsic conditions of an individual life. When we analyze the theories to their ultimate essence the great contrast between Creationism and Evolutionism does not lie in the fact that the one acknowledges God to be the first cause or ultimate ancestor of every living thing, while the other, in magnifying the agency of the environment in controlling the origin of species, denies all first cause: for, in both cases, some pre-existing power or potency that is quite godlike must be assumed as the necessary antecedent to the phenomenal appearance of organisms in all their variety upon the earth.

Edward Forbes on Origin of Species and Centres of Creation.—When we ask how did species arise, we find two dominant opinions have existed regarding the nature of the antecedent condition immediately preceding the individual organism in each case. According to the first view, immediate physical ancestry has explained only the repetition and perpetuation of its own morphological characters, and the origin of any particular combination of such morphological characters was not accounted for, except through the agency of a primitive first cause. The sequence of organisms in paleontology was clearly recognized by naturalists at the beginning of the century, but neither ancestry nor environment was deemed competent to explain anything but what were called varietal modifications of species. It was this idea that was in the mind of Edward Forbes* when he described a natural-history province to be "an area within which there is evidence of the special manifestations of the creative power; that is to say, within which there have been called into being the original or protoplasts of animals or plants." And again he says: "The diffusion of the individuals of the characteristic species of a province is found to indicate that the manifestation of the creative energy has not been equal in all parts of the area, but that in some portion of it, and that usually more or less central, the genesis of new beings has been more intensely exerted than elsewhere." This notion led to the use of the terms *centres of creation* and *specific centres*, at which the species was supposed to have originated, and from which it was distributed, or migrated in the course of time.

Reality of Specific Centres Not Questioned; the Fact Variously Interpreted.—It is a well-known fact, and one that Forbes clearly understood, that each natural-history province is such a specific centre for rarely more than one species of each genus of its fauna; or, in other words, each well-defined species is typically developed in some such specific centre and distributed within such a natural-history province. The specific centre may not be geographic. Geography, in general, is the most commonly observed criterion of distribution

---

* Edward Forbes, "The Natural History of the European Seas," 1859.

of organisms, but in the case of insects the root and leaf of the same tree present greater contrasts in conditions of environment than two trees of the same species a thousand miles apart. Geographical distribution, and other terms associated with it, have reference fundamentally to conditions of environment, whether the distribution is on geographical or other lines.

Representative Species, Common Descent, and Migration of Species.—Similar species of the genus in other provinces were called *representative species* by Forbes. Another idea, included in this hypothesis, was that all the individuals of a species had a *common descent*. The idea of common descent was associated with the definition of species, and when the same species was recognized in two distinct provinces, the fact was explained by the theory of diffusion, or *migration of species;* and in defence of the theory of the specific centres Forbes held that provinces, to be understood, must be traced back, like species, to their history and origin in past time; and again, that "species have a definite existence, and a centralization in geological time as well as in geographical space, and that no species is repeated in time."

Darwin did not deny the Facts, but explained them differently from Forbes.—Darwin, who gave a different interpretation of the facts, recognized the truth of the proposition set forth by Forbes. In his famous "Origin of Species" he says (in reply to the question "whether species have been created at one or more points of the earth's surface," and after some discussion of the topic): "Hence, it seems to me, as it has to many of the naturalists, that the view of each species having been produced in one area alone, and having subsequently migrated from that area as far as its power of migration and subsistence under past and present conditions permitted, is the most probable."

Forbes' Explanation of the Origin of Species.—In Forbes' notion of "specific centres" is included the idea that ancestry is responsible for the "specific characters" of the individual. "Every true species presents in its individuals," he says, "certain features, specific characters, which distinguish it from every other species; as if the Creator had set an exclusive mark or seal on each living type." And in the distribution of fossil as well as living species was seen evidence of "relationship of descent" and of "the derivation from an original protoplast." But descent was supposed by him and his school to be *without modification;* it was the transmission without change of the ancestral characters to their offspring. Whatever modification might appear was considered an irregularity of individual growth, the cause of which was looked for in idiosyncrasies of the individual or in accidents of environment. Forbes was not ignorant of the paleontologic succession of species. Ancestry determined the specific characters, but it was supposed to determine their likeness, and not their differences. All the evolving of new forms was traced to antecessory causes and conditions, but the immediate ancestors, it was believed, were capable of transmitting only the characters which they received from their ancestors. There is nothing wrong with "geographical distribution," or "specific centres," or "specific characters," as used by the older naturalists; the new light has come into the interpretation of descent and the nature of species.

The Meaning of Evolution by Descent.—It is important to distinguish between the names of things and their explanation. The term *evolution by descent* is in this respect faulty, for it means both more and less than is intended. More, in that the most important factor brought forward in explanation of evolution to-day, that of natural selection, is among the extrinsic rather than the intrinsic forces, when the conditions of environment are strictly discriminated; while descent, or ancestry, can be applied only to those forces or conditions which are intrinsic. It expresses less than is intended in that it is not meant that descent alone determines the steps of evolution.

Distinction between Evolution and Development.—Huxley's definition, "evolution, or development, is, in fact, at present employed in biology as a general name for the history of the steps by which any living being has acquired the morphological and the physiological characters which distinguish it," is defective in that it includes a definition of both evolution and development. Development of the individual organism, from the germ to the adult, is a very different thing from the history of the steps by which the same individual acquired the differences which distinguish it from other species of the same genus, which is the particular meaning of evolution. Evolution is the process of *modification of specific* characters, and development is the process of *formation of individual* characters. There are also conditions incident to these processes—conditions which are both outside of and exist before each step of these processes. When these conditions are essentially connected with the preparatory organic functions by which the processes are carried on, they are intrinsic, and they are defined under the general term *ancestry;* when they are accidental to the time or place when and where the processes are acting, they are extrinsic, and are called the conditions of environment.

Immutability or Mutability of Species.—The fundamental difference between the old and new schools of naturalists is found in their opinions regarding the origin of specific differences: the old school held the doctrine of the *immutability of species*, the new holds the doctrine of the *mutability of species.* The result of the change of view has not invalidated the observations of the earlier naturalists, but it has produced a complete revolution in the methods of interpretation of natural history.

In this conception, defined by Forbes, we see that among the contributions which ancestry brings to the actually known individual there are what he called the "specific characters" which distinguish it from every other species, and the possession of these "specific characters" was taken to support the notion of derivation from the original protoplast.

Descent was recognized as without modification; that is, the law of descent was the perpetuation of the ancestral "specific characters" in the offspring. There was in the definition no consideration of the origin of such specific characters. Whatever modifications occurred in the offspring were defined as irregularities of growth, whose cause was located in the idiosyncrasies of the individual, or in what is above called environment, but they were not supposed to be perpetuated.

This school, which Forbes represents, assigned all the steps of progress observed in the history of organisms to causes entirely antecedent to each individual's birth. The explanation was confined to ancestry in its abstract sense of "antecessor" (as the Latin original has it): all cause of changes of a specific rank was entirely antecedent to the organic individual expressing them. The fundamental characteristic of this view is found in the doctrine of the "immutability of species," as contrasted with the doctrine of "mutability of species" of the new school.

Mutability of Species the Central Thought in the New Theory of the Origin of Species.—Nothing that has occurred in the present century has so stimulated investigation of the facts of nature, and has so pervaded the whole realm of philosophical thought, as that which has centred about this question as to the nature and origin of the organic species. Darwin's famous work "The Origin of Species," first published in November, 1859, struck the key-note of the present age of the science. He clearly announced the opinion that *species are mutable*, and as the whole science of natural history was built on the idea of their immutability, a complete readjustment of the science to the new conception has resulted. The importance of a clear conception of the meaning of species is thus apparent, and it will be discussed in detail in a following chapter. The idea of immutability of species obstructed the way to the clear comprehension of the evolution of organisms, very much as the catastrophe theory of the end of the last century prevented geologists from reaching a clear understanding of the agencies and methods by which the earth reached its present condition. Uniformitarianism played much the same rôle for Geology which evolutionism is working for the science of Biology.

Two Extremes of Opinion Regarding the Mode of Origin of Species by Evolution.—Among those to-day who adopt evolution as the explanation of the mode of origin of the different forms of organisms, there are two extremes of opinion with many intermediate compromises.

All will agree in recognizing ancestry and environment as each taking some part in the evolution; but the extreme school, on the one hand, holds that *environment* is the chief factor determining the direction and extent of the modifications, which heredity tends to perpetuate, and that ancestry plays only the part of holding and preserving, in its offspring, what it gets from the agency of environment.

The other extreme is the opinion that *ancestry* is the more efficient factor in bringing about the evolution; that in what is called *variability* there is working out, not a mere accidental reflex of environment upon the plastic organism, but a fundamental property or force of organisms, ever tending from homogeneity to heterogeneity, and resulting in the specialization of functions and the differentiation of organic structure always; the line of evolution followed out by any particular race being influenced little by environment,—the adjustments being active and not passive,—the successful organisms seeking and adopting conditions favorable for their existence if out of them, dying out if the conditions favorable are not within reach, or if crowded out of them. Natural selection, to this school of opinion, plays rather an eliminating rôle than one of causation, and explains rather why there are gaps in the series of organisms than why the characters assumed in the modified forms are what they are. In this latter view the successive steps of modification of a race are as much controlled by the ancestry as are the successive steps of development in the growth of the individual.

In the former view there is the replacement of the theory of immutability of species by that of the mutability of species, but the process of reproduction is still looked upon as immutable, reproducing the characters of the parents in the offspring without change: in the second view reproduction itself takes a part in evolution and normally accomplishes modification of form, either slowly or suddenly, but progressively, and evolution is an intrinsic law of organism.

An Unknown Cause assumed to explain Origins by both Forbes and Lamarck.—The naturalists of Forbes' school, with the fundamental notion of immutability of species, had no other way to explain the series of successive forms which they knew from paleontological research than to call in the resources of a first cause; but they were not ignorant of the series.

Lamarck, who looked upon species as mutable, still found his ignorance impelling him to use the theory of spontaneous generation to start his series. However much they may seem to be independent of a first cause, no scientific theory even of evolution is complete without recognizing the *potency* of the things as existing before their *appearance*.

Conclusions.—It will be apparent now that the discussion of the relation of organisms to environment, or geographical distribution, touches the fundamental problems of natural history. Forbes was of the Linnæan school, who with Cuvier and all that earlier school of naturalists held to the conception of a species immutable; but his studies of distribution were among the more important agencies in clearing the way for the abandonment of that conception of species. The explanation he gave of the origin of species was the most rational one so long as the species was supposed to be immutable. We often imagine that evolution, which has been made the watchword of the new view, is a newly discovered truth; not so. The processes of evolution have been elaborately investigated by the new school, but evolution of organisms, in the abstract sense, had been promulgated almost from the beginning of philosophy, as already stated.

Darwin, in his "Origin of Species," frequently, and with apparently no more hesitation than he had for the use of species, spoke of Creation; he adopted, too, Forbes' term "Centres of Creation." Haeckel, one of the most radical defenders of the new views, entitled one of his most important books "The History of Creation."* These illustrations show that the attempt to explain the process and cause of Evolution is quite distinct from the recognition of the facts of Evolution, and we may conclude that mutability of organic species and the evolution of organisms in geological time are established facts, in the accomplishment of which both ancestry and the conditions of environment have played a part.

---

* "Natürliche Schöpfungsgeschichte." Berlin, 1868.

CHAPTER VII.

GEOGRAPHICAL DISTRIBUTION: SPECIAL CONSIDERATION: THE ADJUSTMENT OF ORGANISMS TO ENVIRONMENT.

Résumé.—In the case of the Madreporarian corals it was observed that as geological time progressed new genera actually were initiated, and the succession of genera and the rate of their increase was seen to be definitely associated with succession of time. Likeness of structure and likeness of time, dissimilarity of form and separation in time, slowness or rapidity of initiation of new genera, and a particular geological period of time for each family, order, or class, are interpreted to mean that there is a definite relationship existing between differentiation of structure and passage of time. This we assume to be a law of the order of events, and we infer the general hypothesis that the form and structure of organisms of one geological period are in some measure determined by the form and structure of the organisms of the period immediately preceding.

This hypothesis involves two particular propositions:

(1) *That each organism is genetically related to some preexisting ancestor whose form and structure were not exactly like its own.*

(2) *That the process of organic reproduction is not a stereotype process of repeating in the offspring the exact characters of the ancestry, but that the production of differences between the parent and offspring is a normal factor in the reproductive process, either continuously or occasionally in operation.*

There is, however, another fact to be noted: the innumerable differences in the conditions of environment are more or less distinctly expressed by differences in the kinds of organisms associated with them. All kinds of animals are not found in every place or condition, but in each particular kind of environment particular kinds of animals are found, and their living is more or less dependent upon those conditions.

Hence we infer another general hypothesis:

(3) *That the conditions of environment do in some measure determine the particular form and structure of each organism.*

The Gastropoda Illustrate the Law of the Relationship between Organisms and Environment.—In order to show more particularly how the differences of form (expressed by different species, genera, and families in scientific classification) are related to differences in the conditions of environment, a class of the Mollusca, the Gastropoda, may be examined in detail. This group of organisms is convenient for the purpose because of the full statistics already accumulated regarding the geographical distribution of its species.

Meaning of the Classification of Organisms.—Without defining the morphological characters indicated by the classification, it is important to remember that zoological classifications are fundamentally based upon morphological differences, that organisms of two distinct classes present greater morphological difference than those of a single class, that lesser divergence in form is expressed by division of the class into subclasses, and that the animals of the same order present greater resemblance to each other than to those of different orders. Families are again subdivisions of the orders, and each family includes two or more genera, and the species of each genus are alike in their general form, differing only in some of the more minute details. Hence when we describe the peculiarities of the distribution of genera, we are expressing the law of association between the generic form and the conditions of environment indicated by the geographical distribution. Thus, the common sea-whelk, *Buccinum undatum* (Fig. 33), represents the class Gastropoda as contrasted with the Dentalium (Fig. 37), belonging to the class Scaphopoda, Hyloea, a Pteropoda or Chiton (Fig. 36), a representative of the class Placophora. The Gastropoda, Scaphopoda, Pteropoda, and Placophora together constitute that division of Mollusca called Glossophora, being alike in the possession of a more or less distinct head-portion of the body, and of a well-developed tongue (*radula*), which is generally armed with minute denticles set in rows (Fig. 34). The other types

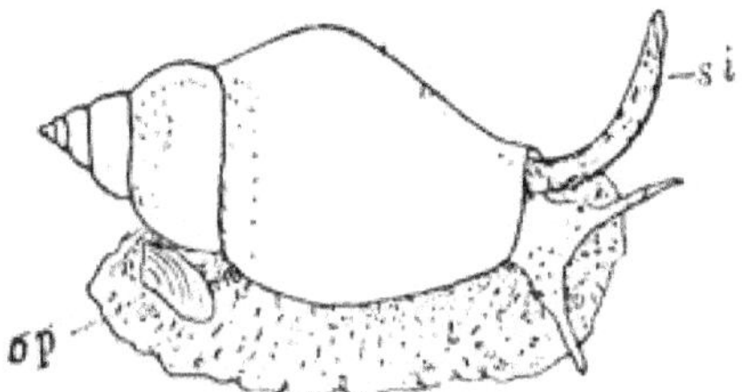

Fig. 33.—A Gastropod, the common whelk, *Buccinum undatum*, showing the spiral shell on the back of the animal, its large flattened foot, distinct head with two tentacles, at the base of which are the eyes. The siphon *si* and the operculum *op* are special parts not found in all Gastropods.

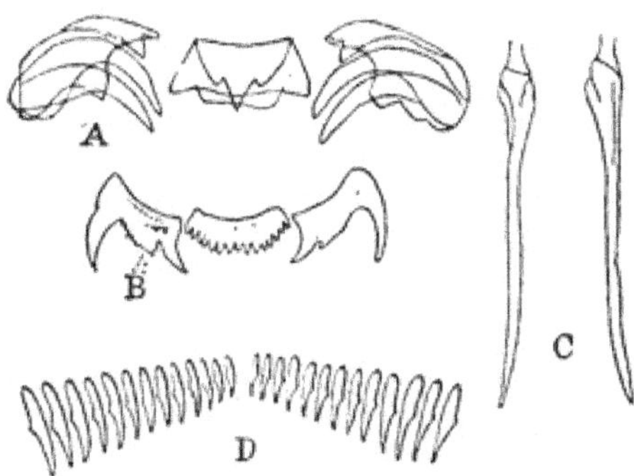

Fig. 34.—Examples of the dentition of Gastropoda, single transverse rows of the denticles of the lingual ribbon (*radula*), greatly magnified, of (*A*) *Natica*, (*B*) *Nassa*, (*C*) *Pleurotoma*, (*D*) *Scalaria*.

of Glossophora are adjusted to various conditions of environment, but for our purpose it will be better to confine our attention at present to the single type of the class Gastropoda.

Distinguishing Characters of the Class Gastropoda.—The common external characters of all Gastropods are these, viz.: Head and sense organs well developed, the former often bearing tentacles; a ventral muscular foot and undivided mantle, which frequently secretes a plate-shaped, or spirally twisted shell. The paleontologist knows Gastropods by their calcareous, more or less spirally twisted, univalve shells. These Gastropods, of which several tens of thousands of species are described, are specifically adjusted to all kinds of conditions of environment, and are distributed from the bottom of the ocean to the tops of the mountains.

Zones of Environment in which Gastropods are Distributed.—If we arrange their environmental conditions in tabular order we have the following series, viz.:

1st. Abyss of the ocean, or an abysmal zone, extending from 500 metres, or 250 fathoms, to the lowest known depths of the ocean.

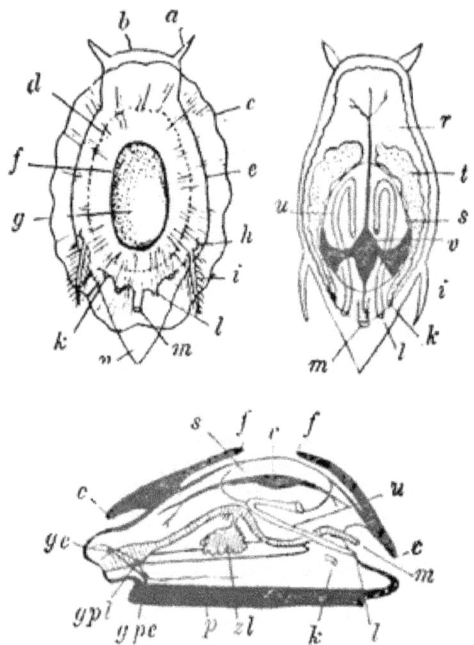

Fig. [illegible].—Schematic Mollusk. (After Lankester.) *a*, tentacle; *b*, head; *c*, margin of mantle; *d*, margin of shell; *e*, edge of body; *f*, edge of shell depression; *g*, shell; *gc*, cerebral ganglion; *gpe*, pedal ganglion; *gpl*, pleural ganglion; *h*, osphradium; *i*, ctenidium; *k*, reproductive pore; *l*, nephridial pore; *m*, anus; *n* and *p*, foot; *r*, cœlom; *s*, pericardium; *t*, testis; *u*, nephridium; *v*, ventricle of heart; *zl*, liver.

2d. Zone of Brachiopods, or of deep-sea corals (72–500 metres, 30–250 fathoms).

3d. Zone of Nullipores, or of Corallines (27–72 metres, 15–50 fathoms).

4th. Laminarian Zone (low tide to 27 metres, 1–15 fathoms).

5th. Littoral Zone (between low and high tides).

6th. Brackish water, sea-shores above tide, where fresh and brackish waters are mixed, and where the surface may be exposed to the air part of the time.

7th. Fresh water, as in rivers and lakes.

8th. Amphibious conditions, fresh water and land.

9th. Land, the surface of the land, or in the air.

These are *zones of environment*, which express a series of varying conditions of light, of oxygen, of air, of moisture, of degrees of temperature, of pressure of the medium, of depth, of height.

Reasons for Selecting the Gastropods.—The Gastropoda are selected because of the wide range of adaptation expressed in their distribution, and because the statistics are particularly full. The classification found in Zittel's Handbuch is adopted, so far as nomenclature and inclusion of genera are concerned; but Gastropoda will be spoken of as of the rank of a class, the more common usage of zoologists,* and the morphologically specialized forms, the Chitons (Placophora, Fig. 36) and the Dentalia (Scaphopoda, Fig. 37), will be omitted from the true Gastropoda, as is done by Zittel, following Ihring and Lacaze Duthiers; the Pteropoda will also be omitted.

Peculiarity of the Divisions of the Gastropods as to Range of Adaptation.—Ranking Gastropoda as a class, with the restrictions above mentioned, it will include the following four orders, viz.: Prosobranchia, Opisthobranchia, Pulmonata, and Nucleobranchiata (or Heteropoda). The whole of the *Heteropoda* are specialized in structure and restricted in distribution to the surface and upper parts of the ocean water, and structurally they may be ranked with the monotocardian Prosobranchs. Six living genera with about 50 species are known, and a few fossil genera are referred to this order. The *Pulmonata* (Fig. 38, 38*a*) are air-breathers, and (with the exception of the Siphonariidæ) are restricted in distribution to land and fresh water. Six thousand (6000) living and 700

* Lankester's classification is (Encycl. Brit., art. "Mollusca," p. 633).

Phylum mollusca:

Branch A, *Glossophora*.

Class 1. Gastropoda.

Br. *a*. Isopleura.

Br. *b*. Anisopleura.

Class 2. Scaphopoda.

Class 3. Cephalopoda.

Br. *a*. Pteropoda

Br. *b*. Siphonopoda.

Branch B, *Lipocephala* (= Acephala, Cuvier).

Class 1. Lamellibranchia (syn. Conchifera).

fossil species are described. The *Opisthobranchia* (Fig. 39) are all sea-snails, and appear to be restricted in distribution to the coastal waters, near the land, and near the line of contact between salt and brackish water habitats; about 1200 species

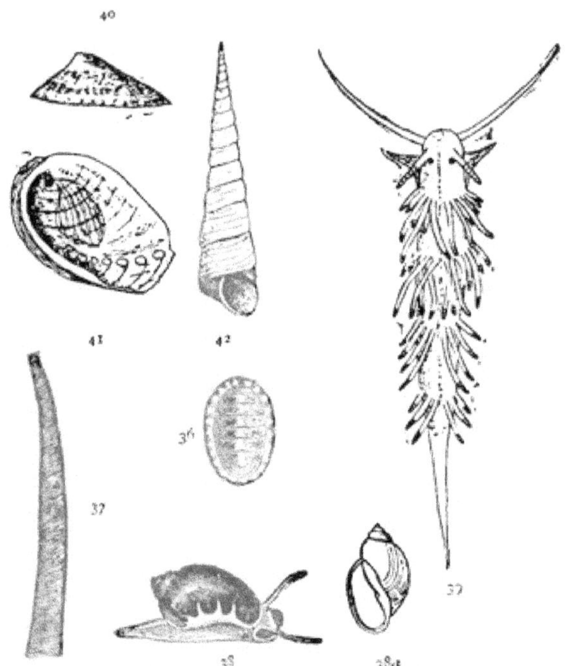

FIGS. 36–42.—Gastropoda. Illustrations of the chief types: 36, Placophora, *Chiton ruber*; 37, Scaphopoda, *Dentalium Indianorum*; 38, 38a, Pulmonata, *Physa heterostropha*; 39, Opisthobranchia (Nudibranchia), *Eolis pilata*; 40, 41, 42, Prosobranchia, 40, Cyclobranchina, *Achmæa testudinalis*; 41, Aspidobranchina, *Haliotis sp.*; 42, Ctenobranchina, *Turritella sp.* (After Packard and McMurrich.)

are described, including fossil forms; the gills are behind the heart. The remaining order, the *Prosobranchia* (Figs. 40, 41, 42), includes mainly marine species, which are adapted to a great variety of marine conditions; there are known some 14,000 species. They are divided into three suborders, separated primarily upon the differences in their breathing organs, viz.: *A*, Cyclobranchina; *B*, Aspidobranchina; *C*, Ctenobranchina, or better known as the Pectinibranchia of Cuvier. In all the Prosobranchs the gills are in front of the heart, that is, the branchial vein enters from the front. They are diœcious (while the Opisthobranchs and Pulmonates are hermaphrodite).

The *mode of existence of the Glossophora* is compactly summarized as follows by Zittel (translation from "Handbuch der Palæontologie," vol. II. p. 161): The greater number of Glossophora are aquatic animals, and the majority marine. The Pteropods, Placophors, Heteropods, Opisthobranchs, live exclusively in the sea. The great order of Prosobranchs comprise also a majority of marine forms. There are a certain number which live in brackish water near the discharge of lakes (Potamides, Neritinas, Rissoas, Hydrobias), and others in fresh water (Paludinidæ, Melaniidæ, Valvatidæ). The Pulmonate genera, furnished with gills, are adapted to a terrestrial life (Cyclostomidæ, Helicinidæ). The Pteropods and Heteropods are pelagic animals, free swimmers, inhabiting the open sea; the great part of the other Glossophora are coast animals, crawling upon plants, rocks, and shore débris. Some Prosobranchs are amphibious (Littorina, Truncatella, Patella, Nerita), and are able to live a long time dry, without water; they then retire within their shell, close the operculum, and breathe the water which they have retained with them. The Ampullarians have the advantage of two different kinds of respiratory organs, and can live alike on land and in water. Some Prosobranchs bore in the sand and mud like Lamellibranchs (Oliva, Mitra, Natica, Buccinum); others inhabit coral reefs, or live as parasites in other animals (Entoconcha, Stylifer). The shells of fresh-water Gastropods are generally covered with a greenish-olive or brown epidermis, their apices are often broken or absorbed; their shell is thin and horny (Limnaeus). Many Gastropods subsist on fresh flesh or carrion; there are some which perforate shells of other mollusks with their tongue, and devour them through the little hole thus perforated (Natica, Murex, Buccinum); the majority of Glossophora (almost all the Pulmonates, and the holostomate Prosobranchs) live upon vegetable food.

Their geographical distribution is little known except for the littoral, fluviatile, and terrestrial species; it is known, however, that the Pteropods and Heteropods, being pelagic, have a very extended distribution; the Scaphopods and the Placophors are equally found in all seas and all latitudes. There are only a few pelagic forms among the Opisthobranchs and the Prosobranchs. The geographical distribution of the marine Glossophora, besides the influence of centres of origin, is determined greatly by the character of the bottom, the form of the coast, the flux and reflux of tides, the currents, and the saltness and depth of the water. The sandy shores are little favorable to Gastropods; they prefer rocky shores, where algæ flourish. The shores much cut up furnish great variety of conditions of habitat, and accordingly have a richer fauna than great estuaries. The movements of the tide produce changes and bring in food, and thus favor life. There are currents, also, which greatly affect geographical distribution. Most of the marine Glossophora die as soon as they are transported into fresh water. There are, however, some which have the faculty of adapting themselves to change of medium. Such notably are certain species of the genera Patella, Rissoa, Trochus, Purpura, Littorina, and Cerithium. Some fresh-water species, conversely, are able to live in salt water (Limnaeus, Planorbis, Melania, Melanopsis, Physa, Neritina). It is probable that all the actual terrestrial or fluviatile species are traceable to a common origin, and that they descended from marine types of the geological epochs, modified by adaptation.

The temperature has a great influence upon the development of the Glossophora: heat is favorable to them, and they are much more abundant in the seas and lands of tropical regions than in temperate or polar regions. The marine bathymetric zones, as the hypsometric zones on land, exercise their influence upon the Glossophora, as well as upon other animals. The study of their conditions has a particular interest to the paleontologist, since he is thus able to account for the conditions under which the fossils lived, and the mode of formation of marine sediments. One knows, in a general way, that the temperature of the ocean goes on diminishing from the surface to the bottom, and that it attains a temperature approximately constant of 4° to 5° Cent. at the depth of 500 feet; it descends scarcely to zero (32° F.) at great depths; the conditions of submarine existence are thus approximately constant in abysmal regions, while they present the greatest range of variation in the shore regions of slight depth in the tropics.

The bathymetric distribution of Mollusca was studied in 1830 by Andouin and Milne Edwards; and later, upon new data, by Sars in Norway (1835) and Ed. Forbes in the Ægean Sea and in England. The most important results in this direction have been attained by the expeditions of the Porcupine (1869-70), of the Challenger (1873-76), of the Gazelle (1874-76), of the Tuscarora (1874-76), of the Blake (1877-78), of the Voraigen (1876-78), of the Vorailles (1880).

The Zonal Distribution of the Ctenobranchina.—Restricting our attention to the families of *Ctenobranchina*, and using for the purpose the classification into families of F. Barnard,* which are 44, we are able to see some evidence of the particular connection between form and bathymetric distribution. Of these families three have land species, and two of the families are restricted to a land habitat (Cyclophoridæ and Cyclostomidæ). There are five families of which the species are all fresh-water species (Paludinidæ, Ampullaridæ, Bithyniidæ, Valvatidæ, and Melaniidæ). One family, Hydrobiidæ, has both littoral and brackish water species. The remaining thirty-four families are all marine; of them many of the littoral species are able to endure exposure to the air and some contamination of the water, but the normal habitat of all is marine. Some of the families are limited in downward distribution: such are the families Truncatellidæ, Hydrobiidæ, Janthinidæ (a pelagic type), Cypræidæ, Solariidæ, Purpuridæ, and Terebridæ. Others reach downward to the abysmal depths, as Littorinidæ, Rissoidæ, Cerithiidæ, Naticidæ, Scalaridæ, Pyramidellidæ, Eulimidæ, Muricidæ, Pleurotomidæ; and it is interesting to note that of these families, having a bathymetric distribution from the abysmal depth to the littoral zone, several are also the most ancient in geological range; the Littorinidæ, the Naticidæ, and the Pyramidellidæ are reported from as early as the Silurian era. The second section,

* "Éléments de Paleontologie," 1893.

Tænioglossa (Zittel's classification), contains twenty-six families; of these, four families contain strictly fresh-water species, a few genera of which are amphibious. One of the families is made up of land species; the remainder are marine forms. Species of, at least, three families have been taken from the abysmal zone. If we consider only the genera characteristic of the several zones, we find them distributed among different families. Three of these are represented in the 1st, or abysmal zone; four in the 2d, or deep-sea Coral zone; five in the 3d, or Nullipore zone; five in the 4th, or Laminarian zone; five in the 5th, or littoral zone.

Genera of the Ctenobranchina characteristic of the Several Bathymetric Zones.—The genera which have already been found to characterize the several zones have been tabulated (from lists derived from various sources) by Fischer.*

It will be noticed that some genera are restricted to single zones, and others characterize the faunas of more than one bathymetric zone. Examination of these lists shows the following genera of Ctenobranchia to characterize the faunas of the respective zones.

(1) *The Littoral Zone.*—From high water to a depth of 12 metres, species of the genera Littorina, Hydrobia, Assiminea, Rissoa, Truncatella, Cerithium, Natica, Pyramidella, Nassa, Purpura, Murex, Conus.

(2) *The Laminarian Zone.*—From low tide to 15 fathoms; a zone characterized by species mostly phytophagous, of the following genera, viz.: Phasianella, Xenophora, Triforis, Rissoa, Aclis, Daphnella, Lacuna, Terebellum, Pterocera, Marginella, Mitra, Nassa, Phos, Drillia, Pleurotoma.

(3) *The Nullipore Zone.*—From 15 to 20 fathoms; the zone of calcareous algæ. The characteristic species are mainly carnivorous, and of the following genera, viz.: Bela, Buccinum, Cassis, Cassidaria, Chenopus, Eulima, Fossarus, Fusus, Nassa, Natica, Pleurotoma, Trichotropis, Tritonium, Trophon, Velutina.

(4) *The Brachiopod Zone*, or that of deep-sea corals, extending from a depth of 50 to 100 fathoms, has for its Cteno-

* "Manuel de Conchyliologie," Paris, 1887.

branch fauna species of the genera Bela, Eglesia, Fossarus, Mangelia, Murex, Odostomia, Pleurotoma, Rissoa, Triforis, and Turritella.

(5) *The Abyssal Zone.*—300 metres, or 100 fathoms, or more in depth, down to the profound depths, supports species of the genera Aclis, Acirsa, Cerithium, Chenopus, Defranchi, Eulima, Fusus, Hela, Natica, Odostomia, Pleurotoma, Rissoa, Taranis, and Trophon.

Evidence of Adjustment of the Morphological Character to the Environment.—An examination in the like manner of the distribution of species shows an adaptation of each species to much more restricted bathymetric conditions, and to restricted geographical areas or provinces. This fact might, however, be accounted for by migration and sorting out of species from choice, or the selection of environmental conditions; but in the case before us, where not only genera, but whole families, —families whose representatives are found in all parts of the globe,—are restricted to special conditions of environment, it seems impossible to account for the fact except by the supposition that the morphological characters of the organisms are adjusted to the environment.

When we examine animals whose structure is more strongly contrasted, as in the case of the fish swimming in water, the beast walking on land, and the bird flying in the air, we are not impressed so much by the morphological adjustment as by the physiological necessity of the restriction to a particular environment; but in the case of the Gastropods, where the differences in form are relatively of small physiological significance, the finding of a close correlation existing between the specific, generic, and even family form, and the particular conditions of environment seen in the zones of the ocean, and climatal differences of land, impresses one vividly with *the immediate connection between differences of form and differences of environment.*

Law of the Adjustment of Organisms to Conditions of Environment.—We learn from these statistics that the morphological differences, which are the basis of the classification of the various species of the ctenobranch Gastropods into genera and families, are intimately connected with the differences in temperature, depth, pressure, medium, aad, in general, conditions of environment in which they are distributed. And at the same time we learn (*a*) that this is the case for a small group of organisms whose general structure is alike, all secreting a spiral shell, all having substantially the same organs, arranged in much the same manner, and (*b*) that the range of the differences of environment concerned—viz., in temperature, in depth, from the abysses to the tops of mountains, in mediums, from high-pressure salt water to rarefied air—is almost as complete as it would be possibie to reach in habitable regions of the globe. The differences in form and structure of the organisms as units are, therefore, not at all in proportion to the differences of conditions of environment. Organisms very much alike, in the same genus even, are found living under conditions of environment as strongly contrasted, almost, as can be found; and organisms of extreme difference in structure are associated together in the same conditions of environment. The conclusion we draw is, that *condition of environment is a fundamental cause in determining differences of form, but that whatever the structure or organization of an organism may be, there have been, and are constantly going on, adjustments to changed habitats, and that the morphological changes resulting in these adjustments to environment have been mainly of low order, i.e., varietal or specific, and rarely are of higher than generic importance.*

This is in strong constrast to the law observed regarding relation of differences of form to time; amount of time-separation being co-ordinate with degree of difference in the whole structure, and not merely in specific and generic characters.

Summary.—In what has been said above the relations of form to general conditions of environment have been discussed. Geographical distribution, in the particular use of the term, is concerned with the association of like forms (the same species or varieties) in areas presenting like conditions of environment, and the distinguishing of different areas by the different faunas and floras inhabiting them. It is supposed that the adjustment, by various processes, of the species to their changed environment may explain their differences of form. What we have been illustrating is the fact that the species of a genus, or the genera of a family, are found adapted to different kinds of environment, and that the adaptation is expressed by modification of the form of the organism. In geographical distribution proper the fact is emphasized that likeness of characters of form is associated with continuity of like environmental conditions; viz., that the same variety is restricted to a particular geographical area or province.

Geographical distribution emphasizes the fact that environment, by the law of adaptation, has the effect of confining the descendants of common parents within boundaries, and thus tends to the continuance of like characters. Bathymetric distribution emphasizes the fact of adaptation itself, by showing that the morphological differences distinguishing the several species of a common genus, or the several genera of a common order, are directly associated with differences in the environment. The two groups of facts together point to a most important biological law: that *divergence of morphological characters is in some way associated with changing of environmental conditions.*

Distribution implies migration, and when we observe that migration is accompanied with modification and adjustment to new environment, we discover this second of the fundamental laws of evolution.

Relation between Zonal Adaptation and Geographical Range. —An analysis of the classification of the Mollusca shows that the Gastropod structure is adapted to all kinds of environment, because we find genera of Gastropoda in each of the several zones expressing the full range of environmental differences on the earth, from the abysses of the sea to the top of the dry land.

Three of the orders of Gastropoda are somewhat specialized in adaptation to environment: the order Heteropoda are pelagic forms; the Opisthobranchia are all marine, living in the zones from Littoral down to the Nullipore zone. The Pulmonata are restricted in adjustment to the high littoral only of the marine zones, and to brackish, fresh-water, and land conditions above the tide-level. The order Prosobranchia has genera in every zone distinguished in our list.

This order is distinguished by the following characters, viz.: diœcious, branchiate, shell-bearing, gills in front of heart; from the latter character the name of the group is derived. The *Cyclobranchina* (Fig. 40) and the *Aspidobranchina* (Fig. 41), two suborders, are all marine forms, but under the suborder *Ctenobranchina* (Figs. 33 and 42)—a division in which all are so far specialized as to possess "a large cervical gill of pectinate form on the left side, with small olfactory organ (so-called rudimentary gill); a spiral shell is very generally present; the male possesses a penis on the right side; most are carnivorous, and possess a protrusible proboscis" (Claus and Sedgwick)—there are genera adapted to each of the different kinds of environment, from the abysmal to dry-land zones. Some of the genera of this suborder are restricted in distribution, and constitute subdivisions of higher than family value. The *Ptenoglossa* are all pelagic. The *Rachioglossa*, the *Toxiglossa*, and the *Rhipidoglossa* are all marine; but the genera included under the *Tænioglossa* are adapted to differing zones of environment from one extreme to the other. This group is still further specialized, and in each transverse row of the elongated radula of the tongue-like rasping organ of the mouth, there are usually seven plates, and two small jaws are usually found at the mouth-entrance.

There are two divisions of the Tænioglossa, the *Siphonostomata*, in which the opening of the shell is canaliculated for the protrusion of a proboscis-like extension of the mouth; in the other, the *Holostomata*, the opening is entire. But when we examine the genera grouped together by possession of such likeness of structure, still we find in the former group, of which most of the families contain marine species, that the Ampullaridæ are restricted to fresh-water habitat. In the *holostomatous* division the Cyclostomidæ, Cyclophoridæ, and Truncatellidæ are air-breathers and live on land. The Paludinidæ, the Valvatidæ, the Melaniidæ are all fresh-water species; while the Littorinidæ are marine forms, but have species in the deepest part of the ocean, and others living between tides; and many other of the families of the latter group are distributed through several zones. The forms of this division, Tænioglossa, which are constructed to breathe air, and thus are restricted to a habitat above sea-level, are included in the five families Cyclophoridæ, Cyclostomidæ, Aciculidæ, Truncatellidæ, and Assimineidæ.

Of the eighteen genera of the first family, twelve are restricted to Southern and Eastern Asia and neighboring islands; one genus (Pomatias) is distributed over North Africa and South Europe; another (Craspedopoma) over the Canary, Madeira, and Azores Islands. Another genus (Megalomostoma) is found in the Antilles and Guatemala, and Aperostoma in Central and South America and Mexico.

The genera of Cyclostomidæ have a similar distribution, mainly in the lands bordering the Indian Ocean, and a couple of genera (Choanopoma and Cistula) in the corresponding lands bordering on the Gulf of Mexico.

The other three families, Aciculidæ, Truncatellidæ, and Assimineidæ, are all found within the same areas.

Families whose Genera have a very wide Range of Adaptation, and Restricted Adjustment only among the Species.—If we pursue the analysis still further, we find that there are some families, like Cerithiidæ, in which for some of its genera there is still a very wide adaptation to conditions of environment; species of the genus Cerithium are living now between tide, and have also been dredged from the abysmal zone. In such families the zonal adaptation can be found only among the species.

Great Difference in the Closeness of Adjustment of the Characters of different Taxonomic Rank.—It is hence evident that there is great difference in the extent to which organisms are adjusted to restricted conditions of environment. In some organisms their class characters are strictly adjusted to a particular group of environmental conditions, as is the case with the insect whose mature structure with tracheal breathing restricts it to a habitat in which air is accessible; but even among insects there are cases of adaptation to life in water. In other cases one order is adapted to one mode of life and another to a different condition of environment—as among the reptiles there are aquatic Saurians, the Enaliosauria, and the true lizards, Lacertilia, adapted to live on land. In such cases as we have been considering, though there are in each group some cases of restriction of adjustment to particular conditions of environment, within the same group, be it order, suborder, family, or genus, there are also those having the same structure which are not so closely adjusted to the environmental conditions.

Species Generally Closely Adjusted to Particular Conditions.—This is the case until we reach the species. Species do appear to be closely adjusted to some particular set of physical conditions. Each one is so constructed that one environment is at least most favorable, and to remove it from such condition is either impossible without killing it, or leads to some adjustment of its habits, and, it may be, structure and form to adapt it to the changes. The adaptation can only be varietal for a single individual; hence it is only among the specific characters that we find the evidence of immediate change of form to adapt the organism to changed conditions.

Fresh-water Families; Restriction in their Distribution.—The following families are made up of fresh-water species: Paludinidæ, Ampullaridæ, Valvatidæ, Melaniidæ, and Hydrobiidæ; the latter two families containing a few brackish-water species.

Such species are by their specialized structure restricted, therefore, to continental or island habitat.

The Paludinidæ and the Valvatidæ are restricted to the northern hemisphere, are mainly in temperate zones, and are not known south of the equator.

The Ampullaridæ are from Central and South America, Eastern Africa, Madagascar, S. Asia, Malaysia, the Philippines, Australia, and vicinity.

The Melaniidæ are chiefly intertropical species, being most abundant in India, Indo-China, Malaysia, the Philippines, Oceanica, Africa, Central America, South America, running from Central America up into Mexico, and from North Africa to Spain and Asia Minor.

The Hydrobiidæ, which, according to Fischer, have been distributed under eighty genera, are scattered over almost all the lands between the temperate zones of the northern and southern hemispheres.

Two Closely Allied Families, Separated in their Distribution.—The Strombidæ and Chenopodidæ illustrate this law. The shells of both these families are heavy, and more or less specialized in their form, developing elongations, spines, and processes giving them peculiar shapes. It is probable, therefore, that in their life they are not capable of much change of local habitation.

The Strombus is confined to warm seas,—the Pacific, the Indian, and the mid-Atlantic, including the Caribbean seas and Mexican Gulf. The other recent genera of Strombidæ are from the Indian and Pacific warm seas. The genus Chenopus of the second family is a North Atlantic form, and is not associated with the Strombidæ in habitat.

But representatives of both families are as old as the Jurassic, and there are also several genera in each family from Cretaceous rocks.

It is evident from this set of facts that the distinction between the two family types of structure was initiated in the Mesozoic, and that there was adjustment to particular conditions of environment very early—an adjustment which change of time did not modify.

The following table will graphically illustrate this fact:

TABLE OF THE GEOLOGICAL RANGE OF THE FAMILIES STROMBIDÆ AND CHENOPODIDÆ.

	Jurassic.	Cretaceous.	Tertiary.	Recent.
Strombidæ:				
Strombus (warm seas, Pac., Ind., Med., and Ant.)		*	*	*
Pereiræa..........			*	
Pterocera (Ind. and Pacif. O.)..........	*	*	*	*
Rostellaria (Ind. O., Red Sea, and China)........		*	*	*
? Mitrafusus..........			*	
Pterodonta..........		*		
Terebellum (Ind. O.)..........			*	*
Chenopodidæ:				
Chenopus (N. Atlantic)..........	*	*	*	*
Diarthema..........	*			
? Malaptera..........	*	*		
? Harpagodes..........	*	*		
Alaria..........	*	*		

The Relation of Antiquity to Distribution.—The distribution of the genera in the family of Cerithiidæ illustrates another law: viz., old genera are widely distributed, while younger genera are more closely restricted in distribution.

Thus Cerithium is a genus of which species are known from the Triassic, Jurassic, Cretaceous, Tertiary, Quaternary and Recent Periods. It is known from all seas, warm and temperate; and a species of the genus has been dredged from the abysmal zone, and other species are known up to the littoral zone of the ocean. Fastigiella, on the other hand, a genus known no farther back than the Tertiary, is confined to the Antilles, as present knowledge goes.

Rissoa, of the family Rissoidæ, has a similar history; it is known from the Jurassic up, and it is distributed in all seas. In the accompanying table these facts are graphically represented.

TABLE OF THE GEOLOGICAL RANGE OF THE FAMILIES CERITHIIDÆ AND RISSOIDÆ.

	Triassic.	Jurassic.	Cretaceous.	Tertiary.	Recent.
Cerithiidæ:					
Triforis (Antil., Ind. and Pac. O.)				*	*
Fastigiella (Antilles)				*	*
Cerithium (all seas, warm and temp.)	*	*	*	*	*
Bittium (all seas)	*	*	*	*	*
Potamides (Ind. O., Afric. coasts, and Cal.)				*	*
Diastoma				*	
? Sandbergeria				*	
? Mesostoma				*	
? Exelissa		*			
? Pterostoma				*	
Cerithiopsis (all seas)			*	*	*
Ceritella	*	*	*		
Eustoma		*			
? Brachytrema		*			
Rissoidæ:					
Rissoa (all the seas)		*	*	*	*
Scaliola (Japan, New Caled., and Red Sea)				*	*
Rissoina (warm and temp. seas, Antil., Med., and Pac.)			*	*	*
? Barlæia (Europ. seas)					*
Paryphostoma				*	

Distribution in Relation to Temperature of the Waters.—Two families may be selected to illustrate this law. In the Lamellariidæ the genus Lamellaria is distributed in all three oceans Atlantic, Pacific, and Indian. The three genera Velutina, Marsenina, Oncidiopsis are confined to boreal seas, and the fifth genus, Caledoniella, is from New Caledonia alone.

In another family, the Cypræidæ, all the genera are adjusted to warm and temperate seas; the principal genus, Cypræa, of which more than a hundred and fifty species have been described, is confined entirely to warm seas; the majority of the species are from the Indian Ocean and the Australian and Polynesian oceans. This genus also dates from as early as the Middle Mesozoic, species having been found in the Jurassic, Cretaceous, and Tertiary rocks. Other genera of the family live in the Mediterranean waters, and extend across to the shores of the West India islands and Central America, and are also seen on the west coast of America.

To select another family, outside the immediate suborder we are now considering, in the Buccinidæ we find as much of an adaptation to cold waters as in the last case there was to warm waters.

The genera Buccinum and Siphonalia are distributed in both boreal and austral seas; Chrysodomus has a circumpolar distribution; Liomesus is only found in arctic and boreal seas. Other genera of the same family are distributed in the intermediate seas, both Atlantic and Pacific; and several genera which are associated by their structure in the same family, as Phos Hindsia and Dipsaceus, are restricted to the warmer seas about the Philippines, Indian, China, and Caribbean shores, or the corresponding warmer western coasts of America.

This family dates back to as early as the Cretaceous era.

Tabulation of the Facts.—The following table expresses in graphic form a summary for all Gastropods, of the facts regarding the actual present adjustment of the form and structure of these organisms (as expressed in the different classes, orders, and families into which they are classified), and the various conditions of environment (ranging from the abysses of the ocean to the extremities of leaves of trees in the open air) in which they find their normal life habitat.

TABLE EXPRESSING THE RELATIONS BETWEEN THE DIFFERENCES IN STRUCTURE OF THE GASTROPODA AND DIFFERENT CONDITIONS OF ENVIRONMENT.

	Abyssal zone.	Brachiopod zone.	Nullipore zone.	Laminarian zone.	Littoral zone.	Brackish.	Fresh Water.	Amphibious.	Land.
	Marine.					Above ocean surface.			Aerial
Glossophora	*	*	*	*	*	*	*	*	*
Class: 1. Scaphopoda	*	*	*	*	*				
2. Placophora			*	*	*				
3. Gastropoda	*	*	*	*	*	*	*	*	*
4. Pteropoda	(Pelagic)				*				
Class Gastropoda.									
Order: 1. Prosobranchia	*	*	*	*	*	*	*	*	*
2. Heteropoda	(Pelagic)				*				
3. Opisthobranchia			*	*	*				
4. Palmonata					?	*	*	*	*
Order Prosobranchia.									
Suborder: *a.* Cyclobranchina		*	*	*	*				
b. Aspidobranchina	*	*	*	*	*				
c. Ctenobranchina	*	*	*	*	*	*	*	*	*
Suborder Ctenobranchina.									
Section: 1. Ptenoglossa	(Pelagic)				*				
2. Tænioglossa	*	*	*	*	*	*	*	*	*
3. Rachioglossa	*	*	*	*	*				
4. Toxoglossa	*	*	*	*	*				
Section Tænioglossa.									
Div. 1. Siphonostomata		Marine & Fresh Water.							
2. Holostomata	*	*	*	*	*	*	*	*	*
Family: Cyclostomidæ									*
Cyclophoridæ									*
Truncatellidæ									*
Aciculidæ									*
Assimneidæ									*
Paludinidæ							*		
Hydrobiidæ					*	*	*		
Valvatidæ							*		
Melaniidæ						*	*	*	
Littorinidæ	*	*	*	*	*				
Rissoidæ	*	*		*	*				
Cerithiidæ	*	*		*	*				
Naticidæ	*	*	*	*	*				

NOTE.—The stars opposite the name of each group indicate the kind of environment to which the genera of the group are adapted.

Summary of Results.—This analysis of the distribution of the various types of Gastropods may be summarized as follows:

The organic structure of Gastropods is such that it is capable of adjustment to all the conditions of environment found inhabited by living things on the face of the earth, or in the waters under the earth.

There are examples in the class Gastropoda of orders, all the members of which are restricted to a narrow and particular set of environmental conditions, as the Heteropods (pelagic), and the Pulmonata (above tide-level).

There are other cases in which the structural adaptation is of subordinal rank, as the Ptenoglossa (pelagic) among the Ctenobranchia; and still others in which members of the sub-order are found under all kinds of environmental conditions; but certain families are restricted in distribution, as among the Taenioglossa the family Paludinidæ are all fresh-water species, the Strombidæ all marine, the Cyclostomidæ all are air-breathing and land forms.

Again, among the members of a family there are genera which are restricted in their distribution to particular conditions of environment, and other genera distributed over a wider range of conditions, as in the Buccinidæ Buccinum is distributed in cold waters, and thus about the northern and southern poles; and Phos is restricted to warm seas, and thus near the equatorial zone.

And, to proceed one step further, particular species, of a genus which is known to be distributed in all oceans, are generally restricted to living in a narrow range of environmental conditions, to a particular limit of depth, to a particular zone of temperature, and often to a particular geographical position along one side of a continent or along the shores of a particular sea or gulf or island.

While, however, there is this great variation in the closeness of adaptation of structure to conditions of environment, it is a general law that *the higher the taxonomic rank of a group of animals the greater is the range of environmental differences to which its members are adjusted;* viz., the members of a family, as a rule, are distributed more widely and under more diverse conditions of environment than the members of some particular genus of the family, or than a particular species of the genus.

CHAPTER VIII.

WHAT IS A SPECIES?—VARIOUS DEFINITIONS AND OPINIONS.

What are Species?—Their Numbers and Importance.—In the previous chapter reference is made to the great importance of the idea of species to the study of natural history, and in the following chapter an attempt will be made to answer the question, "What are species?"

Bronn, in 1849, published a list of all the then known fossil species.* The list comprised 2050 names of plants, 24,300 names of animals. When Zittel wrote his Paleontology † he quoted Günther's estimate of 320,000 species of living animals, and 25,000 fossil animals, already described. Of this 350,000 species of animal organisms, now known to science, what is it in each case which the naturalist observes, and names and enumerates as a species?

Ernst Heinrich Haeckel, in his "History of Creation," insists upon the importance of the idea of species, as follows: "Even now all the important fundamental questions as to the history of creation turn finally upon the decision of the very remote and unimportant question, 'What really are kinds or species?' The idea of organic species may be termed the central point of the whole question of creation, the disputed centre, about the different conceptions of which Darwinists and anti-Darwinists fight." ‡

Linné held that there are as many different species as there

* H. G. Bronn, "Index Palæontologicus," etc. 3 vols. Stuttgart, 1848-49.

† K. A. Zittel, "Handbuch der Palæontologie," vol. 1. München, 1876.

‡ E. H. Haeckel, "The History of Creation; or, The development of the earth and its inhabitants by the action of natural causes; a popular exposition of the doctrine of evolution in general, and of that of Darwin, Goethe, and Lamarck in particular; the translation revised by E. R. Lankester," 2 vols. New York, 1883. Vol. I. p. 42.

were different forms created in the beginning by the infinite Being, and his binary nomenclature, in which each species is given a specific and a generic name, is the foundation of modern Natural History.

Definitions of Species.—TOURNEFORT (1656–1708) defined *genus* of plants to be "the assemblage of plants which *resemble* each other in structure," and *species* as "the collection of plants which are *distinguished* by some particular characters."

LINNÉ (1707–1778) said that we count as species what has been created of diverse form at its origin, and later Linné considered that all the species of a genus were originally a single species.

BUFFON (1707–1788) described species as a "continuous succession of similar individuals which reproduce themselves, and the characteristic of the species is *continuous fecundity*."

DE CANDOLLE (1778–1841), the celebrated botanist (as translated by Wallace in his book on "Darwinism")* defined the term thus: "A species is a collection of all the individuals which resemble each other more than they resemble anything else, which can by mutual fecundation produce fertile individuals, and which reproduce themselves by generation in such a manner that we may from analogy suppose them all to have sprung from one single individual."

CUVIER (1769–1832) gave what is probably the standard definition of this school: "*L'espèce est la collection de tous les corps organisés nés les uns des autres, ou de parents communs et de ceux qui leur ressemblent autant qu'ils se ressemblent entre eux.*" In 1821 the first clause of the definition was changed to "*comprend les individus qui descendent les uns des autres.*" This definition may be regarded as the foundation principle of the school of naturalists of which Cuvier was, probably, the most distinguished teacher.

ZITTEL.—In his treatise on Paleontologie, Zittel† says of species: The single species was considered, by the great classification naturalists, Linné and Cuvier, as having a real existence and fixed invariable value; this opinion was almost

* A. R. Wallace, "Darwinism; an exposition of the theory of natural selection, with some of its applications." London, 1889.

† K. A. Zittel, "Handbuch der Palæontologie," vol. I. pp. 45. 46.

universally admitted by all naturalists until Darwin came to show that this category was also variable, changing, and without fixity. The partisans of the first theory accorded to the species a sum of particular immutable characters; it had always been such as we see it, *(species tot sunt diversæ quot diversæ formæ sunt creatæ)*. . . . The partisans of the theory of Transmutation believe that species have appeared slowly, the one after the other, and by successive transformations. . . . In order to limit living species, the better criterion is furnished by their direct descent. According to Cuvier, one should refer to the same species all the individuals which were born, the one from the other, or of common parents, and which resemble each other as much as they resemble their parents; the individuals of separate species are incapable of fertile union, or produce generally only infertile progeny. In paleontology it is impossible to control real consanguinity by physiological observation, and consequently we are deprived of this criterion in the study of fossil species. . . . One ought to recognize, moreover, that the value of this criterion is not more absolute in the determination of living botanical or zoological species, as numerous species are capable of reproduction without sexual union, (as hermaphrodites, the products of scissiparity, budding, alternate generation, parthenogenesis), and there are other species, recognized as good species, the crossing of which produces fertile hybrids. . . . If, then, the delineation of species is difficult in botany and zoology, it is evident that it will be more so in paleontology. The paleontologist is limited to a knowledge of the exterior forms of fossils, and these, moreover, often incomplete, the better characters having been frequently destroyed by fossilization. . . . In general, there are referred in paleontology to the same species *all the individuals, or all the fragments, which present certain common characters, and form a circumscribed group, independent of geological range or geographical distribution; they can, nevertheless, be associated with neighboring groups by a small number of intermediate forms.*

The Theory of Mutability of Species and Evolution.—BONNET (1720–1793) advanced the idea that diversity of climate, nourishment, etc., might produce new species, and the term evolution, in its general sense, appears to have been first proposed by him.

LAMARCK (1744–1829) definitely adopted this view, and under the name of *Mutability of Species*. The term *Development* was also used by him to express the formation of new species from pre-existing species by gradual modification, and the theory was elaborately expounded by him.*

The restricted use of the word *Evolution* (as adopted in this treatise) meaning the gradual and progressive change in the form of species, as distinct from the *development* of the individual organism from the embryo upward, to which Lamarck also likened it, was first adopted, it is believed, by ÉTIENNE GEOFFROY ST. HILAIRE in 1825, in a report of his travels in Egypt; and the idea was finally elaborated in a book published in 1831, entitled "Mémoire sur le dégré d'influence du monde ambient pour modifier les formes animales." He maintained the principles of mutability of species, common descent of individual species from common primary forms, and the unity of their organization, or unity of plan of structure.

LAMARCK was prominent for the promulgation of the theory of the mutability of species, and there was warm discussion between the Lamarckian and Cuvierian schools long before Darwin produced the "Origin of Species." But before either of these great naturalists the philosophical notion of mutation of organic forms had been definitely announced. In the Ionian school ANAXIMANDER (611–547 B.C.) expressed the view as a philosophical conception. In describing the origin of things he gave utterance to the theory that out of the vague indeterminate first principle by successive transmutation man and animals have sprung.

Philosophical Importance of the Transmutation Theory of the Ionians.—Thus it is seen that as early as the beginning of Greek philosophy the Ionian school of physicists (Thales and Anaximander) recognized the principle of change in nature. Without the idea of change cause has little meaning, and from a philosophical point of view modern science traces back

* Introduction to "Histoire des animaux sans Vertèbres," etc., published in 1815–1820.

its origin to this old school of philosophy, which recognized the difference between the all and the parts, and found the parts necessarily the changed forms of the all. The notion of change of form, or Metamorphism, led to the seeking an explanation of it: and the whole idea of evolution, or the unfolding of things from that which they were not, grew up as men thought on this subject.

Antiquity of the Notion of Evolution. — As Schurman pointed out in a chapter on Evolution and Darwinism in his recent book on the "Ethics of Darwinism": "Like most of the fundamental conceptions of our knowledge, and our science, the essential elements of the theory [of evolution] are as old as human reflection; and among the Greeks we find these five constituent elements of the modern evolution hypothesis: The belief in the immeasurable antiquity of man, the conception of a progressive movement in the life of nature, the notion of a survival of the fittest, and the twofold assumption that any thing, or any animal, may become another, since all things are at bottom the same." *

Reality of Species Logically Antecedent to the Notion of Specific Mutability.—But that particular form of the conception which is formulated in the term *mutability of species* was first clearly expressed in the latter part of the last century, and for its expression it was essential that first there should be a formulated idea of the *reality of species*. The idea of organic species had to be conceived of as a fundamental entity at the foundation of the science of organisms, before any explanation of its origin, or of the laws governing its existence, could arise.

The Idea of Species as Immutable.—The school of Linné and Cuvier developed the idea of organic species, and in giving expression to the idea, which was abstract in itself it became necessary to find concrete delimitation of the species. This idea of species is as essential to the biologist as the ideas of atom, of molecule, of force, of energy, are to the physicist and chemist; and in the order of development of ideas, it was natural that the primary definition of species should include the idea of stability, and it was fully scientific too; for, as

* J. G. Schurman, "Ethics of Darwinism," pp. 43. 48.

we have seen, the species, so far as superficial and even very careful observation goes to-day, when expressed by so acute an observer as Huxley, is fundamentally a group of like individuals, alike for space, alike for time duration. "A species in the strictly morphological sense, is simply an assemblage of individuals which agree with one another, and differ from the rest of the living world, in the sum of their morphological characters." *

A Mutable Species necessarily Temporary.—The idea of "mutability," which was added to the conception of the reality of species by the modern school of naturalists, is intimately associated with the idea that the morphological form of organisms, which constitutes their specific characters, is temporary, and thus is distinguished from the characters of atoms which are conceived of as continuing the same throughout all time. The theory that the species is immutable was associated with the conception of a primary principle underlying each form which was supposed to exist from the beginning with persistent integrity.

The Question of Mutability of Species entirely Distinct from that of the Origin of Species.—This discussion of species is also a thoroughly legitimate process for the scientific investigator, and the two views alike call for an explanation of their origin. The Lamarckian school was not less free from the unscientific cutting short of investigation by referring this origin to an unknown cause. Cuvier and his school argued, We know the species, but the first cause is a sufficient cause of its origin; here it is, and we do not know how it came to be. Lamarck alike believed in scientific ignorance as to its origin when he followed Aristotle in calling in spontaneous generation as the explanation of its existence. According to Lamarck, Life is purely a physical phenomenon. All the phenomena of life depend on mechanical, physical, and chemical causes, which are inherent in the nature of matter itself. The simplest animals, and the simplest plants, which stand at the lowest point in the scale of organization have originated, and do originate, by spontaneous generation. In the first beginning

* T. H. Huxley, "The Crayfish," etc., p. 29.

only the very simplest and lowest animals came into existence; those of a more complex organization only at a later period.

The Fundamental Tenet of the Mutability School.—Thus we find that the fundamental difference between the hypothesis of the "immutability of species" of Cuvier, and that of the "mutability of species" of Geoffroy St. Hilaire, Lamarck, and Darwin, consists in the assumption by the more modern school that *the specific morphological characters of organisms are temporary; are constantly undergoing slight modification from generation to generation; and, finally, that separate species are not such from the beginning, but take their place in an orderly sequence of phenomena; that which constitutes the specific character for each case having an explanation in what preceded it, and bearing the relation of cause, or taking a part in determining what shall follow.*

The removal of "special creation" from the one theory and "spontaneous generation" from the other was the natural result of the progress of ideas,—an opening of the laws of organic evolution to scientific investigation. These two hypotheses were the natural recourses of ignorance, and the present form of our philosophy is no less obliged to find an unobservable origin for the things of whose existence we have observable evidence.

State of Opinions when Darwin began his Investigation of the Origin of Species.—This brings us to the stage in the history of opinions when Darwin began his investigations. The mutability of species had been announced and strongly supported by able advocates. The general principle of evolution had been formulated centuries before, but was rather in the stage of speculative opinion than applied hypothesis; the facts supporting and illustrating it were not greatly accumulated. Linné, Cuvier, and their schools had already defined a great number of species of plants and animals, had classified them, and had erected an elaborate systematic botany and a systematic zoology on the theory of immutability of species. The new theory seemed to shake the foundation of the science of Natural History. If there is no fixity to the idea of species, the query arose, what can we talk about? What is there left for us to investigate? But in fact, while the mutability of species was received and advocated, the idea of species was still retained, as evidenced by the title of Darwin's famous book, "The Origin of Species."

New Conception of the Nature of Species.—The change was a philosophical one; no longer was the species considered to be a permanent entity with definite boundaries, but in the definition of organic species its time-relations and its geographical distribution were elements added to those of its morphology and physiology. This was a great advance. The organism came to be recognized as not a mere concrete being independent and standing by itself, constituted at the beginning what it is and remaining so during its existence, but as a very dependent part of a greater organism, nature itself, and related intimately to its surroundings or environment, to the organisms which preceded it or its ancestry, and to those which are to follow or its descendants, as a sensitive, slowly changing reflex of all that has been and is. In the new conception there is the dim outlining of the idea (an old idea, but one which is day by day growing more distinct and of fuller comprehension) that nature itself is a greater organism in which the species is but one of the organs.

Remarkable Revolution of Thought started by Darwin's "Origin of Species."—Darwinism, although not pure evolutionism, but only one phase of it, has done more than anything else to bring about these changed views of nature. Darwin took up the general theory of evolution, and attempted to give an account of the method of its working. The title of his work clearly sets forth the essential scope of his theory: "On the Origin of Species by Means of Natural Selection, or the Preservation of Favored Races in the Struggle for Life." This definition of the origin of species implies two fundamental propositions, viz.: (1) That the species naturally varies in its characters, for the natural selection is selection among characters that differ; this is the idea of "mutation"; and (2) that the reason why one character rather than another is preserved is its better adaptation to conditions of environment; this is the idea of "natural selection."

Darwin brought out prominently the fact, that what we

call species, i.e., the descendants of common parents, vary among themselves, and that the variability is substantially universal. This was elaborated by study of the variation of plants and animals, and particularly of pigeons under domestication. The selection which man makes in his stock-breeding suggested to Darwin the idea that the very conditions of environment would act in the course of ages as selecting agencies, favoring the growth and transmission of certain peculiarities of structure or habit, and working against other varietal characters, thus eventually perpetuating the favorable varieties, and causing ill-adapted characters to become lost. Undoubtedly his observations, when a boy, of the results of stock-breeding among Leicester sheep and the ideas of Mr. Bakewell, with whom he was acquainted, impressed themselves upon his memory and were the foundation of the theory, the elaboration of which made him famous.

The Evolution Theory of Biology and the Uniformitarian Theory of Geology.—Darwin's "Origin of Species" brought the world to a vivid appreciation of the universal mutability of all organic things, and the theory which bound together the mutability of organisms was evolutionism. It is interesting, from a philosophical point of view, to note that about fifty years before, a like step of progress was reached through the uniformitarian theory of Hutton, which set forth the principle, that during all geological time, there has been no essential change in the character of geological events; but uniformity of law and conservation of force are perfectly consistent with the mutability in the results and the incessant evolution of present life out of the dying past.

Evolution and Development Contrasted.—In its general sense I find no better definition of evolution than that given by Huxley: "*Evolution or development,*" he says, "*is, in fact, at present employed in biology as a general name for the history of the steps by which any living being has acquired the morphological and the physiological characters which distinguish it.*" Evolution, as has been already noted, in this sense confuses two processes which may co-operate in the result, but which may be distinguished in their exhibition in actual facts of the history. They are technically separated under the two categories of Ontogeny and Phylogeny. *Ontogeny*, or *Ontogenesis*, is the technical term for the "history of the individual development of the organized being." *Phylogeny* is applied to the history of the genealogical development. Phylogeny, as Haeckel used it, is associated with the theory* that the steps of phylogenesis, or of ancestral development, may be deduced from the observed history of ontogenesis or the development of the individual. In order to free the term from any theory of accounting for the history, it is proposed to restrict the use of the term evolution to that part of the history of organisms which is seen upon comparing the organisms of one geological epoch with those most closely similar in the preceding geological epochs, and to restrict the use of the term development to the history of those changes which are observed on comparing the successive stages of growth of the individual organism with each other, or the history of a single cycle of organic growth.

Evolution the History of the Steps by which Variation is Acquired, not Transmitted.—It is evident from this analysis that in the case of any particular organism the steps by which it acquires the characters which were possessed by its parents are steps in the development of the individual; but the steps by which it acquires any characters not possessed by its ancestors are steps of evolution. The latter characters in every case are the varietal characters.

It is the acquirement of variation, not its transmission, that constitutes whatever there is of evolution in the history of organisms. The terms thus restricted furnish us with names which can be used independently of any theory. The facts, or series of facts, may be scientifically observed, recorded, and defined, and an explanation sought for them.

A Definition of Darwinism.—For the meaning of Darwinism we may adopt the excellent definition of the Century Dictionary. "That which is specially and properly Darwinian, in

---

* The Recapitulation theory. See, for a clear statement of the principal features of this theory, the President's address to the Biological Section of the British Association, delivered at Leeds, September 1890, by Arthur Milnes Marshall, entitled "The Recapitulation Theory," and republished in "Biological Lectures and Addresses," 1894, pp. 259–363.

the general theory of Evolution, relates to the manner, or methods, or means by which living organisms are developed, or evolved, from one another: namely, the inherent susceptibility and tendency to variation according to conditions of environment; the preservation and perfection of organs best suited to the needs of the individual in its struggle for existence; the perpetuation of the more favorably organized beings, and the destruction of those less gifted to survive; the operation of natural selection, in which sexual selection is an important factor; and the general proposition that at any given time any given organism represents the result of the foregoing factors, acting in opposition to the hereditary tendency to adhere to the type, or 'breed true'"

The Lamarckian Theory of Evolution.—"The portion of the theory of Development [Evolution] which maintains the common descent of all species of animals and plants from the simplest common original forms might, therefore, in honor of its eminent founder, and with full justice, be called *Lamarckian;* on the other hand, the theory of Selection, or breeding, might be justly called *Darwinism*, being that portion of the theory of Development [Evolution] which shows us in what way, and why, the different species of organisms have developed from those simplest primary forms."*

Phylogenetic Evolution.—We may quote again from the Century Dictionary the definition of *Phylogenetic Evolution:* "It is the name for that form of the doctrine of Evolution which insists upon the direct derivation of all forms of life from other antecedent forms, in no other way than as, in ontogeny, offspring are derived from parents, and consequently grades all actual affinities according to propinquity, or remoteness of genetic succession. It presumes, as a rule, such derivation or descent, with modification, is from the more simple to the more complex forms, from low to high in organization, and from the more generalized to the more specialized in structure and function; but it also recognizes retrograde development, degeneration or degradation."

The law of Evolution is put in a terse form by Huxley, who

* Haeckel, "Hist. of Creation," etc., vol. 1. p. 150.

expands the Latin phrase of Harvey "*omne vivum ex ovo*" into "*omnium vivum ex vivo*," and carries the evolution idea still further in the phrase "*omnis cellulæ cellula*."

The Fact of Evolution Established Beyond Controversy: the Real Nature of Evolution to be Learned only by a Study of the History of Organisms.—The followers of Cuvier, with their "immutability of species," recognized the principle of "development" in the sense above defined, but they believed that the metamorphoses, which are called evolution, are the results of independent originating force, or they discarded the belief altogether. The more modern school, represented by the idea of the "mutability of species," fully accepts both development and evolution as established facts in the order of nature. This principle of evolution is so far-reaching in its application, and so dominates the speculations of our times, that typical illustrations of it as exhibited in the history of organisms are worthy of special study in order that these applications to other things may be correctly made, for only by understanding precisely what evolution is in nature can one apply the term correctly in discussing the philosophical application of it to other things.

What is an Individual?—When we push the analysis of organic nature farther, we meet the question, What is the individual? A very superficial consideration of the problem shows us that the organic individual is not merely the sum of the matter constituting the body of the individual at any particular time. The matter of the individual is not made in the course of the growth, but it is only organized. The matter in the case is the food, which before was not a part of the individual. So that it is true to say that an organic individual develops, but the matter it uses is not in any sense characteristic of the individual, nor is the particular structure of the cells or tissues, for this is common to other individuals, but each individual differs from others in the mode and purpose of its activities, and in the results of such activities as expressed in its morphological characters.

In other words, the organic individual is what it is in each case, not by virtue of the chemical or physical materials of which it is composed, but by virtue of the form, structure, and activity of the whole as constructed. Thus the likeness in form and function, which leads to the classification of organisms as of the same species, does not arise by virtue of likeness of the matter involved in its construction, but by virtue of likeness of the agency by which the particular construction is brought about. To put this proposition in concrete form, a particular cat has the form and function it possesses, not by virtue of any qualities inherent in the bones or muscles or tissues of which it is composed, or in the cells or in the ultimate chemical elements of which it is composed, but its individual characteristics are altogether determined by the fact that it developed from a cat which was its mother.

Descent is the explanation of the particular characteristics of each individual. In dealing with such characteristics, we are dealing with the phenomena of life which are continuous, so far as our experience tells, and depend for their expression not alone upon the immediate surroundings of the individuals, but upon pre-existing living organic individuals, its ancestors.

CHAPTER IX.

WHAT IS AN ORGANISM?—THE CHARACTERISTICS OF THE INDIVIDUAL AND ITS MODE OF DEVELOPMENT.

Mutability of Organisms a Foundation Principle of all Evolution.—In an analysis of the meaning of evolution, it is essential to remember, at the outset, that the evolution takes place only in respect of mutable things. The species is said to be mutable, but it is the *organic* species as contrasted with everything else. The mutability, therefore, is respecting organisms only. I have shown how the organic "species," which one school of naturalists calls "mutable," is in one sense a mere abstract idea but in another it stands for an aggregate of real existing individual organisms. Such an earnest advocate of mutability of species as Oskar Schmidt says, "The retention of species is, moreover, scientifically justifiable and necessary, if only the determining impulses be taken into account and the definition reduced to harmony with reality;" and the definition he gives is, "While we regard species as absolutely mutable, and only relatively stable, we will term it, with Haeckel, '*the sum of all cycles of reproduction which, under similar conditions of existence, exhibit similar forms.*' "*

Morphological Similarity the Characteristic of Species.—The essential notion in species is *similarity of form.* The fact recorded in the term species is the occurrence in nature of numerous organisms of almost identical form and structure—individuals which seem, in general, to live and grow separately, but are also organically associated together. In order to explain this community of form among the individuals of the same species, we must examine into the laws by which

* "The Doctrine of Descent and Darwinism," p. 103, New York, 1878.

the individual attains its form, and to this end we must analyze the characteristics of an organism.

The Definition of an Organism.—Organism may be defined in two ways: we may point to a concrete example and say, "*That cat is an organism*," and then take away all those characteristics which are peculiar to the particular example, as its hair, its limbs, its eyes, its teeth, in fact, all its special organs and parts, and come down to a fully abstract definition of an organism, of which the cat is a concrete example; or we may take the philosophical definition, and with Kant define the organism to be "*that whose every part is at once the means and end of all the rest.*" For our purposes it is better to combine the two methods, and say, *An organism is a living being whose every part is at once the means and end of all the rest;* for it should be insisted that, whatever its full meaning may be, *living* is an essential quality of any organism that either develops or evolves, and the idea of organism includes the necessary *relationship of the parts to each other and to the whole.*

Living and Performance of Physiological Functions are Essential Parts of the Definition of an Organism. — "Under one aspect," says Huxley, "the result of the search after the rationale of animal structure thus set apart is Teleology, or the doctrine of adaptation to purpose; under another aspect it is Physiology." *

Inversely, then, a dead animal is not an organism. It is only a mass of organic matter which some organism has constructed. So much are we engaged in handling dead animals and plants that we are apt to overlook this important distinction. Too often the modern naturalist conceives of the organism as only an aggregate of matter having a definite form and structure of parts, as a house might be defined as a building made of mortar and bricks.

A Zoological Specimen in the Museum as much a Vestige of Organism as a Fossil.—The animals we see in the zoological museums and dissect in the laboratories are as much remains or vestiges of organisms as are fossils; growth and structure

---

* Thomas Henry Huxley, "An Introduction to the study of Zoology illustrated by the Crayfish," p. 47, New York, 1884.

are in intimate association in the organism, and the instant the organism ceases those changes incident to growth there remains the inert result of the living, that is, the dead, animal.

Living Implies Change, and Change is Incessant in a Living Organism.—Living implies change in the organism, and incessant change. This change is what makes growth possible. The organism at any particular stage is only the morphological result of the previous growth, and what we recognize as the adult form of the individual is as truly mutable as the species itself. The individual organism, if exactly defined, is not precisely the same for any two days or moments of its existence, but one of its fundamental characteristics is that it grows, i.e., it has development. Almost the same might be said of any of the parts or organs: so long as they are acting they are undergoing waste, and repair, and incessant change; as soon as this process ceases they cease their organic function, decay, and return to their material elements. In the organ, in the individual, in the species, or in the whole organic kingdom, the morphological form and the physiological function are of a temporary nature, and thus essentially differ from the physical or chemical properties of matter.

An Organism is an Aggregate of Cells.—An analysis of a plant or animal demonstrates it to be composed of "cells." Each individual organism is morphologically an aggregate of cells; these cells are not all alike, nor are they combined in the same manner. Another proposition may be accepted without further examination: every animal or plant begins its "existence as a simple cell, fundamentally identical with the less modified cells which are found in the tissues of the adult."

The Organic Cell the Morphological Unit.—The simplest form of the cell, or, as Huxley calls it, a "morphological unit," may be conceived of as a mere mass of protoplasm devoid of cell-wall and nucleus. He sets forth as fundamental propositions that, 1. "For the whole living world the morphological unit, the primary and fundamental form of life, is merely an individual mass of protoplasm, in which no further structure is discernible; 2. That independent living forms may present but little advance on this structure; and 3. That all the higher forms of life are aggregates of such morphological units, or cells, variously modified." *

The primitive form of the organic individual is the simple cell of microscopic size, globular in shape and with no distinguishable differences in the structure of its contained protoplasm. If higher powers of microscope could be brought to bear, it is not improbable that, like the nebulæ of the macrocosm, this amorphous unit of the organic microcosm might be resolved into complexity; but, as we know them, cells are found almost universally to possess three elements of structure: (1) the protoplasmic substance of the cell, (2) a cell-wall or marginal sheath, and (3) a nucleus within.

The Three Ways by which Cell-modification is Accomplished.—There are three ways by which diversity of form is attained by the cell:

(1) By *movement of the cell itself*, exhibited in change of shape of its exterior form, or of the cell-wall. This is seen in the Amœba (Fig. 51), which, by drawing in one part and extending another, assumes various forms, temporarily, but remains in the simple cell state.

(2) The second method of attaining diversity of form is by *cell-division*, which is the common method by which growth is effected. *Reproduction* of a new cell is accomplished by such division of the original cell, separation of one part from the other, and completion of its outlines by each part until division into two distinct cells takes place. The Protozoa are characterized by this mode of development, and by the necessary failure to attain complexity of structure of the individual, which reaches no higher stage of diversity than the unicellular stage.

(3) The third method of attaining diversity of form is by *cell-multiplication* within the individual.

Metazoa Characterized by Histogenesis, or the Formation of Tissues.—All the animals of the classes higher than Protozoa are ranked together under the general name Metazoa, and are distinguished from them by this differentiation of the substance of the body into cells. This, which is the second

* Huxley, Ency. Brit., 9th ed., vol. III. p. 652.

method of organic development, is called *histogeny*, or *histogenesis*—the origination or development of tissues; and the terms *cryptogeny* and *cryptogenesis* may be used to distinguish from it the first method of organic development, which ends in the reproduction of cellular units, and is confined to simple enlargement of the cell, as in the Protozoa.

Histogenesis, Cryptogenesis, and Phylogenesis.—In *histogenesis* the organic unit is enlarged by the division of the initial cell into many separate cells forming a compound organism known as the metazoal individual. In *cryptogenesis* the organic unit is a simple cell. As histogenesis begins with cryptogenesis, and is an enlargement of the scope of organic growth, so we may conceive of *phylogenesis* as an enlargement of histogenesis, in which the unit is the organic species, and the progress is in terms of specific forms, new species arising by evolution of the old and modification and expansion of the ancestral types in their descendants. The growth is growth of the race, and the specialization is in functions of the individuals, first seen in the production of sex; this specialization is further developed in the co-ordination and co-operation of the members of a family, and is still more highly elaborated in the community or the race.

Analogy between the Cell and Organism and the Molecules, Elements, and Minerals of Inorganic Matter.—The results of these several modes of growth of the organism are analogous to the categories used in chemical nomenclature. There are physical units which are called molecules, which may be compared to cells, the organic units. The chemical element is a molecule, or mass of molecules, exhibiting uniform properties, or chemical reactions. A mineral is a combination, or it may be a simple element, exhibiting definite and uniform chemical composition, and physical characters of weight, hardness, crystalline structure, etc. As the molecule is resolvable into imagined atomic constituents, so the organic cell is resolvable into its protoplasm, and according to the theories of some into innumerable *pangenes* or *ids*, each having its personal characteristics.

The Individuality of the Organism.—On the other hand, as any particular mineral exists only temporarily and under special conditions, so the organic species may be looked upon as a temporary thing made up of a certain number of actual individuals, living at a particular time and under particular circumstances, the individuality perpetuating itself by the process of generation. But here the analogies cease, as is explained elsewhere; the incessant changing of the organic form and function of the living organism distinguishes it fundamentally from matter in any other condition.

Growth and Reproduction of the Protozoa and of the Metazoa Contrasted.—As will be seen from the above remarks, the function of reproduction in the Metazoa is a specialization of the simpler function of growth of the unicellular Protozoa. Growth in the Protozoa seems to be limited by what may be called the capacity of the organic cell, and reproduction then consists merely in producing new cells, or in the multiplication of unicellular organisms.

Generation the Fundamental Function of an Organism.—In the Metazoa the growth capacity is enlarged, and in these higher animals reproduction or generation is no longer the function of the whole organism, but is specialized off as a part of its activity; and in the structure of the organism special parts, tissues, or organs are set apart or differentiated for the execution of this special function. The remaining activities are spent in the development of the individual. Individual development, and all auxiliary activities, have to do with actually existing conditions of life, but generation looks forward in its very essence to conditions that have not yet appeared. *Generation is, therefore, at the foundation of all organic life and history, and in the process of generation organs are constructed before they act, and independently of the external environment to which they must be adjusted when they act.*

Summary of the Steps of Progress in Organic Development.—To summarize the steps of progress in the organic development, we find, first, simple growth; the cell increases by absorbing matter from outside, accumulating it, and thereby augmenting, both as to physical size and to the amount of its organic force, whatever that may include. This process goes on until the cell reaches the limit of its individual capacity, until growth ceases.

Secondly. Some sort of division or fission sets in which begins with the cell-nucleus; if fission becomes complete, it is unicellular reproduction and the organism is protozoal. This process, repeated over and over again, is what may be called cell-genesis, or cryptogenesis. This is unspecialized growth, and the cell, when considered as carrying on independent existence, may be called an undifferentiated organism.

Thirdly. When the fission of the developing cell is incomplete within the walls of the cell, the process goes on until there is repeated cell-division, or segmentation, and the dependent cells are more or less specialized and combine to form *tissues*.

Fourthly. The tissues develop into separate organs, capable of carrying on special functions, and we have a metazoal animal, in which the several parts act for the interest of the whole body. The product is a complex *organism* with organs made of specialized cells performing special functions.

Growth, strictly speaking, is thus a function of the cell, which culminates in cell-multiplication by fission, or partial fission, augmenting the mass and force of the individual.

Development is that kind of growth which takes place in a multicellular organism when, by generation, a nucleated cell is set apart, protected, nourished, and by division and differentiation is elaborated into a complex organism, without regard to the growth of the parent—even at its expense, and when fully constructed is set free to begin independent life for itself.

Evolution is that kind of growth which is expressed in the specialization of functions and differentiation of organic structure in some members of a species, enabling them to exceed the capacity of their ancestors, and to adapt themselves to conditions beyond or other than those to which the parent form was adapted. The evolution is exhibited in a series of forms, succeeding one another, in which varietal, and ultimately specific, differences distinguish the later from the earlier members of the series. Such a series is called a *race*, and the representatives of a race which are alike are called a *species*.

Embryology.—The development of the individual is particularly discussed under the name of Embryology, and the student is referred to special treatises on this subject for information regarding the details of the process, but a few general statements may be of use in forming a correct notion of the nature of organisms in general.

The typical cell is composed of a mass of *protoplasm* with a more or less distinct *cell-wall*, and, generally, a *nucleus*, very minute in size and escaping resolution into its elements, but giving evidence of performing some very important functions in the cell when examined with the highest powers of the microscope.

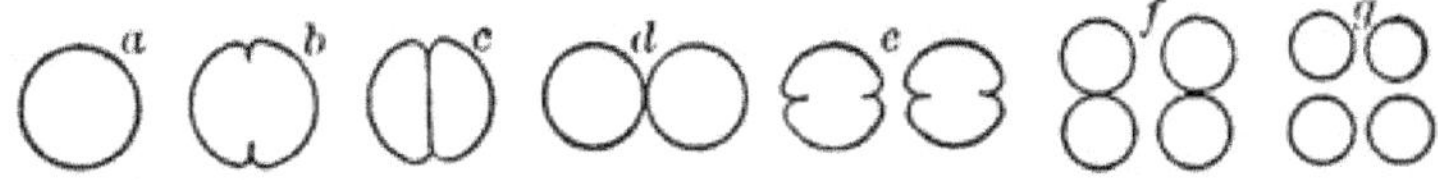

FIG. 43.—Agamogenesis by fission. *a–g* = the several steps in the process of generation from the parent form *a* to the production of four separate individuals *g*.

The Functions of a Metazoal Organism; Generation.—In the Metazoa there are three groups of functions, viz., sustentation, generation, and correlation. *Generation* is the name of the function by which organic individuals are produced, or, as is commonly said, reproduced.

Agamogenesis.—There are three (or, including alternate generation, four) modes of generation. *Agamogenesis*, of two

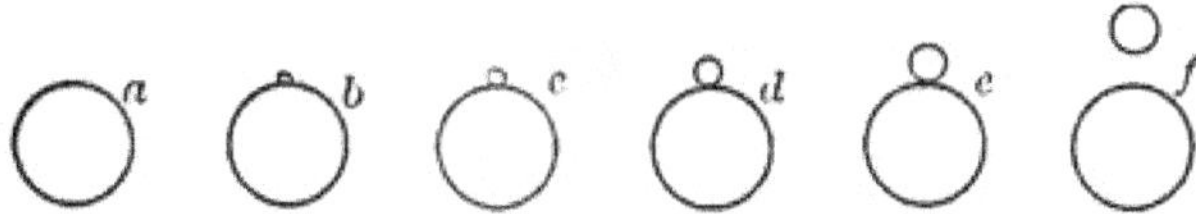

FIG. 44.—Agamogenesis by budding. Generation in which the parent individual retains its integrity, sending off a young but relatively immature offspring (*f*) as an external bud.

kinds, by fission (1) and by budding (2). This mode may be represented diagrammatically by the following series:

I. In this series the simple parent individual (*a*) by subdivision into sub-equal parts becomes four separate individuals (*g*), each capable of independent existence (Fig. 43).

II. The second mode of agamogenesis may be represented by the above diagram (Fig. 44).

Here a modified fission takes place, the original individual retaining its integrity and sending off a bud, which, after partial development, is separated, completely or partially, from the parent: this is called *budding*, from its similarity to the mode of budding in vegetable growth.

III. This budding process may proceed within the parent individual when separation takes place by an act of expulsion,

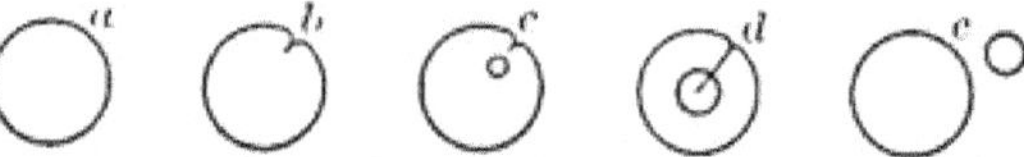

Fig. 45.—Agamogenesis by internal budding, in which the young germ is formed within the body cavity of the parent, and when complete is suddenly expelled as a free individual.

suddenly instead of gradually, which gives a third type of agamogenesis, as may be illustrated by the diagram Fig. 45.

In this case the offspring comes forth immature in development, but complete in organization. All three of these modes

Fig. 46.—Monœcious gamogenesis. Sex differentiation, represented by the symbols ∧ male, + female, taking place within the parent individual (a), the several steps consisting of union of the two elements (b), development of the germ (c), its discharge (d e) and becoming a free individual, the parent retaining its integrity (f).

of generation are called agamogenesis, because there is generation without sex differentiation.

Gamogenesis.—*Gamogenesis* is that mode of generation in which distinction of sex is accomplished in the individuals be-

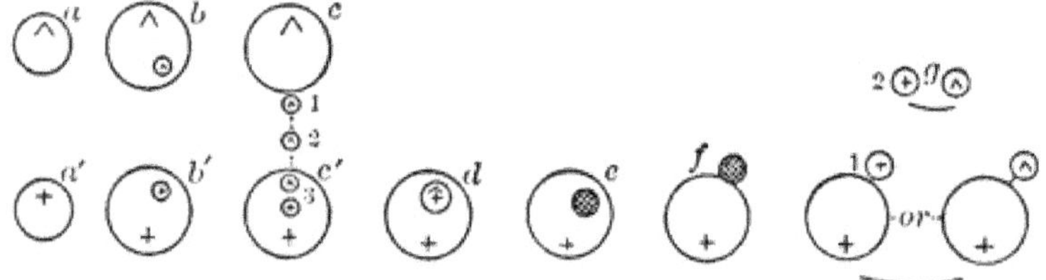

Fig. 47.—Diœcious gamogenesis. In this mode of generation sex differentiation has taken place before the individual is complete, and co-operation of two distinct individuals is essential to each act of generation (a a'). Separate organic elements are developed in the sex individuals (b b'); the spermule is extruded from the male individual (c 1 2), is brought into contact with the ovule (c'), the two elements unite (d), segmentation and development of the ovum (e f) take place, the ovum is developed as a dependent individual until it is capable of independent existence, when it is extruded and set free (g 1 and 2) either as a male or as a female individual (g 2).

fore generation begins. Gamogenesis may (IV) be *monœcious*, in which case the sex differentiation has proceeded only so far as to differentiate the organs within the body of the individual organism, each individual developing both of the sexual elements. This mode of genesis may be represented by Fig. 46.

In this case the generation is sexual, but *hermaphrodite*, and the product of generation is set free after being developed sufficiently to carry on independently the functions of life.

The other mode of gamogenesis is *diæcious* (V), in which the differentiation of sex has proceeded so far as to affect individual life, and to require the co-operation of two different individuals for the accomplishment of the function. This is the more frequent mode of generation in the animal kingdom. It may be represented diagrammatically by Fig. 47.

The Several Stages of Development in the Higher Organisms.—In this series there are several stages of development which it is

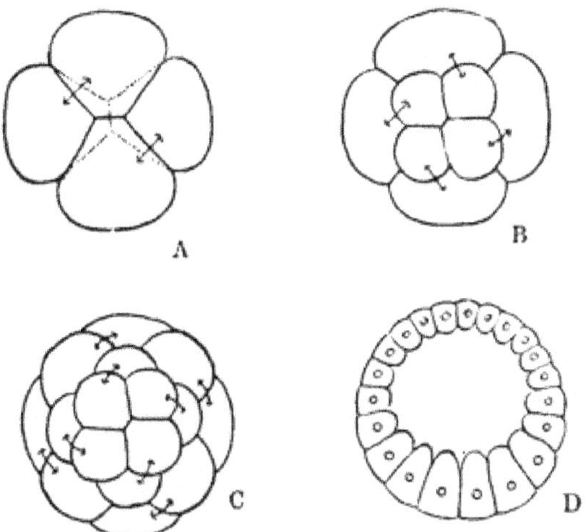

FIG. 48.—Segmentation of the ovum. *A*, *B*, *C*, various stages of segmentation. *D*, blastula. (After McMurrich.)

important to note. There is, first, the stage of *sex differentiation* in the individual, the one being called male, the other female. This appears early in the life of the individual, but in its earliest stages there appears no discernible difference of form in the organs of the two sexes.

Second. This distinction is carried on independently in the growth and development of each kind of individual; organs are specialized and differently formed, and finally result in the production of specialized cells, called in the one case *Spermule* (*Spermatozoan*), and *Ovule* (*Ovum*) in the other.

Third. The *conjunction* of the spermule and ovule, formed

within the organism of separate individuals, is the next essential step in the process, and the ovule, thus *fertilized* (as the result of this conjunction is called) proceeds under proper conditions to further develop, and when sufficiently developed for independent life is thrust out of the parental organism, is separated, and becomes a separate individual, as represented by the stages, *d–g*, Fig. 47. The distinction of sex is again represented in the new-born individual which is born already differentiated (2 *g*, Fig. 47) in this respect, and as it matures develops the organization of either a male or a female individual, and only as thus differentiated is the continuation of the process of reproduction possible.

With the third stage of cell-development, above described, begin the processes of cellular differentiation, or histogenesis, within the walls of the cellular organism. The segmentation of the contents of the interior of the ovule is the first step in this process, and results in the formation of innumerable spherules. The cell in this condition is called a *blastula*, or *morula* (according to the extent and mode of its segmentation). The *blastula* results when the segmentation affects only a part of the cell-contents, and a hollow ball-like cell is formed; in the *morula* the whole cell-contents are segmented, or, at least, the unaffected part is relatively very small, and the result is a solid ball of cellules (Fig. 48).

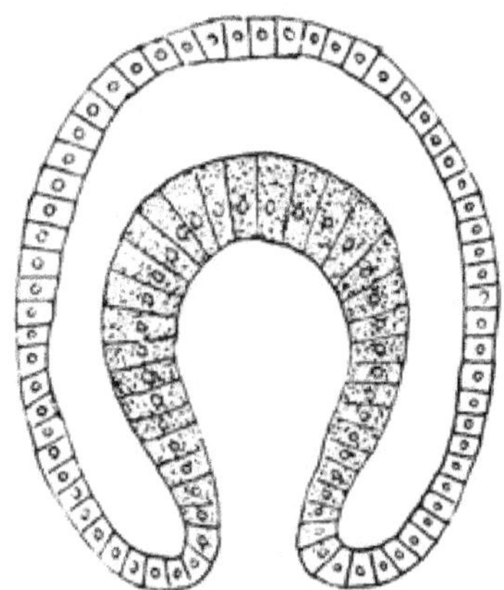

FIG. 4) —Gastrula stage of the ovum. (After McMurrich.)

Fourth. The next stage of development is the formation into a *gastrula*, in which specialization of the secondary spherules or cellules take place, and an outer and an inner layer are formed. The typical gastrula is formed by the dimpling in of the hollow sphere of the blastosphere to form a two-layered cell (Fig. 49).

The Primitive Tissues, Endoderm, Ectoderm, and Mesoderm.— The *Ectoderm* and the *Endoderm* are the primitive undifferentiated tissues from which develop, as growth proceeds, the special organs. There is also formed very early in the development of most of the higher animals, the Metazoa, an intermediate layer called the *Mesoderm*.

These several stages of histogenic development distinguish the Metazoa from the Protozoa, and the distinction might be stated by describing the Protozoan as a *cellular animal*, the Metazoan, as a *tissue-bearing animal*.

The Special Organs Arising from Primitive Tissue Layers.—This is not the place to go into further details regarding the mode of development of organisms, but, as illustrative of the degree of specialization of function already outlined in the distinction of the tissues of the gastrula into Ectoderm, Endoderm and Mesoderm, the following summary of the organs which develope in the Vertebrate from each of the primitive tissue layers is given.

1. From the *Ectoderm* arise the epidermis, the nervous system, and the infoldings at each end of the intestinal cavity.

2. From the *Endoderm* arise the mesenteron and its extensions, the lung, liver, etc., and the notochord (later, the backbone).

3. From the *Mesoderm* arise dermis, muscles, connective tissue, bony skeleton, and probably the reproductive organs.

The Embryo Stage, characterized by Dependence and Passivity, is not subject to Individual Struggle for Existence.—Fifth. The stages of development, enumerated under the preceding section, take place either within the cavity of the parent body or within a food-holding case provided by the parent; in other words, the organism is not free, building up its growth by its own energies, but it is still attached and dependent upon the vital conditions and resources of the parent. It is called a *germ*, and the *embryo stage* of development.

In the development of each metazoal animal there is this dependent stage of development, the embryo stage, of greater or less length, in which the young organism is not an independent individual, and therefore *is not subject to the action of struggle for existence*.

The most important fact to note regarding this stage, is, that it is the stage in which all the differentiation of tissues (up to the formation of the completed organs—those, at least, that are essential to independent activity) is carried on with relative passivity of the embryo itself; and the determination of all this development is traceable directly to the parent, and not to the environment of the developing organism. However much the length and extent of this embryo stage may differ in different kinds of animals, it is clear that there is such an embryo stage of development in all metazoa.

The Stage from the Free Existence of the Individual to the Maturing of its Functions.—Sixth. The next step in the development is the setting free of the organism from its embryo stage; its birth marks the beginning of the infantine stage in the higher Vertebrates. The higher the differentiation and the more complex and specialized the organization, so much the longer is the dependent or preparatory stage extended.

In the higher animals, for instance, some of the systems of organs are not completed at birth, particularly the generative system; these gradually mature, and the stage from birth to the perfection of this system of organs is the *infantine, larval*, or *juvenescent stage*. Full maturity is reached only when the whole organism is fully developed and capable of independent life and the execution of all the functions of life.

The Cell an Organism.—From what has already been said the essential elements of the organism may be learned. Recurring to Kant's definition of an organism as "That whose every part is at once the means and end of the whole," we observe that one of the first marks by which we recognize the simplest cell to be an organism is its division into parts, with what we assume to be different functions, because they do play different parts in the history of the cell.

Differentiation of the Cell a Mark of its Organic Nature.—The simplest differentiation of parts which we are able to observe is that expressed by the *cell-wall*. This is a differentiation of the superficies as a protective shelter for the interior. If, in contrast, we break open a crystal there is no essential difference between the outer and inner parts. A further specialization of parts and function is seen in the *nucleus* as a differentiated part of the cell. All cells do not appear to be possessed of special cell-walls, but the lack of them may be due to the imperfection of our vision, or to imperfectly formed cells; although the cells whose existence appears intrinsically dependent upon their own activity possess the nucleus, it is not fully evident what the function is which the nucleus plays. It is sufficient for the present purpose to note that it is a specialization, by the activity of the whole cell, of a part of itself for the execution of some function essential to the existence of the cell as a whole. Morphologically it is a *differentiation* of form and structure; physiologically it is a *specialization*, a division of labor or function, within the cell. When the cell acts in generation the same principle is at work; that is, a partition of material substance, or of morphological characters, with a retention of common interests. So long as the segmentation of the yolk goes on there is the differentiation of parts, but each part is essentially a part of the whole, and the segmentation is but an increasing of parts with the growth of the individual. As the segmented parts arrange themselves into orderly series, and, like soldiers dividing into platoons and companies, they march off to construct themselves into organs and tissues, the same principle of organic growth is expressing itself in the organism—the enlargement of the function of the whole by the increase of the number of active parts.

Differentiation and Specialization the Marks of an Organism.—Differentiation and specialization are intrinsic marks of an organism. They are essentially *processes of increment of parts and functions by division*, and not by addition. The activity of the organism ever tends to increase heterogeneity, or dissimilarity of kind of its parts. The activity of non-organism tends to the decrease of heterogeneity. In gravitation this is illustrated wherever the law of gravitation expresses itself in action; two things tend to approach more nearly to a state of uniformity regarding the law of gravitation, and so the final end of activity of the law of gravity would be a perfectly homogeneous mass, in which the attraction in every direction would be uniform. So chemical action is a process by which the heterogeneity of chemical composition is reduced; the acid and the salt unite to form a more stable compound, each of the heterogeneous chemicals uniting to form a homogeneous compound. The final result of chemical action is the compound with homogeneous properties throughout, theoretically and historically composed of sundry elements, but effectively simple, uniform and homogeneous. So, too, in crystallization the tendency is, in the heterogeneous solution, for the like things to associate according to regular arrangement of particles; from heterogeneity of arrangement the law is toward simplicity and regularity of form.

The Attainment of Heterogeneity.—When these two modes of activity come into conflict the organism expresses its vitality, we say, by overcoming the disintegrating chemical and physical forces about it. *The intrinsic tendency of organism is, then, to attain heterogeneity, or dissimilarity of kind, dissimilarity of form, morphologically, and dissimilarity of function, physiologically.* This we see in the development of the cell, in the construction of tissues and of organs, in the growth of the individual, or technically, in all the stages of *ontogenesis*.

Grand Results of Ontogenesis, or Development of the Individual.—This is not the place to discuss the details of ontogenesis,—in the departments of Histology, Physiology, Zoology, and Botany these details are fully elaborated; but it is important to note what are the general results involved, or the history of the stages by which the individual attains its distinguishing characters. The first analysis of the organism shows us that the two primary characteristics of organism are form and growth, and, in describing any individual organism, to be complete, our description must include an account of both the *morphological* and the *physiological* characters. From the earliest life of the cell this development is a process of division—division of substance or *differentiation*, division of action or function, i.e., *specialization*. The great complexity of the higher organism is accomplished, not by addition and aggregation of new particles from outside, but it is a work of the cells from within, taking in crude physical matter, assimilating and reconstructing it, and then, by subdivision developing the general structure. In the higher organism the result of this elaboration is seen in a great elaboration of structure and differentiation of parts, called organs, and of the specialization of the functions of these organs

Classification of the Functions of a Vertebrate.—Analysis of a highly specialized organism, such as a vertebrate, presents us with three groups of functions, viz.; Sustentation, Generation, and Correlation.

I. Sustentation, or Assimilation, is seen in the various processes of what we are accustomed to call growth, the taking in and digesting of food, and the building up of tissue.

In assimilation two kinds of results are attained. The morphologic effects are technically called metabolic changes; these may be divided into changes of two kinds: Constructive, or *Anabolism*; and Destructive, or *Katabolism*. The destructive process, or katabolism, results in two special functions: *Secretion*, which is the preparation of products necessary to the anabolism, or to the constructive work of the organism; and *Excretion*, or throwing off of useless products of the activities of the organism.

II. Generation, or Reproduction—vegetative multiplication.

III. Correlation, exhibited in higher organisms in two ways; as (*a*) Contractility—seen in muscular action, and (*b*) Irritability—as seen in responses to any exciting cause or sensation.

In the following table is shown the relation of the systems of organs and special tissues to these groups of functions:

I. SUSTENTATION.	
(I^a) Nutritive	Alimentary system: mouth, stomach, intestines, etc.
(I^b) Circulative	Circulatory system: heart, veins, etc.
(I^c) Purificative	Excretory organs: kidneys. Respiratory organs: lungs, etc. Secretory organs: liver, salivary gland, pancreas, etc.
II. GENERATION	Generative organs: ovaries, etc. Auxiliary organs: egg capsules, uterus, mammæ, etc.
III. CORRELATION.	
(III^a) Contraction	Muscles. Skeletal parts: exo- and endo-skeleton, for fixation, support, protection, and offence, as teeth, claws, bones, shell, coral, etc.
(III^b) Irritability	Nerve-ganglia: nerves and brain. Sense organs: eye, ear, etc.

Such are the steps of the growth and development of the individual by which it passes from a condition of homogeneous protoplasm to the elaborate organization of the highest animal.

Are the Laws of Ontogenesis the Same as those of Phylogenesis?—If we are right in stating that this increasing of the heterogeneity is an essential and fundamental law of organism, does it follow that it is also an essential and fundamental law in the processes of phylogenesis, or evolution of species?

The Meaning of Function.—Before answering this question it is necessary to consider that the use of the term function, as applied to an organ or part of an organism, is quite analogous to the use of the term property as applied to a chemical or physical substance. The mineral loses its crystalline properties when it is melted and the morphological arrangement of its particles is destroyed, although it is the same matter as before, and for the reason that the crystalline properties consist in the morphological arrangement of the molecules, not in their chemical composition: so the animal has lost its proper organic function when the physiological processes cease to operate, although the morphological form and constitution of the organic structure still remain. As the crystalline properties are the peculiar marks of the mineral, so the physiological functions are the peculiar marks of the organism, and, teleologically, the structure of the organism is built up for the purpose of these functions. The question thus arises: In case there are hindrances to the accomplishment of the functions of any organism as it develops, is it not according to analogy in the other fields of nature to expect the organ to adjust itself to the hindrances to the extent of the capacity of the organism to vary its form?

A mineral in crystallizing arranges its particles so that, left free to express its characteristics, a particular crystalline form will appear. If a physical obstruction appears in its way, this form will be imperfect, but the law of crystallization is expressed as far as possible; the whole process of crystallization does not cease because of the hindrance to its perfect action.

If we consider function to exist prior to, and to be the *raison d'être* of organization, it is to be expected that functional activity of growth and development will go on normally at the expense of change of morphological form.

Normal Growth.—This explanation assumes that there is a normal growth, and the determining of what is normal to each individual is found in the ancestry; i.e., at the outset of embryonic growth the normal function of the development of the individual is already determined. This includes the attainment of the morphological and the physiological characters of the class, order, family, genus, and species to which the organism belongs. The egg at the first appearance of the embryo is determined not only to be a vertebrate, but a bird, of the order Rasores, of the suborder Gallinæ, of the family Phasianidæ, of the genus Gallus, and of one of the many varieties of the species *Gallus domesticus*. Such is the normal development for that particular embryo. The laws of the development in its every step may be studied, and have been very fully traced in this particular case, and the knowledge of the law is based upon the observed order of these steps in the development; the inference which we naturally draw is that every new development of a similar egg will be the same.

Natural Selection, as an explanation of the changes which transpire in phylogenesis, assumes that the slight adjustments of the morphological characters, which take place in the ontogenesis of the individual, are added to the determining factors of development for the next generation; that adjustments which are very slight in each case, by accumulation from generation to generation, bring about the differences which distinguish the various species, genera, families and orders of the classes of the animal kingdom. And this is what is meant by "descent with modification." Instead of the idea of descent along a uniform line, in which the offspring differs in only unimportant and strictly variable characters from any of its ancestors, the school of Darwin holds that the slight variations observed (between the offspring and parent, or among the offspring of a common parentage) do not tend to become less in succeeding generations, but that the variations have unequal values in relation to the advantage of the individual; and in the struggle of individuals for life, those individuals possessing the slightest advantage over their fellows will, in the long-run, survive them in the race, and they will increase and prevail while the others will drop out and be lost.

Definition of Ontogeny and Phylogeny.—In the analysis of Huxley's definition of evolution (or development) the two-fold division of the history is adopted, which is expressed in part by the terms Ontogenesis and Phylogenesis, introduced by Haeckel. Haeckel briefly defined these terms, as follows: *Ontogeny*, or *Ontogenesis:* The history of the development of the individual (including Embryology and Metamorphology); *Phylogeny*, or *Phylogenesis:* The paleontological history of the development of the ancestors of a living form. It is proposed to restrict the term development to the meaning expressed by Ontogenesis, and to restrict the use of evolution to Phylogenesis. In Ontogeny we find the individual organism beginning with a great majority of its lines of development or steps of metamorphism already determined for it. Take, as an example, the crayfish, which Huxley has so interestingly dissected and described,* and of it we can say at the first stage of the embryo that in case it lives at all, whatever the conditions of environment may be, it will develop all the characters of the branch, class, order, family, genus, and species to which it belongs. Its name, *Astacus fluviatilis*, applies to it in all stages of its development from the embryo up (Fig. 50).

The Main Features of Development Predetermined before they Begin.—We can predict before any trace of the characters appear (with as great a degree of certainty as we can predict the result of combining a given acid with a grain of chemical salt) what the path of development will be which the embryo will take if it continues to grow. It will surely develop a jointed body, with the articulated limbs and chitinous crust of the Arthropoda. It will surely develop a breathing apparatus of gills situated on the maxilliped and legs of the Crustacea. The appendages of the cephalothorax will certainly be antennæ, and the specialized biting mouth parts of the sub-class Neocaridæ, not the simple legs of the more ancient sub-class

* "The Crayfish, an Introduction to the Study of Zoology" (Appletons, 1880).

Paleocaridæ; it will have the twenty segments, the specialized carapace, the pair of mandibles, the two pairs of locked maxillæ, and other characters of the order Decapoda, and all the peculiarities of the family Potamobiidæ will be strictly carried out. These concern the whole of the morphology, but in some characters of still less importance the certainty is not so great. This individual will develop on the first somite or ring of the abdomen small appendages,—certainly if it be a male, and exceptionally if it be a female,—whereas, if its

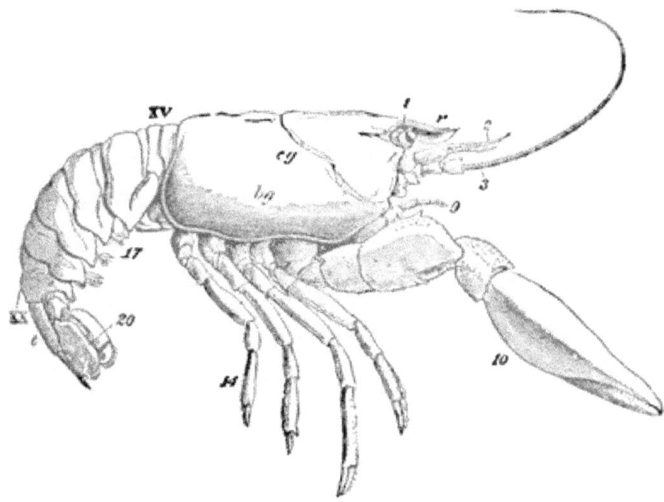

Fig. 50.—*Astacus fluviatilis.* Side view of a male specimen (nat. size). *bg*, branchiostegite; *cg*, cervical groove; *r*, rostrum; *t*, telson; 1, eye-stalk; 2, antennule; 3, antenna; 9, external maxillipede; 10, forceps; 14 last ambulatory leg; 17, third abdominal appendage; *xv*, the first and *xx* the last abdominal somite. (After Huxley.)

ancestors had been the closely allied Parastacidæ, no appendages would be developed. Again, in all the details of structure of parts it will be a true Astacus, and not a Cambarus, a closely allied genus; and finally, if it were taken to California, and placed under identical conditions with the native *Astacus nigrescens*, it would still differ in all its specific characters from that species—characters which consist mainly in differences of form and proportion of the parts, which are in number, structure, and function the same for the two species.

Slight Possible Effect of Environment.—Environment might produce slight modification in some of its very insignificant characters, but otherwise its total anatomy and physiology were predetermined when it began its development. So it is with all organisms. It was this fact, of the perfect repetition of all the essential characters of the ancestors in the new individual, that seemed, in the minds of the early naturalists, so absolutely to fortify the belief in the immutability of species. The slight modifications in unimportant details appeared as mere accidental imperfections of the individual. But it was in these slight variations that Darwin found the secret of evolution.

CHAPTER X.

WHAT IS THE ORIGIN OF SPECIES?—THE PROBLEM AND ITS EXPLANATION.

WE have seen that there are organic individuals; that they all, however complex their organization, begin as simple cells, and pass through, in each case, definite stages of development, assuming by degrees greater and greater differentiation of the cell. The chief stages of this development are the cellular segmentation, the formation of tissues in embryonic growth, and the attainment of maturity by steps of modification which are in almost every observable particular the exact repetition of steps of modification which their immediate parents passed through in attaining their maturity.

Variation and Mutability Essential Presumptions in the Discussion of Origin of Species.—The differences which the individual presents, when closely compared with its parents, are called *variations*, or varietal characters. The characters which each individual possesses in common with its parents are classified according to their importance and permanence, and arranged in order from lesser to greater, as specific, generic, family, ordinal, class, or branch characters.

It is a generally accepted belief that the assumption by the individual of all of the characters which it bears in common with its immediate ancestors is sufficiently accounted for by what are called the natural laws of reproduction: that the slight departure from exact repetition is an insignificant and indeterminate accident of all organisms, or that it is an expression of the imperfection with which the process of reproduction acts.

The theory of zoologists of the first half of the century was that the species were immutable; that variations were not cumulative, but were always simply variations, the species continuing so long as the race continued to reproduce in its original integrity. With this theory there was no way to account for species, except by assuming that the difference between two species is intrinsic, and is not to be accounted for by the natural laws of reproduction.

The problem of the origin of species came to be a question for scientific investigation and speculation at the time when the idea of fixity of those characters was replaced by the theory that variability belonged to the specific as well as to the so-called varietal characters. In other words, in discussing the origin of species we assume that reproduction is not a process of exact, but of inexact repetition of characters, or of imperfect reproduction of ancestral characters in the offspring.

Variability an Inherent Characteristic of all Organisms.—Variability is thus assumed to be an inherent characteristic of all organisms, and origin of species has primarily to consider how comparative permanency of characters, and of different sets of characters in different lines of descent, is brought about.

The Origin of Form, not of Matter.—The origin of organic matter takes us back to the earliest stage of the universe, and as to a choice between a spontaneous origin from inorganic matter, or an eternal existence of the two kinds of matter, theories may differ, and for our purposes it is useless to inquire. Our search is for the origin of forms expressed by organisms, and since our studies of paleontology present us with an orderly procession of changing forms, it is quite legitimate for us to seek among fossil forms for a scientific explanation of the origin of the separate forms, which we call species.

Definition of Species whose Origin is Sought.—The definition of species, quoted from Huxley, will suffice for the present stage of this study of science: "The species regarded as the sum of the morphological characters in question, and nothing else, does not exist in nature; but it is an abstraction, obtained by separating the structural characters in which the actual existences, the individual crayfishes, agree, from those in which they differ, and neglecting the latter."

But again: "Species, in the strictly morphological sense,

is simply an assemblage of individuals which agree with one another, and differ from the rest of the living world in the sum of their morphological characters;" and further, "in the physiological sense, a species means, firstly, a group of animals the members of which are capable of completely fertile union with one another, but not with the members of any other group; and, secondly, it means all the descendants of a primitive ancestor, or ancestors, supposed to have originated otherwise than by ordinary generation."*

Meaning of "Origin of Species."—What, then, do we really mean when we ask, What is the origin of species? It is not the sum of morphological characters, which Huxley says does not exist, but the morphological characters themselves, which concern us. It is not the assemblage of individuals which agree or differ one from another, or the group of animals which have certain capabilities and have certain common ancestors, whose origin we are seeking; it is *the origin of those differences and agreements in morphological characters which are the marks of the morphological species, and of the capabilities and disabilities which constitute the characteristics of the physiological species, that is meant by the phrase "origin of species."*

Development of Individual Characters Known and Observed.—The naturalist is familiar with the development of the individual; he knows very well that the adult differs by well-marked morphological characters from the infant, and more so from the embryo; and he further knows that the stages of development are brought about by successive minute changes of form. The difference existing between the gamecock, with its complex physical organization and high qualities of courage, skill, and determination, shown while fighting its fellow to the death, and the motionless and apparently homogeneous yolk suspended in its bed of albumen, are differences brought about by the processes of ontogenesis in a very short space of time.

The Law of Development.—The origin of the individual organism with all its complexity, both morphological and phys-

* *Loc. cit.*, pp. 242, 291, 296.

iological, is not explained by simply calling it development. Development is the history of the steps by which these characters are attained. It is the term by which we express the law of this history; and so long as the idea of the immutability of species prevailed there was supposed to be a particular law of development for each species. This law of development was alike for all the descendants of a common ancestor. By the expression law of development is meant a regularity in the order of changes or in the sequence of steps by which the results seen in the mature individual are attained. Every descendant of a common parentage was thought of as passing through the same stages of growth in reaching its maturity.

No Analogy between the Origin and Development of an Immutable Species.—The origin of the species from this point of view was explained, necessarily too, in some other way than by natural development; the reverent were satisfied with considering it a special act of the Creator; others preferred to explain it by the fortuitous concourse of atoms. Neither found any explanation in the natural laws of either generation or development. That there are different species and that new species have arisen were accepted facts; but the idea that different species could be explained by any laws noted in the development of the individual was not maintained. It was believed that the characters were specifically distinct for each species, and that this difference was in itself original.

Inorganic Properties and Organic Characters Compared.—The case was somewhat analogous to our idea of two kinds of mineral substances, as gold and iron; as to seeking an explanation for their origin, we do not attempt it: we either say they were created so in the beginning, or they appeared spontaneously concurrent with cooling of the solar system. Their differences we conceive of as their intrinsic properties. So with the idea of immutability of species logically there was associated the other idea, that the characters both morphological and physiological are essential properties or qualities of the species, and it was no more to be expected that one would ask why do birds have feathers and dogs have hair, than why is gold yellow and iron gray. The sufficient answer in each case was, they are the natural properties of the species.

The Idea of Mutability at the Foundation of the Discussion of the Origin of Species.—Thus we see that the attachment of the idea of mutability to organic species naturally led to the inquiry as to the origin of the properties or distinguishing marks of different species; and still further, it led to the dissociation of the characters from the species, causing them to be considered separately. The difference in point of view is a radical one, and the great amount of dispute and controversy which has resulted may be traced in great measure to the radical difference of meaning which the two schools attached to the word species. To a naturalist of the older ideas it was as absurd to speak of the origin of species as to speak of the origin of gold; both of these were supposed to occur in the world naturally, and that was enough.

What is Mutable?—When we speak of mutability, then, we ask, "What is it that is mutable?" Physiologically, the mutable element about species is the *steps of the development;* that is, there is not a perfect fixity of the law of development of the offspring when it starts upon its individual career as an embryo. Morphologically, the mutable characters of the species are among the most unimportant of the characters it assumes; for each individual of the species they are called its varietal characters.

A Concrete Example; Its Characters Symbolically Represented.—In order to fully answer the question what is mutable, and therefore what is it that is evolved in the course of the evolution of a new species, we are obliged to consider a concrete case. We must take an actual individual specimen of a particular species, and ask, What is it about this specimen organism which is mutable and has arisen by the evolutional, as distinct from the developmental, processes of the individual growth?

Such an example, whatever it be, has numerous characters which are recognized by the systematic zoölogist, and are defined by him under separate heads arranged in the order of rank, the whole constituting the taxonomic definition of the particular species. To express the relation of these characters to each other and to the individual it is not necessary to describe them, but symbols may be chosen to stand for them, and by examining the symbols we may arrive directly at the meaning of the expression "origination of characters and species."

If we then express the morphological and physiological characters by symbols, using the letters B for the characters of the branch, C for those of the class, O for those of the order, F for those of the family, G for those of the genus, S for those of the species, V for the varietal characters, and the numerals 1, 2, 3, 4, etc., for the different types of each category, we may combine these symbols in such a way as to express the sum of the characters of a particular individual organism.

Spirifer striatus Martin, var. S. Logani Hall, taken as the Example.—The example chosen for examination is a well-known fossil, specifically recognized in each of the continents in limestones of Eocarboniferous age, *Spirifer striatus* Martin. The variety which is found in the Keokuk limestones of the Mississippi valley is called *Spirifer Logani* Hall in the "Iowa Geological Report" (vol. I. pt. 2, Pl. XXI., p. 647). In order to fully define this specimen and assign its place in the classification of organisms we must refer it to the branch *Molluscoidea* (B 6) of the *Animal Kingdom*,* to the class *Brachiopoda* Dumeril (C 2) and subclass *Arthropomata* Owen, to the order *Telotremata* (O 4) of Beecher,† to the suborder *Helicopegmata* Waagen,‡ family *Spiriferidæ* King § (F 4), subfamily *Trigonotretaria* Schuchert,‖ genus *Spirifer* Sowerby ¶ (G 10), species *striatus* Martin,** and variety (so-called

* Claus and Sedgwick, "Elementary Text-book of Zoology," translated and edited by Adam Sedgwick, with the assistance of F. G. Heathcote, part II. p. 71. London and New York, 1884.

† *Am. Jour. Sci.*, ser. III. vol. XLI. p. 355.

‡ *Palæontologia Indica*, ser. XIII., "Salt Range Fossils," by William Waagen, Pt. I., "Productus-limestone fossils," IV. p. 447. Calcutta, 1883.

§ Thos. Davidson, "British Fossil Brachiopoda," vol. I. p. 51. *Paleontograph. Soc.*, London, 1853, etc.

‖ *Am. Geol.*, vol. II. p. 156.

¶ Schuchert's list, *Am. Geol.*, vol. II. p. 156.

** See Davidson's "Brachiopoda," vol. II Pt. V. p. 19, Pl. II. and III.

species) *Logani* Hall.* Taking Miller's "American Paleozoic Fossils," † we count this as species 115, and variety 2.

Thus, to express it symbolically, we should have the formula for the characters developed in this single particular species = B 6 + C 2 + O 4 + F 4 + G 10 + S 115 + V 2; and all of this is implied in the common scientific name for the species, *Spirifer Logani* Hall.

This expresses the morphological characters of the species arranged in the order of their respective ranks.

New Species Conceived of as Arising by a Process of Variable Characters becoming Permanent.—Thus it is seen that there are various degrees of mutability of the characters expressed by any particular specific individual. The accounting for the repetition of the characters already known in the ancestors is by the natural laws of generation. In the example before us the characters represented by the symbols (B 6), (C 2), etc., to (S 115) are supposed to be relatively fixed characters so far as transmission by generation is concerned, but the characters represented by (V 2) are distinctly mutable in generation, the descendants expressing them with varying degrees of modification from their ancestors. These varietal characters in the course of successive generations either (*a*) drop out by degrees, (*b*) do not reappear at all, or (*c*) continue to reappear in the offspring In case they continue to appear in the offspring, then they become added to the more permanent specific characters, and when so added, in place of (S 115 + V 2) we have (S 116), or a new species, all the other characters remaining the same. Species (S 116) may be supposed to show further variation, and (S 116 + V 3) and (S 116 + V 4) appear, and assume the same relations of repetition by generation, forming species (S 117), (S 118), etc.; but after a time the species (S 116), (S 117), (S 118), (S 119) become dominant. (S 115), (S 114) drop out, and we have a new genus (G 11), composed of the newly arisen species (S 116), (S 117), (S 118), and (S 119), the constancy of what was once a specific character becoming more fixed, and

* "Geol. Survey of Iowa," vol. 1. pt. 2, "Paleontology," by James Hall, p. 647, Plate XXI.

† 3d Ed., p. 374

the characters reaching a greater prominence and constituting the marks of another higher group, and then they constitute distinct generic characters.

This theory of the origin of species accounts for the morphological appearance of the new species by supposing that the future specific characters were first in the state of simple varietal modifications of the parental forms, and became fixed and permanent in the course of regular development in the whole or a part of the members of the descended race. Those members of the race permanently developing the new characters constitute the new species.

The varietal character may be algebraically expressed as either a plus or minus quantity; i.e., the variety may differ from the typical species by the addition of some slight character, or by the absence of some character, possessed by the normal species.

Characters of any Particular Specimen Differ Greatly in Antiquity.—In regard to the antiquity of the characters the following facts are known, as expressed in the following table:

TABLE REPRESENTING THE VARYING ANTIQUITY AND DIFFERENT GEOLOGICAL RANGE OF THE CHARACTERS OF AN EXAMPLE OF THE SPECIES *SPIRIFER LOGANI* HALL.

Symbol.	Taxonomic rank of characters.	C	O	S	D	Cr	T	J	K	Ty	Q.P.
B 6	Molluscoidea	—	—	—	—	—	—	—	—	—	—
C 2	Brachiopoda	—	—	—	—	—	—	—	—	—	—
O 4	Telotremata		—	—	—	—	—	—	—	—	—
F 4	Spiriferidæ			—	—	—	—	—			
G 10	Spirifer			—	—	—					
S 115	S. striatus.........					—					
V 2	V. Logani.........					—					

The varietal characters, expressed by the name *Spirifer Logani* Hall, appeared geologically for the first time in the Keokuk limestone in Middle Eocarboniferous time in North America. The specific characters, represented by the specific name *S. striatus* Martin. are found in all parts of the world and are characteristic of limestone rocks of the Eocarboniferous period.* The generic characters of the specimen named *Spirifer* began to appear in Eosilurian time and continued to appear till the close of Paleozoic time. The family characters, Spiriferidæ, do not date back earlier than the genus, and they continued to appear till the Jurassic era. The ordinal characters, Telotremata, began in the Eoördovician period, and species developing the ordinal characters are living at the present time. The class characters, Brachiopoda, appeared among the earliest Cambrian fossils, and are represented by numerous species and genera in the seas of the present time, and the same may be said of the branch characters, because we have reached the beginning of our record. Thus it is seen that the form of organism, called *Spirifer Logani*, although it has been extinct for millions of years, developed certain characters, described as ordinal and class characters, which are still being repeated in organisms now living; and although the species is characteristic of the Carboniferous era, and did not appear earlier or later, it developed characters (genera and family) which began as early as the beginning of the Silurian, and others which began in the Ordovician, and still others that began as far back as our record goes.

The Majority of the Characters of a so-called New Species have Appeared Before.—When we say, then, that at a particular time in geological history a new species arose, we do not mean that the new species differs *in toto* from its ancestors, but that a form has arisen which, agreeing with them in the great majority of its structural characters, yet differs from them by certain so-called specific characters, their specific rank being indicated by the fact that they are transmitted to their offspring without modification. The fact of their constancy is all that distinguishes these characters from varietal characters; and the generic characters are like specific characters in this particular of being transmitted without observable modification from generation to generation.

Theoretically, however, it is assumed that this perma-

* See the time-scale on page 54.

nency is only relative; that, somehow, the higher characters become modified as well as the lower. Thus it is supposed that by such gradual modification, taking place in the course of genealogical descent, successive individuals arise which differ specifically from their ancestors, later others which attain generic difference, and after a great many generations the family characters are changed; and still later they differ ordinally, and, theoretically, even such radical differences of structure as distinguish one class from another may be thus attained.

Fixed Characters those which are Transmitted Unchanged in Natural Descent.—In ordinary natural development, or ontogenesis, there is a law of constancy regarding all the characters expressed by the symbols B 6, C 2, etc., to S 115. These may be then called the fixed characters of the species at any particular time, and be indicated by the letter M. But, as we have explained, in the course of time among the descendants may appear a new genus, G 11; the point of geological time, or the stage in the history, marking such an event is when the new species assumes dominance in individuals, and the old forms drop out, and leave a gap in the series. The species M may be considered as expanding at this point to include new generic characters, or we may consider the new genus as arising as an offset from the old forms. It will be seen that all the individuals possessing the characters M form a common race, and that divergence of race proceeds from varietal, through specific, generic, family, etc., characters, and in the order here given; and that the series, branch, class, order, etc., are expressive of the natural order in rank of importance of the characters, in their antiquity, and in their fixity.

Rank of Characters, the Precision of their Reproduction, and their Antiquity.—If we arrange the characters in the inverse order, thus: V, S, G, F, etc., we have expressed the characters in the order of their increasing importance, increasing fixity, and constancy of their repetition by generation. There is thus seen to be a law of relation existing between the certainty and accuracy of repetition in reproduction, and the number of times the reproduction cycle has been repeated. This leads us to the further analysis of this process —the *plasticity* or the *permanency of the characters.*

Plasticity of Characters.—In the characters recognized as plastic in the development of the individual there is possible adjustment to changed conditions. So long then as any character is in a plastic, undeterminate condition, it is evidently not essential. All varietal characters may be regarded as in such a condition. The theory of Darwin explains that these tentative characters will necessarily prove of advantage or of disadvantage; it may be extremely slight, but, in a close contest, sufficient to give the possessor greater or less chance of success in the struggle for life; and the perpetuation of such characters will be brought about by the living of the possessor of the favorable variation to perpetuate its kind, and the death of the others.

Origin of Species from the Physiological Point of View.—At this point we need to consider the origin from the physiological standpoint. The name for the process of assuming morphological and physiological characters by the individual is *development*, as has already been explained. *Reproduction* is that process by which one set of individuals initiates the cycle of development for a new individual. The principle determining the repetition of like characters in the parent and offspring is called *Heredity* or *Inheritance*. *Variability* is the principle expressed in the tendency of all vigorous organisms to exceed the mere repetition of ancestral characters by divergences. Darwin's theory of the origin of species was proposed to account for the existence of different species by a physiological process.

Darwin's Theory of the Origin of Species.—The full title of Darwin's work is, "Theory of the Origin of Species by Means of Natural Selection, or the Preservation of Favored Races in the Struggle for Life," and its chief points are the following:

1. *Variability* Darwin found to be a natural law in the development of all plants and animals.

2. *Artificial Selection.*—Darwin observed that men, by selecting, under domestication, plants or animals which already possess particular varietal characters, can, by breeding them together, and by preventing their mixing with other

varieties, perpetuate the varieties, or can cause a race to grow up in which the varietal characters shall become relatively permanent. Numerous facts of this kind are familiar, as our common breeds of horses, cows, pigs, domestic pigeons, flowers, fruits, etc. As illustrative of the extreme modification possible, the greyhound and the pug-dog may be cited.

3. Darwin further observed that *varieties occur under natural conditions ;* that there are doubtful species, or forms which are intermediate between the typical species.

4. He found by an analysis of the plants of twelve countries, and the coleopterous insects of two districts (and this result was confirmed by later study), that *the larger genera present the greater number of varieties* and are the more widely distributed.

5. The natural *increase of organisms* by generation is vastly in excess of the actual number reaching maturity; *increase* is by *geometric ratio*, but the increase of adults is, at best, only a very slight *arithmetical ratio.* Linné showed that from a single plant producing only two seeds, if all the seedlings were to live, in twenty years there would arise a million plants. Darwin estimated that from a single pair of elephants breeding at the age of thirty years, and continuing breeding until ninety years old, producing three pairs of young in the interval, at the end of the fifth century there would arise fifteen million elephants alive at one time descended from the first pair.

6. There are innumerable *checks to increase*, as nature of climate and of food, but particularly mutual checks, as struggling of individuals for the same food, or for the same set of favorable conditions. This is the general law of *struggle for existence.*

7. *Natural Selection.*—Darwin then argued that the conditions of environment; the abundance of food, or lack of it; the favoring climate, or the opposite; the accessibility of food, or difficulties in the way of obtaining it; would all work together and separately, as either favorable or unfavorable conditions for each individual, according as he was more poorly or better adapted to live under them than his fellow; that each of the characters of a varietal nature must have

some slight value in favor of the possessor, or against him, in the struggle: the result would be the extinction of those less well adapted and the preservation of the more favored—i.e., a survival of the fittest. This is the law of *natural selection.*

8. Darwin further added the principle of *sexual selection;* that is, that variations in habit, or even in color, are adapted to cause a selection in pairing, which will lead to a further perpetuation of certain characters and the isolation of varieties into breeds, and thus the formation of species proper, or larger groups of individuals, repeating by reproduction the originally varietal characters of the few.

9. Darwin noticed that *divergence of characters* is produced in animals and plants under domestication, gradually and as the result of continued artificial selection; hence he inferred that the selection acting in nature will also tend to perpetuate more and more markedly the strongly contrasted varieties, the intermediate ones blending with the stronger types; thus, he believed, the differences, or gaps marking species from species, are formed.

There were other laws of variation which he noticed. That use tends to develop, disuse to suppress characters, had already been emphasized by Lamarck. Habit or custom favors certain characters. Correlation of parts in growth tends to cause variation in other parts, as adjustments to changed organic conditions, and many others; and the facts of distribution of organisms were found in line with this theory of origin of species, and paleontological succession is in harmony with it. In his sixth revised edition of "Origin of Species," published in 1888, Darwin says definitely: "I believe that animals are descended from at most only four or five progenitors, and plants from an equal or lesser number. Analogy would lead me one step farther, namely, to the belief that all animals and plants are descended from some one prototype."* And in the closing passage of the book he sums up the essential points of his idea of the origin of species, speaking of the laws by which all animals and plants have been produced, thus: "These laws, taken in the largest

* Vol. II. p. 299.

sense, being *growth* with *reproduction*; *inheritance*, which is almost implied by reproduction; *variability* from the indirect and direct action of the conditions of life, and from use and disuse; a *ratio of increase* so high as to lead to a *struggle for life*, and, as a consequence, to *natural selection*, entailing *divergence of characters*, and the *extinction* of less improved forms. There is a grandeur in this view of life, with its several powers, having been originally breathed by the Creator into a few forms, or into one, and that, whilst this planet has been cycling on, according to the fixed laws of gravity, from so simple a beginning, endless forms most beautiful and most wonderful have been and are being evolved." *

Do Characters become of Higher Rank as they are Transmitted?—The natural and general inference from the Darwinian explanation of the origin of species is that variations, by selection and invariable transmission, become, in the course of generations, fixed and permanent characteristics in the offspring, which removes them from the rank of variations to that of specific characters; by degrees in the course of more generations these same characters are supposed to become of higher rank and constitute the generic characters of their descendants; and in the same way further fixation and repeated inheritance might make them to become still more important, and thus to attain ordinal and finally class rank in classification. The paleontologist may with good reason ask if this be the fact. Are early genera made up of species whose distinguishing specific characters constitute the distinguishing marks of genera of later times? There are those who question the truth of this proposition as a matter of fact.

Evolution of Genera and Acceleration and Retardation.—The opinion was expressed by E. D. Cope† that the evolution of generic characters has proceeded in a different manner from the evolution of specific characters; that the evolution of generic and of specific characters has not been *pari passu*, but independently of each other. He further distinguished two special laws of evolution—the law of acceleration and retardation, and the law of natural selection.

* Pages 305, 306.

† "Origin of the Fittest: Essays on Evolution," p. 43. New York, 1887.

The essential idea set forth by Cope may be found in the following quotation from the chapter "On the Origin of Genera":

"There are, it appears to us, two laws of means and modes of development [evolution]: I. The law of acceleration and retardation. II. The law of natural selection. It is my purpose to show that these propositions are distinct, and not one a part of the other: in brief, that, while natural selection operates by the 'preservation of the fittest,' retardation and acceleration act without any reference to 'fitness' at all; that instead of being controlled by fitness, it is the controller of fitness. Perhaps all the characteristics supposed to mark generalized groups from genera up (excepting, perhaps, families) have been evolved under the first mode, combined with some intervention of the second, and that specific characters or *species* have been evolved by a combination of a lesser degree of the first with a greater degree of the second mode."

Growth-force or Bathmism.—The defenders of this view are called by Wallace, in criticising them, the American school of Evolutionists.* There is assumed to be a special developmental force, called growth-force or "bathmism," which is exhibited in variation itself, and becomes effective, as phylogenetic evolution, through retardation and acceleration, in the same way as the force which is expressed in *natural selection* operates through the death of the unfit and the survival of the fittest toward the evolution of species.

The Origin of Species Still an Open Question.—Many other theories have been advanced to explain the origin of species: the examples above cited are sufficient to explain the nature of the problem; but it is aside from the purpose of this treatise to go into detail in the discussion of theories.

It will be observed from the statements already made that the two great factors in evolution and the origin of species are species and mutations. Species with the repetition of characters and the adjustment to environment are facts which every naturalist is more deeply aware of the fuller his knowledge of organisms becomes. Mutation, or the acquire-

---

* Wallace, "Darwinism," p. 420. This American school is in other places called the Neolamarckian school.

ment of variation, is also a conspicuous fact in nature. To explain the origin of species involves the accounting for the becoming fixed or permanent of variable elements of organization, as well as the accounting for the previous variability of the characters now fixed.

Darwin's theory and those like it are chiefly engaged in accounting for the acquirement of permanency of originally variable elements. The Lamarckians and Neolamarckians are chiefly interested in accounting for the variability. While natural selection is effective when the differences themselves are already on hand, it assumes variability to be a fact without explaining it. It is necessary to account for variation itself, and those who assume, that any structural modification which an organism may acquire during its lifetime may be transmitted to its offspring, necessarily emphasize the effects of use and disuse, the retarding or accelerating of growth, and, in general, all the factors of variation tending toward variation of the individual during its life.

It is in the field of observation rather than in speculation that the solution of these questions is to be found. So soon as we admit the possibility that the transmission of characters from one generation to another may not be absolutely constant, we throw back the whole discussion into the field of the actual laws of progress in generation. If the organisms have varying degrees of the growth-force, if they can in the least degree choose for themselves the course of development of their organization, the whole problem of evolution may be accounted for by the operation of this force—a force which then becomes the most important factor in the case. But before we can reach a final theory of the origin of species we need to know what the facts are. Hence it is that the whole subject of variation, both in living forms and as expressed in the historical series, is of vital importance. Not only is variation an intrinsic law of organic generation, but as has been shown with overwhelming force, the discontinuity which we observe separating the character of one species from those of species next to it in likeness is not a result of natural selection, " nor has it its origin in environment," " nor in any phenomenon of adaptation, but it is in the intrinsic nature of organisms themselves, manifested in the original *discontinuity of variation*." *

It is certain that more light is required upon these fundamental factors of evolution before the final word can be said upon the origin of species. That which distinguishes the species, in contrast to the variety, is the constancy of transmission of its specific characters, but it is evident that constancy here is not absolute constancy—at least it is not known to be absolute.

In variation, the nature, causes, degrees, and rate of variation are the subjects of investigation which now promise to give the true explanation of not only the nature but the origin of species.

* "Materials for the Study of Variation, treated with especial regard to Discontinuity in the origin of species," by William Bateson, London, 1894, p. 567.

CHAPTER XI.

THE PRINCIPLES OF NATURAL HISTORY CLASSIFICATIONS.

ILLUSTRATED BY A STUDY OF THE CLASSIFICATION OF THE ANIMAL KINGDOM.

FOR a clear understanding of the meaning of the origin of species it is essential to consider the nature of the nomenclature of the classification of organisms We have already considered what species are and what the organic individual is, and how development is an appropriate term for the growth and perfection of the individual, and how evolution pertains to the progressive modifications of the successive species of a genus.

Classifications in Natural History.—Classifications and systems of classification in natural history are but methods of expressing, briefly, almost symbolically, the knowledge we already possess of the characters of organisms and their relations to each other. A single word, the name of a class or order, or even the specific name of a species, stands for all the morphological and physiological characters peculiar to that species, order, or class. Hence such terms are highly technical: and though it may not be possible to learn the full meaning of any of them in a brief course of lectures, it will be possible to describe the right manner of using them, so that the knowledge of the details will be arranged in an orderly manner under the proper heads as it is gradually acquired.

Species and Genus of Aristotle.—As the facts of biological science have accumulated it has been found necessary to distribute them in some systematic manner, and for this purpose a number of arbitrary divisions having definite names has been gradually evolved. The use and meaning of these names will be most easily explained by a brief examination of their development from the terms *Species* and *Genus* of Aristotelian logic. *Species*, the translation of the Greek term εἶδος, meant, when applied to organisms, those having a number of like and peculiar characters. *Genus*, the translation of the Greek γένος, in logic was that which can be predicated of things differing in species, and as a biological term it was applied to a group which included several different species.

Scaliger's Terms.—Scaliger expanded the Aristotelian nomenclature: by him *Individual* was used to indicate a single organism (plant or animal), distinguished by having a separate body, and having a separate and independent activity. *Species* was used in the Aristotelian sense, but *Genus* was found of three degrees of importance: the *Genus proximum*, the *Genus medium*, and the *Genus summum*.

The Terms of Linné.—Linné (1735–1766) classified organisms (both plants and animals), retained the name *Genus* for the *Genus proximum* of Scaliger, and proposed the term *Ordo* for *Genus medium* and *Classis* for *Genus summum*.

Cuvier's Perfection of the Nomenclature and the Present Usage.—These names were later adopted by Cuvier, about the beginning of the present century, and he added the term *Embranchment*, or *Branch;* and thus was established the nomenture still in use in Biology, which in English is as follows: *Individual, Species, Genus, Order, Class*, and *Branch* (or Subkingdom, or Phylum, or Type). To illustrate the meaning of these divisions the following examples may be given: A black and a bay horse would be called two individuals of the same species. The horse and the ass are two species of the same genus (*Equus*). A horse, an ass, and an elephant all belong to one order (*Pachydermata*). The horse, ass, elephant, and lion are of the same class (*Mammalia*). All these would be united in the same branch with the alligator (the branch *Vertebrata*). Further subdivision has been very commonly made of the order into suborders or families, viz., the family of *Elephantidæ*, including the elephants and the mastodon, and the family of *Equidæ*, including the horse and the Hipparion.

The Classification of Cuvier.—Linné recognized six classes in the Animal Kingdom (*Mammalia, Aves, Amphibia, Pisces, Insecta, Vermes*). Cuvier made great progress in the distinction of the lower animals. He recognized four branches (*Animalia Vertebrata, Animalia Mollusca, Animalia Articulata, Animalia Radiata*). The first four classes of Linné's system were united to form the first branch of Cuvier. The most prominent character uniting them was the possession of an internal skeleton, bound together by a segmented vertebral column. The second branch of Cuvier, called *Mollusca*, included six classes (*Cephalopoda, Pteropoda, Gastropoda, Acephala, Brachiopoda, Cirrhopoda*), and the conspicuous characters of the Mollusca were the possession of a soft, bag-like body, enclosed more or less completely by a hard exterior shell composed of one, two, or more parts. Cuvier called the third branch *Articulata*, including in it four classes (*Annelida, Crustacea, Arachnida, Insecta*). The chief character in this branch was the segmented external skeleton, composed of joints with lateral articulated appendages. The fourth branch was *Radiata*, and included five classes (*Echinoderms, Intestinal Worms, Acalephæ, Polypi, Infusoria*). The prominent character was the radiate structure, typically exhibited in the Star-fish or Sea-urchin, but ignorance of internal structure led to the association of many unlike forms. Since Cuvier's time great advance has been made in the knowledge of the structural anatomy of animals, especially in the smaller and lower organisms, and many other classifications have been proposed, but the majority of Cuvier's classes have remained. Animals referred to some of the classes by Cuvier, and some newly-discovered animals, have been made the types of other classes, and stricter definitions of the classes already established have been made.

Uniformity of Usage of Specific and Generic Names.—The branches have been considerably remodelled, especially by later zoölogists, according as one or other organ or system of organs has been taken as of chief importance in distinguishing the groups. Of the later classifications those of Leuckhart, Huxley, Claus, Gegenbaur, and Lankester have expressed new points of view in the arrangement of the organisms, but in all the confusion of systems a common usage has grown up in the application of specific and generic names to animals and plants, and these have constituted the standards.

At the present time hardly two standard authors of text-books of Zoölogy or Paleontology will be found to apply the nomenclature of classification in the same way throughout: that is, they will not distribute the genera in the same manner, or will give different value, or will apply different names to orders, families, and classes.

Selection of a Standard Classification.—It becomes necessary to use some standard in the matter of classification, and Zittel's "Manual of Paleontology" may be selected as the standard in the present case. Editions of Zittel are published in both German and French, but at the present time (1895) no English edition has appeared.*

Differences of Opinion regarding the Rank of the Characters.—The difference in usage of the nomenclature of classification is determined by differences of opinion as to the taxonomic value or rank of characters expressed by the organisms rather than by any difference in recognizing the characters as matters of fact. Classifications, therefore, although differing in the hands of different authors, may be used with precision when considered as descriptive of the combination of characters expressed in actual organisms.

There are several standard classifications of more or less common use among paleontologists, three of which may be here referred to: Claus and Sedgwick's, as given in "Elementary Text-book of Zoology," 1884; Zittel's classification in "Handbuch der Palæontologie," vol. I., 1876–1880; Nicholson and Lydekker, "Manual of Paleontology," 3d Ed., 1889.

Claus and Sedgwick's Definitions of the Nine Branches of the Animal Kingdom.—Brief definitions of the nine branches, as given by Claus and Sedgwick, are as follows, viz.:

"*Protozoa.*—Of small size, with differentiations within the sarcode, without cellular organs, with predominating asexual reproduction.

"*Cœlenterata.*—Radiate animals segmented in terms of 2, 4, or 6; mesoderm of connective tissue, often gelatinous;

* A briefer text-book in German has appeared: "Grundzüge der Palæontologie (Palæozoologie)," von Karl A. von Zittel, pp. i-viii, 1–971, and 2048 figures; Munich, 1895. An English translation of this work, with some revision by American paleontologists, is in preparation.

and a central body cavity common to digestion and circulation (gastro-vascular space).

"*Echinodermata.*—Radiating animals, for the most part of pentamerous arrangement; with calcareous dermal skeleton, often bearing spines; with separate alimentary and vascular systems; and with nervous system and ambulacral feet.

"*Vermes.*—Bilateral animals with unsegmented or uniformly (homonomous) segmented body, without jointed appendages (limbs), with paired excretory canals sometimes called water-vascular system.

"*Arthropoda.*—Bilateral animals with heteronomously-segmented bodies and jointed appendages, with brain and ventral chain of ganglia.

"*Molluscoidea.*—Bilateral, unsegmented animals with ciliated circlet of tentacles or spirally rolled buccal arms; either polyp-like and provided with a hard shell-case, or mussel-like with a bivalve shell, the valves being anterior and posterior; with one or more ganglia connected together by a periœsophageal ring.

"*Mollusca.*—Bilateral animals with soft, unsegmented body, without a skeleton serving for purposes of locomotion; usually enclosed in a single or bivalve shell, which is excreted by a fold of the skin (mantle); with brain, pedal-ganglion, and mantle-ganglion.

"*Tunicata.*—Bilateral unsegmented animals with sac-shaped or barrel-shaped bodies, and a large mantle cavity perforated by two openings; simple nervous ganglion, heart, and gills.

"*Vertebrata.*—Bilateral animals with an internal cartilaginous or osseous segmented skeleton (vertebral column) which gives off dorsal processes (the neutral arches) to surround a cavity for the reception of the spinal cord and brain; and ventral processes (the ribs) which bound a cavity for the reception of the vegetative organs; never with more than two pairs of limbs."

Zittel adopts the older Claus classification, in which the fifth branch, Mollusca, includes Molluscoidea, Mollusca, and Tunicata—divisions which are given higher rank in the newer classification.

Nicholson separates the Sponges from Cœlenterata under the branch name Porifera; includes the Vermes and the Arthropoda of Claus in one branch, the Annulosa, making of them three sub-branches: I. Solecida, II. Anarthropoda (these two sub-branches together constitute the branch Vermes of Claus), and III. Arthropoda, which includes the same classes as assigned to that division by Claus.

The Classes of Importance in Paleontology and their Known Range in Geological Time.—Those classes which are of importance to the student of the history of organisms are the following: the names are used uniformly so far as to include the same organisms, but their theoretical relations to each other are not stated alike by different authors. (See next page.)

Species and Genera of Chief Use in Tracing the History of Organisms.—When we come to the actual study of the historical relations of organisms it is specific and generic characters with which we chiefly deal, and the grouping of them into families, orders, classes, and branches is the result of the study rather than a matter of direct observation.

We agree with Zittel* that the systems of classification in biology are only the expression of our actual knowledge of the reciprocal relations of the organisms: they depend directly upon the present state of our knowledge, and are subject therefore to more or less profound modifications.

The higher categories are built up of generalizations derived from comparison of the detailed structure of the individuals. All our systematic categories are artificial abstractions which rest upon the greater or less resemblance of form in the individuals. The historical relations between the characters marking these larger categories are not matters of observation, but only of speculation. The history is to be observed in series of successive species, and the study of classifications becomes of importance in restricting our attention to the field within which all the evidence to be had must be found. The actual evidence of the history, which the paleontologist may see and examine, is presented in the specific and varietal characters of the fossil remains preserved in the rocks.

* See "Handbuch der Palæontologie," vol. I. p. 39, etc.

THE CLASSES OF THE ANIMAL KINGDOM AND THEIR GEOLOGICAL RANGE, GROUPED IN BRANCHES ACCORDING TO CLAUS AND SEDGWICK.

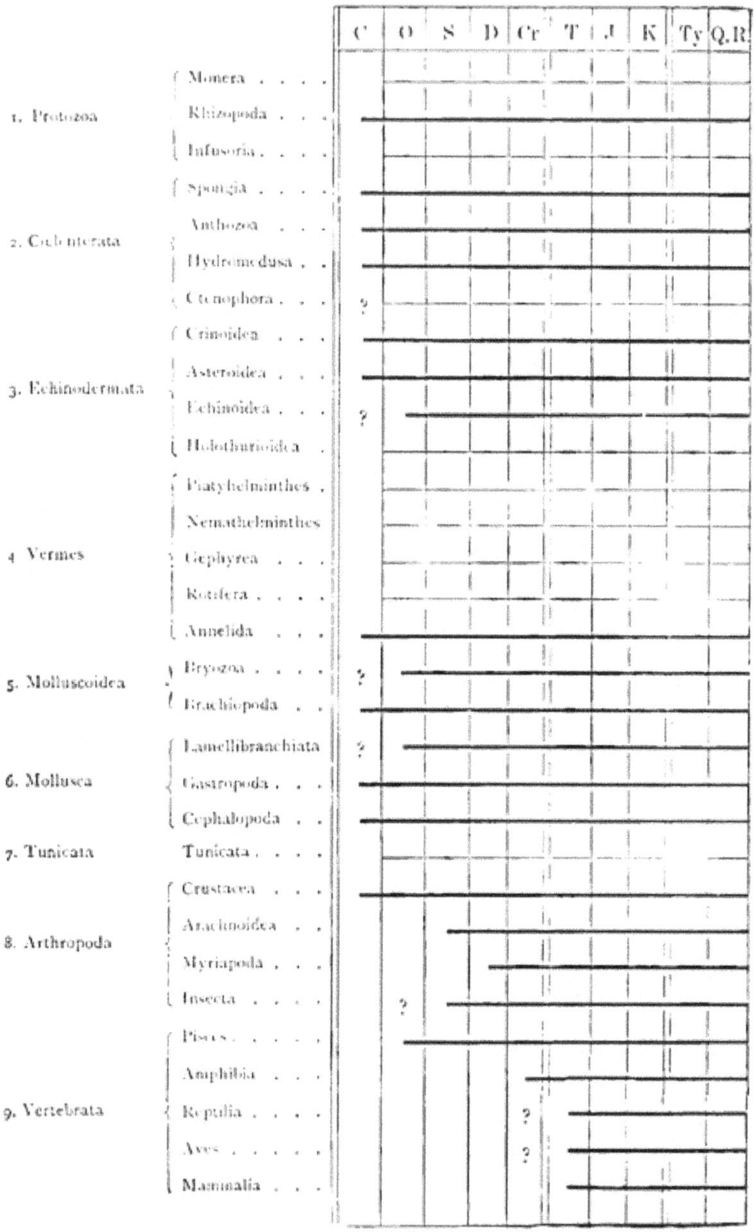

Branch	Class	C	O	S	D	Cr	T	J	K	Ty	Q.R.
1. Protozoa	Monera										
	Rhizopoda . . .										
	Infusoria										
2. Cœlenterata	Spongia										
	Anthozoa . . .										
	Hydromedusa . .										
	Ctenophora . . .	?									
3. Echinodermata	Crinoidea . . .										
	Asteroidea . . .										
	Echinoidea . . .	?									
	Holothurioidea .										
4. Vermes	Platyhelminthes .										
	Nemathelminthes										
	Gephyrea . . .										
	Rotifera										
	Annelida . . .										
5. Molluscoidea	Bryozoa	?									
	Brachiopoda . .										
6. Mollusca	Lamellibranchiata	?									
	Gastropoda . . .										
	Cephalopoda . .										
7. Tunicata	Tunicata										
8. Arthropoda	Crustacea . . .										
	Arachnoidea . .										
	Myriapoda . . .										
	Insecta		?								
9. Vertebrata	Pisces										
	Amphibia . . .										
	Reptilia					?					
	Aves					?					
	Mammalia . . .										

Species of the Paleontologist.—We have already considered the philosophical notion of species, but the real species which we deal with in Paleontology is, as defined by Zittel, all those individuals or all fragments which present certain common characters and form a circumscribed group, independent of geological range or geographical distribution, and which may be linked with allied species by a small number of intermediate forms. If in the same species certain individuals possess some peculiar characters which are more or less conspicuous, they constitute *varieties* or *races* of the naturalists. The varieties maintain in some cases the same habitat with the stock form, in other cases they live in different regions (*representative varieties*). It is more difficult for the paleontologist than for the zoölogist to distinguish species from varieties. It often happens that there are in two contiguous formations fossils of the same genus, presenting differences, very slight but constant, in which case they should be distinguished as separate species. Fossil species are not always restricted to either a single geological horizon or bed, nor are they confined to the same geographical region.

Varieties.—The same fact applies in some measure to varieties. Those slight differences, observed upon comparing the representatives of a species coming from different strata or from different regions, are considered to be varietal, and not specific, in case the differences consist in unequal degrees of modification of the same part or parts, so that the several specimens may be arranged in a continuous series connecting the extremes by intermediate forms. When such a series of forms of one species exhibits the differences in connection with geographical distribution only, the degrees of modification are defined as varietal, and those prominent in a particular locality may then be called distinct *varieties*.

Mutations.—When the modification of form is observed to be associated with succession of their appearance, the differences are called *mutations*—a term proposed by Waagen. Thus modifications of specific form, when contemporaneous, are called *varieties* or *variations;* when successive in time they are called *mutations*.

The History of Organisms; the Two Methods of its Study.—The history of organisms may be examined from either of two points of view. (*a*) We may examine the embryonic and ontogenetic course of differentiation of the individual, and, adding the theory of descent with modification, apply the laws of individual development to the building of a theoretical phylogenesis for the whole series of organisms. This is the method of Zoölogy. (*b*) Or, we may examine the fossil remains of organisms which have appeared in geological history, and by comparative study of their characters, arrange them in series according to their resemblances and differences, and thus reconstruct the history of the organisms from the observed order of their appearance on the globe. This is the paleontological method.

Embryos or Fossils; the Imperfection of the Evidence.—In the first case the chief criteria upon which the history is built are the changes taking place in the growing embryo, minute and, generally, microscopic, and of great difficulty of study. This method requires great use of imagination in the interpretation of rudimentary traces of characters, is based necessarily upon few examples, and those seen mainly by single observers. The results are of necessity highly theoretical, and, like all hypotheses, should be regarded as of no value in the face of facts to the contrary.

In the second case the chief criteria are fossils, which are the remains of the hard parts and, in most cases, of adult forms, imperfectly preserved, presenting a very small percentage of the total variety of forms that must have lived. In this method the imperfection of the evidence and the fragmentary nature of the fossils are the chief sources of imperfect judgment. The hypothetical series erected may be proven by the actual sequence of the forms themselves. The species may be arranged in the wrong race, but actual succession is always distinctly indicated, and the filling of gaps is readily known to be theoretical. The known affinities of living organisms are also in evidence here, to prevent wild hypotheses based upon rare and imperfect fossils.

From either point of view the possibilities of error are enormous, and the proportion of theory to knowledge is large; but at the same time it must be said that the two methods agree in the general results; and while there is a vast amount to learn, to which future theories must adjust, the general facts in the case, which alone we are considering in these lectures, are already fairly well established.

Mature Individuals, not Embryos, used by the Paleontologist.—The chief difference between the two points of view, as they concern us, is that the paleontological method deals essentially with the matured results of individual development. It is remains of the mature organisms that he investigates, and he examines the differences between the mature individuals of the successive periods; while in the other method it is the rudimentary conditions of individuals that carry the evidence of the affinity.

Differentiation attained during the First or Cambrian Era.—The paleontologist asks, To what extent has differentiation proceeded in the individuals of any particular geological epoch, and on comparing the fossils of successive epochs, in what respects and at what rate has differentiation proceeded? In carrying out this method of study we inquire, first, To what extent has morphological differentiation reached in the first geological age of which we have record, i.e., the Cambrian? In reply the answer may be briefly given in terms of abstract scientific nomenclature, by stating the numerical relation existing between the number of the branches and of the classes of the Animal Kingdom which are known to have lived in Cambrian time and the total known number in each category.

On page 206 is given a table of the branches and classes of the Animal Kingdom of which record is preserved in the rocks, with their known geological range. In this summary we may omit from consideration the branches Tunicata and Vertebrata, of which we have no evidence in Cambrian time; and the Protozoa may be omitted from the consideration because, although it is altogether probable that they were well represented, traces of them are almost entirely wanting on account of the minuteness and simplicity of their forms. We may also omit the consideration of such classes as the Holothurioidea, of which no evidence is found in a fossil state. And, finally, taking all the other branches, classes, and orders, known in fossil condition, the answer to the question is as follows: Of the six branches of the Animal Kingdom all six were differentiated in the Cambrian era; 13 classes of the 26 were differentiated in the Cambrian: of the 73 orders, 14 are known from the Cambrian, 14 more are first seen in Ordovician time, 4 more in the Silurian; or before the close of the Silurian out of a known 72 fossil orders 32 had already appeared.

Represented in the form of percentages between the numbers represented in the early ages and the number appearing throughout all the geological ages, we find that, of the differentiations of the primary and fundamental nature which distinguish the branches of the Animal Kingdom one from another, 80% of all that has ever taken place was already accomplished before the close of the Cambrian. It may have been still more complete, but this amount we know to have been the fact. Of differences of only second rank in importance, i.e., those which mark the separate classes of the animal kingdom, 13 out of a known 23 fossil classes are already known to have appeared in Cambrian time, or 56% of the differentiations of class rank had been already attained. In the evolution of orders at least 32 of the 72 fossil orders appeared before the close of the Silurian, and 14 orders are represented in the Cambrian era, or 20% in the Cambrian era and about 40% of ordinal differentiation had been accomplished before the close of the Silurian.

It is probably well within the facts to say that six out of the nine known branches were already differentiated in the Cambrian, and that in all probability all the classes of these six branches were already differentiated before the close of the Silurian or third geological era, and probably four fifths of them in the Cambrian era. In respect of ordinal differentiations, it is probably true that, of the total ordinal differentiation known in these six branches, one fourth, and probably more, took place before the close of the Cambrian, and one half before the close of the Silurian. If we recur to the time-scale, described on page 54, bearing in mind that the rocks of the Cambrian system may not and probably do not contain records of the earliest organisms that appeared upon the earth, but only the earliest records we have of distinct organisms, it will be seen that the statistics given above mean that at least three quarters of the total evolution of the grander distinguishing characters of organisms are known to have been completed before the close of the first quarter of their recorded history. The percentage would be much smaller if the generic and, particularly, if the specific characters of all known organisms were to be considered; but to form a correct idea of what the statement means it is necessary to consider that these latter characters are, both from the point of view of importance of the characters in the economy of individual life and from the point of view of the degree of specialization to particular conditions of environment, far less important than those whose differentiation was so rapidly culminated.

Nature and Extent of the Elaborations.—In order to form a definite notion of the extent of the differentiation thus early attained in the evolutionary history of organisms, we may next consider what structures and functions had been elaborated in each of the several branches of the Animal Kingdom in the Cambrian era.

In the Cambrian system are found traces of six, at least, of the nine branches of the Animal Kingdom, and when we are looking at organic form, of either the morphology or physiology of organisms, this means that the characters by which these various branches are distinguished were differentiated before the close of the Cambrian era; and in most cases there is evidence to show that it was before the close of the lower division of the Cambrian. As has been noted, this statement applies also to a remarkably large proportion of those characters by which the different classes and even orders of these branches are distinguished.

Recurrence of Characters accounted for by Descent.—There follows as correlative to the fact that these characters have appeared in the Cambrian, that their reappearance in successive organisms up to the present time is to be explained by the ordinary laws of heredity. Regarding them no evolution is observed. Whatever evolution is necessary to explain their appearance in the world took place prior to the Cambrian era. It is difficult to appreciate how far back in the world's history this shifts the great events of evolutional elaboration, and how little it leaves to be accomplished within even the immense periods of geological time of which we have the least trace of the history of organisms.

Modern Zoology applicable to the Fauna of the Cambrian Era.—During the preparation of these pages the writer took occasion to examine the details of form and structure discussed in the lectures of a well-known professor of Invertebrate Zoölogy. It was found that, so far as the evidence is preserved, the great majority of the differentiations which are considered in such a course of lectures were actually present in the Cambrian era. What has taken place since is differentiation in respect of less important characters. In other words, a preliminary course of lectures on Invertebrate Zoölogy (eliminating the animals adapted to aerial and pure fresh-water environment) would be adapted to the fauna of the Cambrian era. This statement will probably surprise the reader to whom it comes now for the first time. It is certainly a most remarkable fact that the great plan-work of structure of all the invertebrates was so fully elaborated at such an extremely early period, and that since that time, for the millions of years that have followed, the modification in organic forms has been so slight. It is more impressive than the fact that several genera of Brachiopods (Lingula, Discina, etc.) living to-day were represented in the Cambrian by forms separable from them only by the closest scrutiny.

Characters whose Origin is Traced Back to Cambrian Time.—Assuming the correctness of the above statements, the inquiry may be made, What are the characters, expressed continuously up to the present, which made their first appearance in Cambrian time?

First, there is a branch, called *Protozoa*, all the animals of which are relatively minute, some of them truly microscopic; their bodies are composed of a jelly-like substance, called protoplasm, without cellular differentiation, and void of permanent specialization of function. They show great bodily activity, but in no permanent direction. The whole substance of the body seems transiently to be experimenting in the elementary functions of motion, sensation, digestion, and reproduction. The one differentiation, which at least numerous kinds of the Protozoa have accomplished, is shown in the secretions with which they surround themselves, constructed in definite forms, but of almost infinite variety.

Second, the next stage of differentiation is seen in each of the remaining types of animals, inclusively called the *Metazoa.* In all of these animals (1) there is the localization of the digestive functions in the interior of the body, the *gastro-vascular* cavity, (2) a *mouth* leading to this cavity, and (3) the location of the *motory functions* on the outer side of the body. In the second branch of the Animal Kingdom, the *Cœlenterata*, there is little more of specialization of the digestive functions than this, i.e. (1) there are two elementary *tissues* differentiated, and in this simplest type (as in all higher) the tissues are formed in the course of individual development by the segmentation (4*a*) of the primitive cell (4*b*), the formation of numerous cells, and then a (4*c*) specialization of some of these cells as tissues for one function, others of them for other functions. This process, which is called development, may be regarded as a specialization of the generalized function of reproduction. In the Protozoa reproduction takes place by simple fission and gemmation. In this lowest branch of the Metazoa, the Cœlenterata, the integrality of the body is continued after the separation into parts, and what constitutes the whole of the reproductive function in the Protozoa here constitutes but the segmentation of the contents of the egg, which differentiates the two layers of tissue—the *Ectoderm* (4*d*), or outside layer, and the inner, or *Endoderm* (4*e*). The fundamental function of the Ectoderm is motory, the primitive function of the Endoderm is digestive and assimilative. In the sponges there is developed between these two layers the *Mesoderm* (4*f*), in which a rudimentary type of skeletal parts, in the form of horny fibres, or silicious or calcareous spicules, is deposited. The sponge has differentiated a digestive or gastro-vascular cavity, but the mouths are several and indefinite, and the cells within, by their ciliary motions, perform the functions of motion as well as digestion, thus not exhibiting the full elaboration seen in the true Cœlenterata, but rather constituting a colony of Protozoa-like individuals. The true Cœlenterata, as illustrated by the corals or Anthozoa, are elaborated a step further: in their gastro-vascular cavity a certain *polarity* (5) of the body is differentiated, of which the mouth is the centre, the polarity is expressed functionally in the direction of the currents inward and outward through the mouth; in the motor system special (6) *motory organs* are developed, radiating from and surrounding the mouth as *tentacles* (6*a*), and the whole of the body, in the higher forms, also expresses this radial arrangement of parts into compartments (6*b*), called *mesenteries*. This radial differentiation is indefinite in the earliest forms, but there are two modes of division that are well expressed later, seen in the tetracoralla (7*a*) and the hexacoralla (7*b*). In the Cambrian only the four-parted type (Tetracoralla) was specialized. These constitute the Rugosa; also, the Medusa (8) appeared in the Cambrian, according to Walcott. In the Mesodermal layer are differentiated both *muscular* (9) and *skeletal* (10) tissues, which take the radiate form of the mesenteries, and in the living forms there is a differentiation of the *sex* (11)—a differentiation we have all reason to believe was existent in Cambrian time. In the ectodermal or outer layer of the body there is differentiation of a set of cells for offensive and defensive action upon other organisms; these are the *thread cells* (12), which are used offensively (12*a*), probably to benumb their prey and thus aid in the attainment of food, and as defensive (12*b*), in the way of protecting themselves from attack of larger animals which might seek them for food.

There is no certain differentiation of sense or nervous organs in the Cœlenterata, and the above points are about all that can be said certainly to apply to the organisms referred to the Cœlenter..:a for the Cambrian era.

The branch *Echinodermata* also was present in the Cambrian. In them the body presents the *radiate* (13) type of structure in the adult, but the parts are normally *five* (13*a*), and there is more or less distinct *bilateral symmetry* (14) exhibited by them in the adult form generally, or only in the embryonic form in some of the living types.

In the adult there is developed a more or less resisting *integument* (15), either in the form of coriaceous (15*a*) integument, with granules (15*b*), or spicules (15*c*), or definitely formed and articulated calcareous plates (15*d*). There is elaborated a peculiar hydrostatic apparatus, called the *ambulacral water-vascular system* (16), which subserves the purposes of locomotion (16*a*) and the conveyance of food particles (16*b*) into the mouth, and may be considered as a special elaboration of the elements which are tentacles in the Cœlenterata. In the Echinodermata there is a considerable elaboration of the alimentary system. There is a *closed gastric cavity* (17), separate from the *somatic* or *vascular cavity* (18); this constitutes a *rudimentary stomach*. In the more perfect type of the Echinoderms, the Echinoids, there is a distinct alimentary canal of several parts, composed of a *mouth* (19), provided with special organs for reducing food, *five teeth* (20), and a differentiated *œsophagus* (21) leading to a *stomach* (22), and a distinct *intestine* (23) terminating in an *anal* (24) opening. There is also a *pulsatory heart* (25), with radiatory vessels, or *blood-vascular system* (26). Thus in this higher type of Echinodermata we find already differentiated organs for mastication, digestion, nutrition, and distribution or circulation. It is not well established that the function of *respiration* is specialized, or that distinct organs are differentiated for this function. The Starfish (Asterioidea) do not have distinct teeth, but the Ophiuroidea do, and the hæmal system is present in both, but the mouth in many cases serves for ejection of fœcal matter. These two types are developed very early—as early as the Ordovician, so that it is evident that all these differentiations of the Echinoderm type of the digestive system were elaborated by the beginning of the Ordovician, and probably in the Cambrian. The Crinoids and Cystoids were Cambrian, the Blastoids appeared later; the digestive functions were less elaborate in them, but the differentiation into a stomach or digestive cavity, as distinct from the nutritive tract or intestines, was present. The *nervous system* was also developed as a *ring* about the mouth, or œsophagus, and sent out *nerves* to the other parts of the body, and there are pigment-cells developed on the upper side of the Echinoids, which are regarded as of the nature of *optic organs*, but it is doubtful if any such organs were differentiated in the Cambrian type. The nervous system in this type of animals at that time probably performed the function of co-ordination of organs. With the differentiation of the alimentary canal there was probably a specialization of cells for the particular function of these several parts of the canal. The *reproductive function* had its special organs differentiated, but they were as numerous as the partitions of the body, and the elaboration of this system in the Cambrian era had not proceeded far.

The *Annelids* (which is in our classification a representative of the branch Vermes, but in Huxley's classification is placed in a branch Annulosa, distinguished in some particulars from the Arthropoda, but only as a sub-branch, the Anarthropoda) are represented in the Cambrian. They are the lowest or less differentiated type of the articulate mode of body development. There is an elongation of the body, and in the adult there is a definite division into *segments* (27) or *metameres* (somites repeated and arranged along a longitudinal axis). A prominent distinction separating the Annelids from the Articulata proper, as representatives of branch or class groups, is the absence of jointed appendages articulated to the somites in this division of Vermes, and Huxley recognized this distinction in applying the name Anarthropoda (meaning without joints) to the class, while the Crustacea, Insects, and allied forms develop *jointed and articulated appendages* to the somites. In this type of structure the differentiation of parts in the first or radiate direction is completed in the strictly bilateral symmetry. The function of motion has specialized into definiteness of relation of the motions to the body—a longitudinal polarity. The direction from which supply of food comes toward the body, or towards which the motor system propels the body, is *anterior* (28*a*); it is distinctly in front of the mouth, while the other parts of the digestive system are arranged definitely *posterior* (28*b*) to the mouth, along the longer axis. The parts about the mouth are reduced to their smallest number, and are determined definitely in relation to a surface upon which progression takes place; a *ventral side* (29*a*) and a *dorsal side* (29*b*) become thus distinguished. The nervous system is present and surrounds the œsophagus (30), and expresses the differentiated bilateral symmetry by consisting of a *double, ventral, gangliated cord* (31), and in some genera there are differentiated distinct *optic organs* (32) and special *organs of touch* (33). The digestive system is differentiated into *mouth*, sometimes armed with distinct *jaws* (34) for mastication, a distinct *œsophagus*, a *stomach* or digestive cavity, an *intestine* or assimilative canal, and the two openings, *mouth* and *anal*, of the digestive canal are permanent. In the Annelids there is a *pseudo-hæmal system* (35), a vascular distributing system, but not so highly developed as the circulating system of the true Arthropoda.

In the true *Arthropoda*, in addition to the elaboration seen in the Vermes, there is differentiated a distinct system of *motor organs* (36), articulated appendages moved by muscle and not by hydrostatic device, and articulated to the segments. The segments are repeated in more *definite numbers*, but in the Cambrian there was not a permanent selection as to number. In the Trilobites there was evidently (37) a *selection* of number of segments in contrast to the indefinite number of the Vermes, which in *Eunice gigantea*, a modern type, has 400 segments. There was a permanent specialization of a (38) chitinous *exo-skeleton*, which is a distinct elaboration of the motor skeletal system, and made possible a number of special differentiations of the motor organs. There was specialization of *appendages* for special functions; that is, for *sense organs* (39), for mouth or *mandibular masticatory organs* (40), for *swimming* or *locomotion* (41), and other sets connected with *respiratory* (42) function.

The definite differentiation of organs for the respiratory function (*gills* or *branchiæ*) (43) is a further elaboration of the alimentary system, but this function was evidently specialized even in the Cambrian representatives, the Trilobites, which were the most highly elaborated organisms of that era. In these Trilobites we find thus a thorough differentiation of special organs for each of the systems of functions characteristic of the highest type of animals, viz., *Correlation*, further elaborated in the two systems, (*a*) motory, illustrated by the muscles and skeletal parts, and (*b*) nervous system, ganglia, nerves and organs of sense; *Sustentation*, as exhibited in organs of alimentation, digestion, nutrition, circulation, and purification; *Reproduction*, with special organs and separation of sex.

Insignificance of Characters of Marine Invertebrates Evolved since Cambrian Time.—When we would speak of evolution of different kinds of organisms, it is not regarding the evolution of the differences above described that the geologist has any evidence; they were present at the beginning of the records. All this had been accomplished when we get the first glimpse of the earliest known relic of an organism. The simplest types of organisms are living to-day, as are the most elaborated types; but when we go back to this earliest page of geological history we find (with the exception of vertebrates) all the grand types of animals already living together. So far as these grander differences of organization are concerned, the millions of years of geological time throw no light upon the way by which they came about.

When we consider that our knowledge is only of marine organisms, and how extremely meagre is the evidence we have of them, it becomes highly probable that for animals adapted to this environment nothing of branch, class, or possibly of ordinal rank has been evolved since Cambrian time.

CHAPTER XII.

THE TYPES OF CONSTRUCTION IN THE ANIMAL KINGDOM.

Records of Evolution expressed chiefly in Generic and Specific Characters.—From what has been said in the previous chapter it will be learned that the grand features and the great majority of the more important details of the structure of any living organism are of extreme antiquity. Not only so, but since very early geological time no new types of structure of as high as ordinal rank have been evolved in the majority of the branches of the Animal Kingdom.

In respect, therefore, to a great number of the more important characters of organisms the development of offspring has resulted in the repetition, without substantial modification, of the characters of the ancestors. This is the law of Heredity—the repetition in the offspring by generation of characters like those of its ancestors. *Evolution has to do with the acquirement by organisms of morphological characters which their ancestors did not possess*; hence we must seek for evidences of evolution chiefly among the characters of less than ordinal rank—those of ordinal and higher rank having been evolved almost at the beginning of the history.

Course of Individual Development supposed to have been Constant.—It is not unreasonable to assume that all the course and the stages of development, of characters of ordinal and higher rank in the development of the individual, are repetitions of what has taken place since their first appearance at the beginning of the geological time-record. In the several types of organisms now living, the laws of individual development, as of the steps by which in each case diversity is elaborated out of simplicity of structure, may reasonably be regarded as applicable to all organisms of which we can study the history. The reason why the course of development has been what it is may be no more evident than the reason why gold is yellow and heavier than sulphur; in a particular case the sufficient reason is that it is like that of its ancestors.

Beginning of Individual Life and Development.—In a previous chapter the stages of development of the individual are described. It is there shown how the simple cell is without distinction of parts, other than as *protoplasm* with *cell-walls*; a *cell-nucleus*, which is of great importance, and regarding which recent investigations with high power of the microscope are bringing out wonderful characters and functions; and a *vacuole*, often present, but the function of which is unknown. From such a cell the individual grows to the state of a complex, independent organism, such as the living Vertebrate, seen in its highest representative, Man.

Hypotheses regarding the Phylogenetic Evolution of Races.—The term *Ontogeny* has been applied to this development, and to distinguish it therefrom, *Phylogeny*, or race-development, has been proposed to indicate the analogous passage from the simplest undifferentiated Protozoan, the Amœba, or Monera, through the several stages of increasing complexity of organization to the most highly differentiated Vertebrate. Many attempts have been made to construct the history of the whole organic world on this basis, i.e., to construct phylogenetic trees of the ancestors of beings now living on the earth. Haeckel's "History of Creation" is one of the earlier and most elaborate, and perhaps most artificial, of such treatises; for as science has developed, our knowledge of the true genetic relationship in some particular lines of organisms has greatly increased. When Haeckel's work was published (1868), the new methods of investigation, so greatly stimulated by the appearance of Darwin's "Origin of Species," had only begun to affect the students of fossil remains; and it is mainly since that date that the classification of organisms has been revised on the basis of genetic affinities determined by comparative studies of structure.

The analysis of organic structure, from the phylogenetic point of view, is very instructive and suggestive, if it be not overdone. It helps us to attain general notions of organization, or what we may call the principles of construction of the Animal Kingdom.

The Undifferentiated Cell.—From this point of view the primitive living organism is assumed to be an undifferentiated cell, having no tissues, no organs, no permanently specialized functions. If it moves, the motion is spontaneous, irregular, temporary motion; if it takes food, it is by attaching the food to itself; and in a sense such a protozoal cell is all mouth, all stomach, all everything necessary to living, but nothing particular in any part of itself is permanently different from any other part: it is an undifferentiated organism. The amœba comes nearest to fulfilling these homogeneous conditions, but even there appear the nucleus and the contractile vacuoles, which are differentiated, and perform some, though not well understood, special functions.

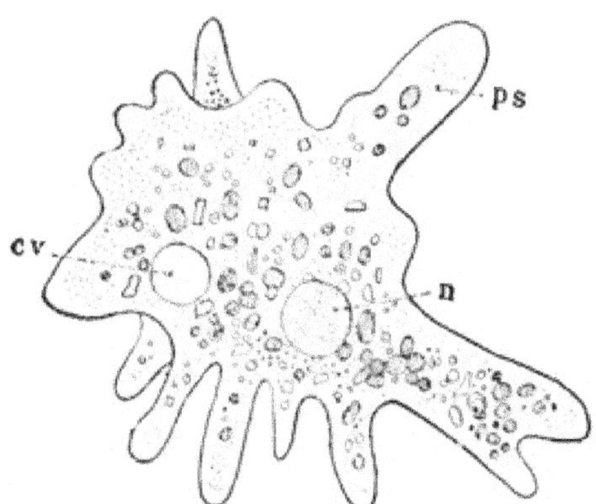

FIG. 51.—*Amœba proteus* (after Gelüber), greatly enlarged. *cv* = contractile vacuole, *n* = nucleus, *ps* = pseudopodium.

In the simplest form of the metazoal cell very considerable complexity is found at the earliest stage in which the cell is observed. The steps by which the cell reaches the organic structure which is characteristic of any of the metazoa when adult is explained in works on embryology and animal morphology.*

When we look at the progress more rapidly, and note the steps of progress in function rather than in structural mor-

* See McMurrich, "Text-book of Invertebrate Morphology," chapter IX, Subkingdom Metazoa.

phology, we observe that in attaining differentiation from this simple state several systematic groups of differences are expressed. The first is concerned with general direction of motion, expressing itself in the arrangement of the body shape, or in its development.

Polarity.—If we imagine the primitive form to be a globe, its motion is expressed by assuming polarity of direction—a definite *anteriority*, or direction toward which motion approaches, and the opposite, *posteriority*, from which it goes. Every living animal having reached the first stage of differentiation (seen in the Metazoa, as the Cœlenterata, for instance) expresses some degree of polarity. The longitudinal axis of the body, in the Metazoa of this simplest form, is clearly expressed, and the anterior end is primarily determined by the position of invagination in the growth of the embryo forming the gastrula. Thus the simple coral polyp is a mature animal representing the *Gastrula stage* of embryonic development of higher animals.

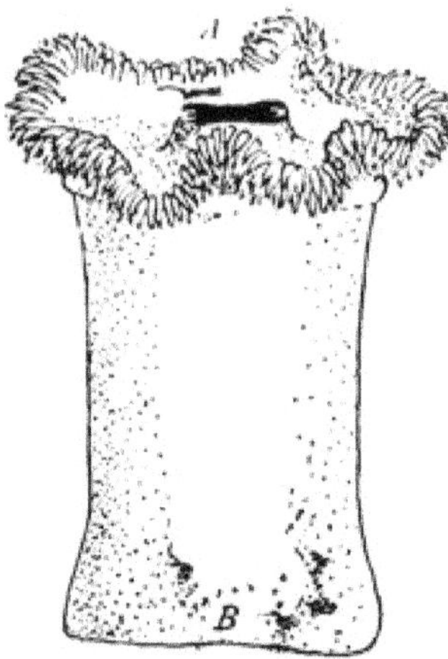

FIG. 52.—A simple coral polyp (*Meridium marginatum* Les.), representing the gastrula stage of differentiation, in which the posterior end of the body *B* is attached and the anterior end *A* is free.

In Fig. 52 the anterior end of the axis of the body, *AB*, is at *A*, which is the mouth or oral end of the enteron or digestive cavity. This is the centre of the free end of the body, and the opposite end, *B*, is in mature stage often fixed.

Antimeres and Metameres.—As such an organism is supposed to develop parts by differentiation, these parts are arranged in one of the following three ways: radially, or around the axis, when they are called *Antimeres;* or one after the other in the direction of the long axis, when they are called *Metameres;* or, third, without repetition of parts, except to express bilateral symmetry and a dorso-ventral opposition of parts.

Radiate Structure, Bilateral Symmetry, and Actinimeres.—The *primary axis* (*AB* in Fig. 52) is the one which is longitudinal to the body, and the secondary axis is at right angles or transverse to this. In the course of growth *repetition of parts* is first noticed as evidence of elevation of rank, and the organism which has no duplication, or multiplication of parts, is lower in the scale, because less differentiated, than one in which there is multiplication of parts. Where there is multiplication of parts the simplest mode of arrangement is around the longitudinal axis. When each of the parts about the axis is alike there is *radiate* structure (see the tentacles, *t*, of Fig. 53). This is the case in the coral animal, or in the starfish, and the separate parts are called *antimeres;* thus the tentacles of the coral, or the arms of the starfish (Fig. 54), are antimeres, or opposed parts. When there is difference

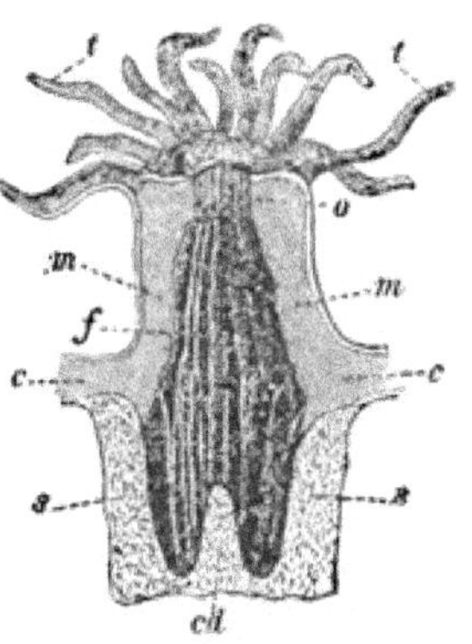

FIG. 53.—Coral animal and its calcareous base, *Asteroides calycularis* Lmk. Longitudinal section. *s*, *cd* = calcareous skeletal base, the "coral;" *o* = mouth; *t* = tentacles; *m* = the mesentery folds of the body; *c* = cœnosarc connecting the individual polyps of the colony; *f* = chambers between the mesenteries. (After Steinmann and Döderlein.)

FIG. 54.—A typical radiate, Starfish, *Asterias arenicola.* (After Agassiz.)

among these parts, and there are series of parts opposed to each other, the differentiation has progressed one step higher, and we have *bilateral symmetry*. When there is multiplicate division, whether there is symmetry or not, the rays thus formed are called *actinimeres*, or ray parts. This mode of differentiation is characteristic of the Cœlenterata and Echinodermata (omitting from the former branch the sponges), and suggested to Cuvier the name Radiata (see Figs. 14–19).

Somites, Arthromeres, and Diarthromeres of the Arthropods.—Another large and diverse group of organisms is characterized by repetition of parts in the direction of the longitudinal axis. The technical name for body without its parts is *soma*; the repeated parts which are longitudinally multiplied are called *metameres*, somites, or segments (see Fig. 55).

The annelids represent the simple metameric type, without appendages to the separate metameres or segments. In the higher class, the Arthropoda, including the Crustacea, Myriapods, Insects, etc., the somites are provided with lateral appendages which are jointed in regular manner (see Fig. 56, also Fig. 50).

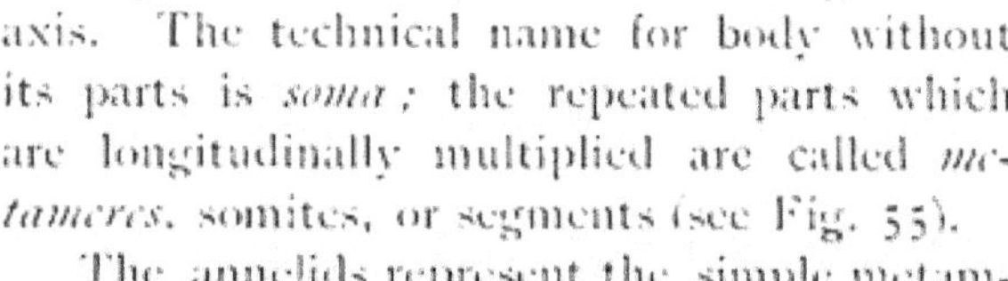

FIG. 55.—The metameric type, a diagram of a typical annelid: *m* = mouth; *ce* = cerebral ganglion; *n* = ventral nerve-cord; *pr* = head (prostomium); *a* = anus. The body (soma) is composed of twenty-five segments or metameres.

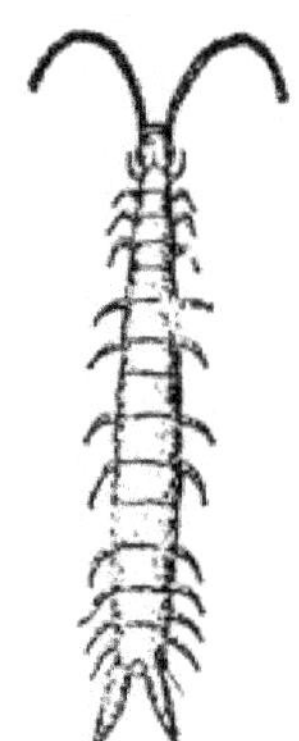

FIG. 56.—A metameric animal with jointed appendages, *Scolopendrella immaculata*. (After Leunis.)

In the Arthropoda, such as the common lobster, and in an insect, these separate somites form a single ring enclosing the interior organs; but in the Vertebrates the somite is double, the special system of correlation lying in the upper arch, the organs of assimilation or auxiliary function lying in the cavity below. To distinguish these two forms of the metameres the first is called a joint part, arthromere; the corresponding part in the vertebrate structure is called a two-joint part, diarthromere. The joints of the appendages of a metameric part, as the joints of the legs of a lobster or the several bones in the limbs of vertebrates, are illustrations of multiplication of parts by division in a transverse direction. The technical name for this mode of repetition of parts is *antimeric*.

Distinctive Characters of the Metazoa.—All the higher tissue-bearing animals, or Metazoa, differ from the Protozoa by the possession of the following characters, viz.:

Metazoa.—Reproduce by developing egg, or ovum, which passes through the stages of (*a*) nucleated cell, (*b*) segmentation, (*c*) blastosphere or morula, (*d*) gastrula; tissues differentiated into (*e*) ectoderm, (*f*) endoderm, and (*g*) mesoderm; (*h*) alimentary cavity, or enteron permanent, and (*i*) sexual differentiation the rule and almost universal.

Molluscan Type of Structure.—The third type is that in which neither metameric nor antimeric repetition is carried on, but bilateral symmetry and simple antero-posteriority and dorso-ventral polarity are more or less conspicuous. In this type of organisms (the Mollusca) differentiation is expressed in the relative positions of the organs in the body-cavity, and in the relative development or importance of the different organs or regions of the body.

Development of Organs and their Taxonomic Rank and Value.—In the molluscan type is seen in its simplest form that relative development of the several systems of organs which marks the rank of the stage of progress in differentiation in each particular case. Thus of the several systems of organs sustentation is more fundamental, and may be regarded, if prominent in relative development, as indicating primitive or low rank. Organs of correlation, when more specialized and according to the degree of differentiation of the special organs, imply specialization, hence high rank. Thus among the Mollusca those which are simply digestive sacs, with no specialized organs of sense, or of definitive motor organs, are low in rank (the Lamellibranchiata). The specialization of sense-organs anterior to a mouth and of the muscular system for giving definiteness to the motion, indicates higher rank (the Gastropoda and Pteropoda). Special tactile organs, and high development of sense-organs, all in front of the oral opening, show still higher rank (the Cephalopoda).

This principle of differentiation in the development of organs throws light upon the rank of particular organisms in the phylogenetic line of their evolution, and relatively in each line those expressing greater differentiation in the general development, or higher specialization of the more dependent or secondary characters, are necessarily of higher rank, on the theory of acquirement of characters by direct descent only.

The Principle of Cephalization.—The relative development of the organs of correlation, especially of the organs of sense, has been recognized for many years as indicative of grade of rank among animals.

James D. Dana has written much on this subject under the name of Cephalization.

In discussing the principle of Cephalization Dana wrote: "Such growth or progress in the brain and nervous system, the seat of power in the animal, is accordant with, and consequent upon, the great fact that this is the part of the structure which comes into actual contact with outside and inside nature. It is the means in the animal by which communication is had with the outer world, and also with its own inner workings and appetites; that which takes impression, which feels whatever inspires energy, prompts to action, exhilarates, or exalts; the part, therefore, which must grow whenever circumstances favor progress, and, at the same time, fail to grow or dwindle under unfavorable circumstances; which communicates whatever it receives to the being to which it belongs, and in each case to the part or parts responding to its condition; which reaches every part of the system and dominates in all action and growth, and hence must cause an expression of its own condition in some way on the structure; which, moreover, must ordinarily produce correlative changes in correlative parts, if any, because in its own nature and distribution the system of correlation has a full expression" . . . . . "We may, therefore, believe that in all progress in grade, upward or downward, there was involved some change in the animal structure of the kind expressing degree of cephalization."

"Whatever the types of structure in course of development, there was also a general subordination in the changes to the principle of cephalization." *

* *A. J. Sci.*, ser. III., vol. XII., Oct. 1876, pp. 245–251.

Cephalization one of the Expressions of the General Law of Differentiation.—Cephalization may be regarded as but one of the expressions of the general principle of differentiation. Differentiation concerns the whole organism, because increase in the specialization of function of one organ always involves the provision, through the activity of other parts, for the supply to that organ of resources which it fails to supply to itself.

Meaning of Homology and Homologous Parts.—When animals are compared there are some terms which are applied to the relationship noted in the parts compared; a few of the terms are the following:

Homology and *homologous* are applied to the organs or parts of different organisms which correspond in type of structure. Thus the secondary joints of the appendages of Arthropods are homologous parts, and one appendage may be used as a swimmer or a claw, another as a mandible, and therefore be constructed in different form; but the parts, although of different form, are said to be homologous, because modifications of the same element of differentiation (see Fig. 50).

Another example is the case of the forearm of a bird and the forearm of a bear. When the bones are compared they are found to possess corresponding parts—a shoulder-blade, a humerus, a radius, an ulna, a carpus, metacarpus, and finger-bones. Although the arm in one case is adapted to the function of flying in the air, in the other to walking on the ground, and the shape of each bone is different, the several parts are homologous, because bearing the same relation to the structure of the whole, and representing the same typical part of the primitive structure.

Analogy and Analogous Parts.—*Analogy* is used in a different sense. Two parts or organs of different animals are said to be analogous when the likeness has to do with the functions or adapted usage of the parts; and not to either the morphology, or the relationship to other parts of the organic structure of the animal. For instance, the leg of a fly and the leg of a mouse serve the same function—walking or locomotion, but they differ morphologically, i.e., in form; they differ also in their structural relation to the whole organism of which they are parts. Homology may be said to be based upon morphological unity, and analogy is based upon functional or physiological unity.

Differentiation Illustrated in the Case of Motor Organs—To illustrate this mode of analysis of the organic structure from the point of view of extent of differentiation of parts, or in order from homogeneity to heterogeneity of structure, a study may be made of the devices developed for the execution of motion or locomotion, in the various branches of the Animal Kingdom.

Organic motion, in its simplest form, is *contraction*, the bringing together of two ends of a contractile tissue, as muscle, with no hard parts, no specialized organs: this is what is seen in the lowest forms of the Protozoa, and expresses itself in change of form of a globule, drawing in of a part, or pushing out of another part.

Two Directions in which Differentiation Proceeds.—In differentiating the mechanism of motion, elaboration may take place in two directions.

(A) *By subdivision, or multiplication of the moving parts*, and increasing the rapidity of the contracting: this results in *ciliary motion*, and the specialized organs thus elaborated are called *cilia*.

(B) The second is *by concentration, or massing of the parts* of motion, and thus increasing the energy expressed in a single motion: this leads to the construction of *muscular tissue*, and the expression of specialized *muscular motion*.

In (A) the direction of the motion is *indefinite*, in (B) it is *definite* in direction and united in time, or period of action.

Ciliary Motion.—The real function of ciliary motion is seen in an augmented state in the special organs called *tentacles*, which act by muscular methods, but whose function is vibratile. These may add the functions of ingestion and prehension to those of simple ciliary motion. But ciliary motion itself is fundamentally applied for the ingestion of food. This is accomplished in minute organisms either by causing the organism itself to move in its medium towards the food, or by setting up currents in the medium and thus causing the food to flow to the mouth of the organism. The tentacle is an enlarged cilium in so far as the function is concerned. The Cœlenterata exhibit this mode of elaboration of the motor organs in a typical way.

Water-vascular System of Echinoderms.—In the Echinodermata a higher elaboration of this kind of action of muscular tissue is expressed in the *water-vascular system*. This is a peculiar adaptation of simple muscular contraction.

Cilia in Molluscoidea and Mollusca.—The Molluscoidea have a system of ciliary motion drawing the food particles to the mouth-opening by setting up currents in the water. Some of the Mollusca have a similar method of producing currents by means of cilia on the edge of their mantles. In the Gastropod and Cephalopod motion is accomplished by specialization of muscular contraction. Various types of modification of the foot are elaborated in the different classes of these interesting forms.

Skeletal Parts.—In the Arthropoda and the Vertebrates organs of motion are more highly elaborated by the addition of *hard parts* acting as levers, and thus giving special direction and change of direction to the simple contraction of the muscles passing between two articulated parts. The general difference between the motor systems, or the modes of motion, in these two grand divisions of the Animal Kingdom, is seen in the different relation which the contracting part (the muscle) bears to the mechanism, or skeletal part. In the Arthropod the muscles are attached on the inside of hollow skeletal elements. In the Vertebrates the muscles are outside and around the levers which they move, and in these two groups of organisms motion, and both the muscles and the machinery of motion, reach a high degree of elaboration.'

Multiplication of Like Parts Preceding Specialization of their Functions.—The course of differentiation is from simplicity or homogeneity of parts, first, to multiplication of the parts possessing like functions and often uniformity of form, and, second, to the specialization of function of these parts, their division into groups, their consolidation, and, finally, definiteness in number and precision of use or function. Hence division of labor follows the multiplication of parts and does not precede it. Multiplicity of laborers is a condition necessary to the division of labor, and the organic co-operation of separate parts.

Comparison between Embryonic Development and Succession of Ancestors.—Prenatal or embryonic development of higher animals may pass through stages similar to those expressed in the mature form of lower animals which are supposed to be in the line of descent of the former: as an example, the Mammalian embryo develops gill-arches, which are characteristic of the mature stage of fishes: but in the embryo of the mammal this feature appears in the earlier embryonic life, and is lost as development proceeds. Much has been made by embryologists and also by systemists of this embryonic calling back to supposed ancestral characters; but in deriving conclusions from these facts it must be remembered that, since the organs are neither fully co-ordinated nor completed for action in the individual embryo, and that not until the natal stage is past, the likeness of these characters to the mature parts of supposed ancestors is rather a likeness in the plan or course of development than in the results of the development. The course of the development may be alike in two organisms; that is, the steps by which the morphological features may be attained may be according to the same plan, and indicate a fundamental affinity, which is less evident or quite lost in the mature animal. However, it is not a necessary inference that in the embryonic development we will be able to recognize the relationship to an ancestral mature form. Changes, such as abbreviation, or a different course of development, of the embryo, can be assumed to be indicative of phylogeny only in case environment was the determining cause of their original appearance. If there be an evolution in these modes of differentiation, as there is an evolution in the final product, the resultant differences may be determined by other laws.

Muscular Motion or Specialized Motion, and Locomotion.—The preparation for motion of the organism in definite direction is exhibited in the differentiation of the head as the oral end of a moving organism. Next, it is seen in the differentiation of the assimilating cavity into a tube, the enteron, with separate entrance and exit, for the materials of assimilation. A third stage is represented in the elongation of the body in such a manner that it may move in part without any actual locomotion, the one end becoming sessile, or attached, as in the case of the Cœlenterata. In this case no specialized organs of locomotion are developed, but the mouth-parts are moved in relation to the body, and are moved also in relation to the source of food. In Vermes there is locomotion, but no special articulated parts are developed in this lower type. In the Arthropoda articulated organs subserving the function of locomotion are developed; associated with the specialization of motion, as local motion, we find a specialization of the poles of the body into an anterior and posterior end, relative to the direction of the motion. The first mode of differentiation spoken of has not to do with locomotion, but rather with relation to reception of food. An oral end of the alimentary canal was established through which food reached the interior of the organism. The polarity was a polarity between the oral end of the alimentary canal and the excretory end, or rather between the approach of food and the discharge of effete results of digestion.

In the Cœlenterata the oral orifice serves also for the discharge, and therefore the oral and aboral poles are brought together, typically, at the same point. In the Echinodermata, in some cases, the aboral corresponds to the oral pole. In other Echinoderms there is a distinct separation of the two ends of the alimentary canal. With the setting up of the antero-posterior polarity of the chief axis of the body, and of these specializations of locomotion, there was expressed a decided advance by the appearance of sense-organs at the anterior pole.

Differentiation of Nervous System a Concomitant of Locomotion. —Motion, bringing about a change of place, implies the selection of better conditions of environment, and the guidance of the locomotion toward such favorable conditions. Thus the *differentiation of the nervous system* follows, or is intimately associated with, the specialization of motion into *locomotion*. Again, we notice that the head, being thus specialized, is only one of the kinds of differentiation. Thus the metameric mode of development first makes possible heteronomy of parts, i.e., the specialization of functions, along the digestive tract. In such an organism as the lobster, for instance, we find a definite arrangement of specialized functions with differentiated organs, distributed along the line of the axis from the antennæ to the extreme posterior end of the body.

Differentiation Along the Digestive Tract.—Without considering the skeletal parts, but looking at the organism in respect to its digestive tract alone, we find the following series of differentiated parts:

First, the *detection of food.* Provision for this is made by special organs of sense, antennæ, eyes, organs of smell and of taste, and, finally, those of hearing.

Second, the *prehension of food.* For this purpose jaws and teeth and other apparatus are provided.

Third, the breaking or gross *reduction of food* for digestion. For this purpose the teeth and jaws are brought into action.

Fourth, the *digestion of food.* In this process several special organs take part, the most important of which are the *stomach* and the secretions which are furnished at that point in the enteron; but there are, in addition, in higher organisms, numerous specialized glands, secreting digestive fluids with differing properties.

Fifth, the *absorption of food.* For this function the *intestine* and associated organs are differentiated.

Sixth, the *distribution* and *application* of *food-products.* To this group of functions are applied the organs of *circulation,* and auxiliary to them are those of *respiration,* and the corresponding organs.

Seventh, *elimination* of effete matter. The organs for this function are at the anal termination of the enteron, and auxiliary organs are found associated with the circulatory system, as, in the higher animals, renal organs; and even the skin subserves the same function, in part, in perspiration.

Differentiation of the Motory System into Muscular and Skeletal Organs.—This principle of differentiation might be traced in relation to the skeletal framework of the body; but these relations are not fundamental, and the organs are adjusted to them to conserve convenience and compactness of arrangement.

The motor organs, however, express their differentiation in the skeletal parts and in the form of the body. There are two types of differentiation of the motory system resulting in the construction of what may be called muscular and skeletal systems, or parts. Muscle and skeletal parts are correlative to each other. Hard parts of some kind, to which the general name *skeletal* is applied, are essential to specialization of the direction of motion, and contractile muscles are just as essential to the motion of these skeletal parts themselves. The relationship between these two elements of the motory system is as intimate as that between steam and machinery in the steam-engine.

Archetypal Structure.—The further elaboration of this method of analysis of organic structure may be pursued only by tracing the elements of structure to their specific characters in many separate types.

Sufficient may have been said to emphasize the fact that there is a logical foundation for the idea of archetypal structure, so much insisted upon by Agassiz, explained embryologically by Von Baer as early as 1828, and expressed in Cuvier's classification of the Animal Kingdom, in the four general plans upon which the various kinds of animals were constructed. Cuvier wrote in 1812 *

". . . On trouvera qu'il existe quatre formes principales, quatre plans généraux, si l'on peut s'exprimer ainsi, d'après ses quel tous les animaux semblent avoir été modelés." etc.

Cuvier's Classification.—Although, later, more minute studies have produced modification in the systematic classification of the Animal Kingdom Cuvier's division of animals into Radiata, Articulata, Mollusca, Vertebrata, expresses the most profound distinction exhibited by these organisms; and whatever criteria we take as the basis for classification, and with slight modification due to increased knowledge, these grand divisions of the Animal Kingdom stand out as pre-eminently the most important groupings that can be made.

To say so much is not an acceptance of the philosophy of the earlier naturalists as final. That there are a few general

* "Ann. des Mus. d'Hist. Naturelle," vol. XIX., Paris, 1812, quoted by Agassiz in "An Essay on Classification," London, 1859, p. 309.

plans of structure upon which the multitudes of animals were built, does not carry with it any theory as to the reasons for the differences, or as to the mode by which the several types of structure came to be carried out in such multitudinous fashion.

The fact is beyond dispute that there are a few types of construction to which the animals of the whole kingdom conform, and these are expressed in the mature forms as well as in the course of the individual development. Cuvier, when he considered the mature results, found them to be four; no one since has found reason to dispute the validity of three of them. The branches Cœlenterata and Echinodermata, constitute one of them—the "Animalia Radiata;" the Arthropoda typify another—the "Animalia Articulata;" and the Vertebrates include substantially the same organisms classified by Cuvier under the type "Animalia Vertebrata."

Von Baer's Embryological Classification.—Von Baer, from a study of the course of embryologic growth, thought he had found a more intrinsic reason for the types in the modes of their embryonic growth than in the gross result. He defined them as the "peripheric type," with *evolutis radiata* (i.e., the Radiata); 2d, the "massive type" (Mollusca), with "*evolutis contorta*;" 3d, the "longitudinal type" (Articulata), with *evolutis gemina*, or production of symmetrical parts on both sides of an axis; 4th, "doubly symmetrical type" (Verbrata), "with *evolutis bigemina*," i.e., the development producing symmetrical development on both sides of a median axis, and also developing two cavities, one above and one below the central axis.

Fundamental Divisions of Classification discerned by Earlier Naturalists.—If we throw the light of more recent investigations upon the matter, we find it necessary to make expansion of those divisions; but very little alteration in the fundamental classification, thus early recognized, is required in order to express the scientific classification of present usage. Looking upon classification from this point of view, we first divide the *Animal Kingdom* on the fundamental character of cell-growth. When cell-growth proceeds no higher than the multiplication of cells, having no differentiated and complementary functions, the individual is not a compound organism, but is always *cellular*. All such animals are grouped under the one division, *Protozoa*.

When cells divide, so that as cells they are separate, but remain in close association, with division of labor, one function played by some, another function performed by others, the result is *tissue* and *organ*, and the individual is an organized individual; this constitutes the group of *Metazoa*. Classifying these Metazoa on the basis of the direction along which development of the specialized parts of the body proceeds, we find the same grand division appearing prominently before us as types of construction of the complex organism.

The Polymeric Type.—There are two fundamental directions along which their development proceeds. Taking the mouth, the opening or entrance to the enteron, as the centre, multiplication of parts may be around this centre (radiate), or it may be from it in the direction of the axis of the enteron (longitudinal); and besides these two ways there is probably no other direction of multiplication of parts. When the multiplication is indefinitely radial, it produces the antimeres of the coral polyp, having chambers and tentacles distributed about the mouth as tentacles, and extending backward as septæ in the body-cavity. The Cœlenterata express this mode of construction in its most indefinite manner; the Echinodermata express it with definiteness of number of radiations but with a tendency to the following type, and this may be named the *polymeric* type of construction.

The Dimeric and Monomeric Types.—The second path is by the specialization of the polymeric types with limitation of multiplication to repetition in two opposite directions, forming bilateral symmetry. This is seen in some of the Mollusca, as in Fig. 35, representing the idealized primitive Mollusk. This may be called the *dimeric type* of construction.

In a third case there is no full duplication of organs, although the motion and the form are as in the dimeric type. This may be called the *monomeric type*.

The Cephalopoda are in part polymeric, but in the main monomeric. The Molluscoidea are polymeric and monomeric in different parts of the body. Thus the branches Cœlenterata, Echinodermata, Molluscoidea, and Mollusca are all associated together by the fact that their development of separate parts is in a direction radiately, or in circle about the mouth, and hence they are antimeric Metazoa.

The Metameric and Diarthromeric Types.—The Vermes and the Arthropoda are, on the contrary, *metameric;* the development adds parts by repetition longitudinally along the median axis. In the Vertebrates, in which there is added, as Von Baer already saw, the *diarthromeric* separation of a dorsal and ventral cavity, with specialized parts distributed in each, the arrangement of specialized organs is on a monomeric plan, as in some of the Mollusca.

Meaning of Typical Structure and Types in Modern Zoölogy.—It may be remarked that the difference between the old and the more modern use of such classifications, as above made, consists in the theoretical value placed upon them. Cuvier and Agassiz considered such "types" as in the nature of "ideal plans" which all animals for some reason were obliged to conform to, and departure from the "type plan" or "arche type" was an abnormality, or required theoretical adjustment to the plan.

In modern Zoölogy by the "typical" structure of a group is meant a generalized statement of the most conspicuous features observed in the members of the class under consideration, and departure from the type in individual cases is evidence not of aberration in the particular case, but of imperfection of the description. The fact that there is development along one or another line is important: the generalization of the law so as to cover the principle and omit the details aids the formation of clear notions; but the notions are not the things, and the latter have to be constantly rectified to express the increase of knowledge of the former. The four types of Cuvier had their representatives in nature, but all organisms did not stick closely to a particular type of construction expressed by his formulation of characters of the types.

CHAPTER XIII.

PHYLOGENESIS IN CLASSIFICATION.

Principles of Classification Illustrated by the Mollusca and Molluscoidea.—In order to reach a closer view of the meaning of the relationship of organic form to the place in the time scale at which it appears, we must examine more particularly the principles underlying the classification of organisms.

The groups of organisms from which examples will be chosen are the Mollusca and the Molluscoidea; chiefly for the reason that they present hard parts which are abundantly preserved in the rocks, and therefore afford more satisfactory records of their geological history than any furnished by any other class of organisms. A second reason for selecting them is the fact that the statistics, regarding the relation of their forms to conditions of external environment, are so satisfactory as to be at least equal to those regarding any other group of animals.

The Author's Philosophy Reflected in his Classification.—From what has already been said it will have been perceived that form and function are both regarded in the classification of organisms; but hitherto the fact has not been emphasized that the classification of organisms, i.e., the description and orderly arrangement of the characters which one is supposed to see in particular examples of organisms, is affected by the philosophy of the classifier. At best the classification expresses only the author's interpretation of the laws of association of different things; hence if we know the theory by which the association is reached we are better prepared to learn truth from the resulting classification.

Effect of Theories of Phylogenesis upon Classification.—Much is found in the modern literature about phylogenesis as a basis of classification, and it is supposed to supersede quite entirely the classifications which were made by the earliest naturalists who believed in the original creation of all species.

The difference between the two methods is quite simple, and may be explained in a few words. Cuvier and his school observed the morphological characters of organisms, not always knowing the exact physiological function, and compared them together, and then wrote descriptions of the *differences* they observed. They separated organisms into distinct species and genera by the different characters they observed in each, and thus their method of classification is based upon *observed differences in form.* The new school of naturalists is intent, first of all, upon the discovery of the *affinities* of each kind of organism studied. Their point of view is directly the reverse of their predecessors. Their descriptions, and finally their classifications, are based upon the *points of resemblance* which can be detected upon comparing different organisms with each other.

Analytic and Synthetic Method of Classification.—These two schools differ as to the kind of characters which they consider to be of chief importance in classification, and, as a general effect upon classification, the one school is apt to overestimate imagined resemblances, not to be seen by the ordinary observer, and the other may err on the side of making too much of external, often trivial, characters.

Irrespective of the way by which the two methods of classification arose, both methods are now in use and both are useful.

In order to give them names, free from any accidental association connected with their origin or application, the first may be called the *analytic method* of classification, the second may be called the *synthetic method;* and for the purposes of illustration Zittel's classification of the Molluscoidea and Mollusca may be selected as examples of the analytic method, and Lankester's classification of the Mollusca may be taken as an example of a synthetic classification.

Here, as elsewhere in this treatise, the reader must be left to learn the full meaning of the descriptions, only outlined, by a study of the objects themselves. No possible description of natural objects, particularly organisms, can convey to a student impressions which he has never before experienced. And the best way for any one to gain a true notion of the meaning of the distinctions pointed out in these pages is to take a lot of mollusca of different kinds, to be found on the sea-shore, and with the descriptions of the expert zoologist at hand attempt to identify and classify them.

And the naturalist who may possibly look into these pages will appreciate, no more keenly than the author, the great difference between such an introduction as is here attempted, and the earnest investigation of the history of organisms by a study of the organisms themselves.

Mollusca and Brachiopods as Illustrations of Evolutional History. —The Mollusca and the Brachiopods present a peculiar interest, because, having no skeletal parts, the mode of action,—the result of adjustment to environment,—the adjustment of several parts and organs to each other in body structure, and the marks of stages of growth are all concentrated in an external, hardened, and therefore preserved, single, or rarely more than two-parted shell.

To the extent to which such a shell can express the characters of the organism, the perfection of its preservation, and the fact that so much of the individuality of the species and so large a number of individuals are accessible to study, give to this kind of fossil its great value in illustrating the problems of evolutional history.

Zittel's Classification of the Branch Mollusca.*—The classification of the Mollusca proposed by Zittel differs in some respects from that of Lankester, Gegenbaur, and many of the stricter modern zoölogists.

In his branch Mollusca were included as sub-branches—

A. Molluscoidea (which is a branch in Gegenbaur's classification),
with the Classes I. *Bryozoa*;
II. *Tunicata* (which is raised to the rank of a distinct branch by Gegenbaur);
and III. *Brachiopoda*.

(Gegenbaur places Bryozoa with the Worms

* Note: This classification was taken from the "Handbuch," published nearly twenty years ago. In the author's "Grundzüge" (1895), the Mollusca and Molluscoidea are relegated to separate branches, in accordance with present usage. The above passage is left as originally written because it well illustrates the point under discussion. See *Am. Jour. Sci.*, ser. III, vol. L, p. 268.

(Vermes), and treats of Brachiopoda as a distinct class, but allies it on certain accounts with the Vermes.)

B. Mollusca (proper).

Class 1. *Lamellibranchiata.*

2. *Gastropoda.*

3. *Cephalopoda.*

The embryologists make greater point of resemblances observed in the early stages of development, and hence the distribution made of Bryozoa and Brachiopoda next the Worms, and Tunicata next to Vertebrates; but when the mature animals are studied and compared the Brachiopoda are found to possess structures closely resembling in important features the Mollusca proper

The embryological resemblance of Bryozoa and Brachiopoda to worms is lost when the adult stage is reached. Hence, for the geologist particularly, the association of the two is not suggested by any apparent similarity of characters.

Points of View of the Embryologist and of the Morphologist.— In studying the philosophy of natural history it is interesting to note this difference in point of view between the strict embryologist and the pure morphologist. They compare animals on a different basis, and therefore there results in some cases a different classification.

The embryologist classifies animals primarily on the theory of phylogenetic relationship; the student of adult morphology classifies them according to the nature and extent of differentiation attained in the adult. Here, too, the two examples selected will illustrate the differences. The two modes of classification differ much as the classification of houses might, viz. by considering them, either, according to the styles or schools of architecture, as Norman, Roman, Queen Anne, etc., on the one hand, or according to the materials of construction—brick, stone, or wood, on the other.

The line of descent, through which any particular organism has come to be what it is, is all-important if it can be discovered by the study of embryology; but of no less importance is it to distinguish the different and similar structures which have been developed for the accomplishment of the same functions in organisms, whether of near or distant genetic relationship.

Embryological Likeness of Organisms whose Mature Characters are Diverse.—In grouping the Mollusca with the Molluscoidea it is not denied that they may differ in origin—even that in their earliest stages of development Brachiopods may be akin to Worms and Echinoderms; and what animals are not? The adult modes of life and construction of hard parts of the Brachiopods presented greater resemblance to the Mollusca in the Cambrian than they did to Worms or Echinoderms, and it is not ignorance alone which has led the paleontologist to compare them in studying the faunas of geological time. On the other hand, when we go back to the primitive steps of development of the germ it is to be expected not only that two branches will show likeness of development, but if we should go back far enough we shall meet with no visible distinction between the germs of all the Metozoa; in fact all animals, if we go back far enough, may be supposed to present no differences. On the ground of embryology, the Tunicates are akin to the Vertebrates; the Worms are akin to the Echinoderms, the Molluscoidea, and the Vertebrates: but the differentiation took place very far back in geological time.

Evolution not Traceable between Different Classes.—The arrangement into branches, therefore, is from a structural point of view highly artificial; and for purposes of tracing the history, or even from a taxonomic point of view, it is of little importance to deal with characters more ancient or of higher rank than the class characters.

It may be convenient to associate the classes together into larger groups; but to reach the point of real union of their characters, in order to associate two or more classes in a common group, leads us far back into the uncertain mists of the earliest geological time, and into the similar mists of embryonic homogeneity. It is impracticable in the present stage of science to trace the evolutional history of classes.

The Mollusca and Molluscoidea are of particular interest because, lacking internal skeletal parts, and developing a single or two-valved shell, there is concentrated on this shell everything recordable of the characters of the whole organism. These shells from their imperishable character are preserved in the rocks in great numbers, so that variations are found for comparative study. The particular consideration of these hard parts and the study of the marks upon them, which have determined the classifications of the paleontologist, cannot be overlooked when it is an historical study we make of organisms.

Having in view the importance of the characters of these hard parts, of parts which can be examined both in living and fossil condition, Zittel has described and classified them according to the characters which they exhibit in their mature condition after their development and whatever of evolution has taken place in their history, are complete. The following is a translation of Zittel's description of the Mollusca:

General Character of Mollusca.—For paleontologists, and particularly for geologists, the Mollusca present a peculiar interest; for all these classes, except the Tunicates, furnish numerous fossil remains. Principally the shells of the Brachiopods, of the Lamellibranchiates, the Gastropods, and Cephalopods, are so widely distributed in the formations of all the periods of the earth, that one chooses them in preference as characteristic molluscs ("Leitfossilien"), wherever the attempt is made to determine the age of the different sedimentary formations. It is quite evident that it is only the calcareous shells, their moulds in stone, or their imprints which are at the service of the geologist. But as these fossils are ordinarily distinguished by their characteristic form and by their varied ornamentation, as the classification within the several classes is essentially based upon the characters of the shells, there is established a special science, Conchology, which the geologist particularly cultivates. Moreover, although the characters presented by the shells are so insignificant, they are often deceptive; as in the case where the animals of quite different organization (Patella, Ancylus) are able to produce shells absolutely similar: so the classification of shells requires, as in the other divisions of the Animal Kingdom, a firm zoölogical basis, and the determination of species should be made according to zoölogical principles. On account of the relative ease in determination of species in Conchology, the molluscan fossils have always had particular favor with mineralogists and geologists. In any other division of the Animal Kingdom it is impossible to collect, describe, and figure fossil remains in such great abundance; and besides, it can be said that the major part of the bibliography in Geology and Paleontology is devoted to shells—not always, it is true, in an ideal manner. If, indeed, the insufficient knowledge of living Mollusca is a great cause of frequent errors in the determination of genera, so too the determination of species is at present in an almost chaotic condition. As each author, according to his own views, extends or contracts the limits of species, it happens that one rarely finds in the works of different authors the identical fossils of the same fauna described in the same terms in the definition of characters. A chief cause of this unfortunate state of affairs comes from the vertical range of fossil Mollusca. Very frequently in a series of superimposed beds of different age one meets with a characteristic type of which the specimens from each of the different formations (although presenting minor differences) preserve a special facies throughout. In older works all the mutations of such a series of forms were considered as belonging to one and the same species, while more recently the inclination is conspicuous either to raise the smallest differences of their kind to the rank of different species, or to distinguish them from one another by application of trinomial names. Mollusca are in major part aquatic. Of these classes, the Tunicates, the Brachiopods, and the Cephalopods, live exclusively in the sea. The greater part of the Bryozoa, the Lamellibranchs, and the Gastropods are found in salt and in fresh water. The class of Gastropoda alone presents representatives living in salt, in brackish, and in fresh waters, and terrestrial species. All the classes capable of preservation appeared in the Lower Silurian (probably all in the Cambrian). The Brachiopods attain the maxim of their development in the Paleozoic age, the Cephalopods in the Mesozoic; the Lamellibranchs and Gastropods appear to have continued their differentiation and expansion quite up to the Tertiary or recent period.

Mollusca.—An animal presenting bilateral symmetry: soft body, non-segmented: possesses neither internal skeleton nor external skeleton: shows digestive organs very well developed, and a nervous æsophageal collar, with three pairs of ganglia in the highest types. Very many Mollusca secrete in a fold of the skin, called mantle, a calcareous shell with a single or two valves; others are entirely naked, and develop no solid formation. Respiration is mainly effected by gills or branchiæ, more rarely by lungs or folds of the skin. A circulatory system imperfectly closed, with a pulsating organ driving its contents to the periphery, exists in Mollusca, except in the lower types. Reproductive organs, differentiated into sex organs, sometimes hermaphrodite and sometimes separate individuals, and in Bryozoa by budding and formation of colonies, and of various forms. All these animals now called Mollusca were ranked with Worms by Linné.

The Molluscoidea are particularly characterized by a calcareous shell, horny integument, or cellulose tissue. Respiratory organs often in front of mouth, as tentacles or appendages; central nervous ganglion, between mouth and anus; besides sexual reproduction, often also budding. All aquatic, and mostly marine.

Bryozoa.—Small animals, increasing by budding, and united into colonies, branching like moss (hence the name), and forming incrustation, etc. Animals enclosed within membranous or calcareous cellules, and possessing at the anterior extremity of the body a mouth surrounded by tentacles; no heart; intestine; well-developed body: anal opening near mouth; hermaphrodite.

Tunicata. — Sac-like animals, free-swimming or fixed, united into colonies; hermaphrodite: furnished with an envelope (mantle) having the consistence of cartilage or leather, which completely surrounds the body, and presents only two openings. Branchiæ on the internal part of the cavity formed by the mantle; mouth in front of the branchial sac: heart tubuliform [now ranked as a separate branch].

Brachiopoda.—Soft animals, living solitary; furnished with a bivalve symmetrical shell, presenting two free lobes of the mantle, which secretes the two shells; near the mouth two respiratory arms, rolled into a spiral; heart present.

The Mollusca (proper).—The Mollusca (strictly speaking) always multiply by sexual reproduction, never by budding; respiratory organs either branchia or lungs; a central nervous mass (brain) with three pairs of ganglia; body enveloped by a thick mantle, which frequently secretes a shell of one unarticulated or of two articulated valves; mouth with or without maxillary appendages.

Lamellibranchs.—Mollusca, with an unsymmetrical bivalve shell, furnished with a large mantle split into two lateral lobes, upon which the branchial lamellæ are developed equally from one part to another; the two valves of the shell are united by an elastic ligament, and generally by a hinge furnished with teeth and sockets; mouth and arms situated between the branchia in the plane of separation of the two valves; ordinarily there is a muscular foot.

Gastropoda.—Soft animal, creeping, more rarely swimming, with a robust muscular foot; presenting a head more or less distinctly separate from the trunk, and a mantle undivided, which generally secretes an orbicular shell in form of a low cone or shield, or spirally enrolled.

Cephalopoda.—Head pointed, separated from the rest of body; sense organs, especially eyes, attaining high degree of perfection; mouth surrounded by a crown of muscular arms. Body sac-form; 2 or 4 arborescent branchia, placed in a cavity formed by the mantle; shell often spiral, of one or many chambers, sometimes internal, or again entirely wanting.

While the Bryozoa and the Tunicates are scarcely above the Cœlenterates, and are inferior to the Echinodermes, considering the differentiation and perfection of their organs, the Cephalopods should be ranked, without doubt, among the most elevated of invertebrate animals, and in some respects they seem superior to certain vertebrates. *

In contrast to this analytic classification of Zittel, in which the definition and grouping of the organisms is based upon the visible and generally conspicuous characters distinguishing

---

* Zittel, "Handbuch der Palæontologie," vol. 1. pp. 571-575.

the mature individuals, we turn to the synthetic classification of the same organisms as presented by Lankester, in which the distinguishing points are chiefly found in those characters by which resemblance or relationship to some other different organism is traced.

In the first case differences and in the second case resemblances form the chief criteria upon which the classification is based.

Lankester's Classification of the Mollusca.—According to Professor Lankester,* whose classification is one of the most radical and modern, the branch Mollusca includes four classes, divided into two groups: Class 1, Gastropoda: Class 2, Scaphopoda; Class 3, Cephalopoda; Class 4, Lamellibranchia.

The Cœlomata.—Among the Metazoa, to which he applied the name Enterozoa in distinction from Protozoa, Lankester recognized two fundamental divisions: (A) the Cœlentera, in which the enteron or digestive cavity communicates directly and is continuous with the cœlom or body cavity; (B), the Cœlomata, including the Mollusca and higher invertebrates, in which the enteron is separate from the cœlom which surrounds it and with which it communicates through its tissues by osmosis. The products of digestion thus transmitted into the cœlom or body cavity are distributed through a system of canals and caused to circulate by a contractile organ which in its more differentiated condition is the heart. The special advance in differentiation in the Cœlomata consists in the separation of the alimentary cavity into a distinct digestive cavity and an assimilative cavity, the circulative and purificative functions being auxiliary to the general assimilative, as distinct from the digestive, functions.

Description of the Mollusca.—The Mollusca are typically Cœlomata. They have also, in common with the other Cœlomata, a region in front of the mouth developed as the expression of the specialized function of *forward*, as distinct from rotatory, *motion;* and in this region, which Lankester calls the *prostomium*, are differentiated, when present, the chief organs of sense. As to body form, the Mollusca have differ-

* See article Mollusca, "Encyclopædia Britannica," 9th Edition, vol. XVI.

entiated a permanent bilateral symmetry, which may be considered as the final elaboration of the radial type of differentiation which was dominant in the Cœlenterata and Echinodermata. In those branches the antimeres are numerous, or at least five, and the bilateral symmetry is only partially exhibited, while in the Mollusca it is a dominant character in the adults. In the Vermes and Arthropoda the bilaterality of each somite is further differentiated in the longitudinal direction by the division into segments. Although the order of rank exhibited by this mode of differentiation of parts would lead us to look for Cœlentera with multiple radiate structure first, then the differentiation of the Medusa, then the Echinodermata, with symmetrical radiation, then simple bilaterality of the Mollusca, to be followed by Vermes and Arthropoda, we actually find that the Arthropoda are already abundant in the Cambrian, and in the Trilobites constitute the dominant type of organism.

Digestive System in the Mollusca.—As already suggested, in the Mollusca the alimentary system is differentiated into a digestive cavity. The products of digestion, finding their way through the walls of the digestive cavity, are received into a system of canals with a contractile reservoir, which is the circulatory blood-system. In connection with the alimentary system is also developed in the higher Mollusca, and perhaps in all the classes, organs called *nephridia*, which are apparently purificative in function, and are primitive kidneys in differentiation. In all the Cœlomata *gonads* or special reproductive organs are differentiated.

Muscular, Nervous, and Motory Systems of Mollusca.—*Muscular tissue* is distinctly differentiated, and also *nervous tissue*, with the peculiar specialized functions of contractility and sensibility. The nervous system consists of a *gangliated ring* of nerve-fibres around the œsophagus, and in the higher types of Mollusca *special sense-organs* are differentiated for touch and sight. The *motory system* is developed in a characteristic way in the several classes of Mollusca, in a foot which is the most permanent and characteristic feature of the branch. Perhaps the simplest way of expressing the relations of this foot to the structure of the organism and the development of its organs is to compare it with the radiate type of the Cœlenterata—to suppose a Cœlenterate with its row of tentacles, specialized in function and differentiated in structure, so that, first, the contractile function is on the one side, and muscular tissue is developed in place of numerous tentacles, thus forming a mass of fleshy tissue contracting in two directions and accomplishing locomotion; while the correlative function of sensibility is specialized on the other side of the mouth into two tentacles, with, in some of the higher types, distinct and well elaborated eyes for sensation. In the Cœlenterata the tentacles are both sensitive and contractile, they are multiple, and muscular tissue and nervous tissue are not fully differentiated. The *foot* of the Mollusca may be considered as a specialized motor organ, in structure it is differentiated muscular tissue, and marks the side of the body as ventral; while the *tentacles*, as of the snail, are specialized sensory organs expressing the differentiation of nervous tissue for the special function of sensation, and mark the dorsal side. In the Mollusca and in other branches *the gangliation of the nervous cord* is co-ordinate with supply of nerve-fibres to specialized organs; thus as the sensory organs are more elaborated and specialized there is developed a large nerve-ganglion on the corresponding part of the nervous ring, and the foot too has its special nerve-ganglia developed. We find also specialized in the Mollusca *tissues for secreting fluids accessory to the digestive function*, and the differentiated organ for this purpose may be compared to the liver of vertebrates. The *stomach* is not, however, highly differentiated from the digestive tract in the lower types. The Mollusca have specialized organs for respiration. These organs—the *gills*, or *mantle fringes*—are present in all; but in the class Lamellibranchia the function is not entirely respiratory, but is also partly ingestive, or has to do with bringing food to the mouth.

Differentiation of the Nervous System.—The *nervous system* is differentiated to correspond to the differentiation of other organs, and in two directions; at an early stage contractility and sensibility were differentiated. Sensibility, then, has two essential relations: sensibility as *receptive* of impressions from without, or sensation; sensibility as *active*, or excitatory of the functions of organs, reflex or motor action. The differentiation of the nervous system corresponds with this distinction. In the lower Mollusca the distinction between the two functions is little seen; but in Gastropoda, for instance, and still more in Cephalopoda, special organs are differentiated for sensation, and the nervous system is in communication with each of the differentiated sets of organs, stimulating and directing their activity. All this differentiation is associated with the distinction of polarity of motion; the nervous system is essentially co-ordinative, and binds together the activity of organs in the way of compensating for the separation of parts due to their differentiation and development in size. The nervous system compensates for separation of the functional activities of the organism, and the circulatory system compensates for separation of the physical parts of the body of the organism, maintaining unity for the organism co-ordinate with the physiological specialization and the morphological differentiation.

Branches, Classes, and Subclasses of Mollusca. — Lankester's division of the Mollusca as a Phylum is first into two branches:

Branch A, the *Glossophora*, characterized by a "head region more or less prominently developed; always provided with a peculiar rasping tongue—the odontophore—rising from the floor of the buccal cavity;" and

Branch B, *Lipocephala*, of which the characters are "Mollusca with the head region undeveloped. No cephalic eyes are present; the buccal cavity is devoid of biting, rasping, or prehensile organs. The animal is sessile, or endowed with very feeble locomotive powers."

All these latter branch characters are practically negative characters: the Glossophora is a group formed of the Mollusca which possess in common a few important characters, and the Lipocephala are those which do not possess those characters. Only one class is recognized in the Lipocephala, i.e., the Lamellibranchia. The Glossophora comprise the three classes: first, Gastropoda, with two subclasses, (1) the Isopleura and (2) the Anisopleura: second, the class Scaphopoda; third, the class Cephalopoda, with two subclasses, (1) the Pteropoda and (2) the Siphonopoda.

The classes are chiefly distinguished by modifications of the foot, as is beautifully shown in Fig. 57.

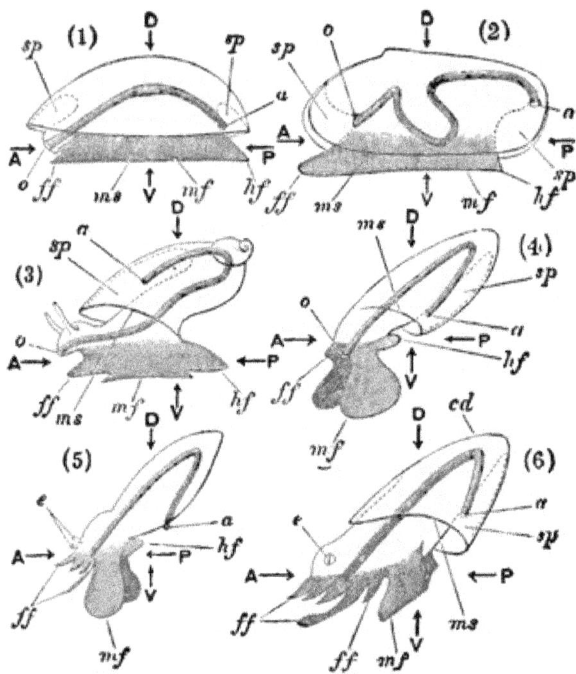

Fig. 57.—Diagrams of a series of Mollusks to show the form of the foot and its regions, and the relation of the visceral hump to the antero-posterior and dorso-ventral axes. (1) A Chiton. (2) A Lamellibranch. (3) An Anisopleurous Gastropod. (4) Thecosomatous Pteropod. (5) A Gymnosomatous Pteropod. (6) A Siphonopod (Cuttle). *A*, *P*, antero-posterior horizontal axis; *D*, *V*, dorso-ventral vertical axis at right angles to *A*, *P*; *o*, mouth; *a*, anus; *ms*, edge of the mantle skirt or flap; *sp*, sub-pallial chamber or space; *ff*, fore-foot; *mf*, mid-foot; *hf*, hind-foot; *e*, cephalic eyes; *cd*, centro-dorsal point (in 6 only). (After Lankester.)

In the Gastropoda the foot is simple, median in position, and flattened so as to form a broad, sole-like surface (No. 3).

In the Scaphopoda the foot is adapted to burrowing life in the sand.

In the Pteropod the mid-foot is developed laterally into paddle-like swimming organs, and the fore-foot may be specialized into tentacles (Nos. 4 and 5).

In the Cephalopoda the fore, middle, and hind foot parts are separately specialized, the fore-foot merging with the head part and developing into arm-like processes, in some cases beset with hooks or suckers, and the mid-foot is developed into a tube either closed or with lapping edges (No. 6).

Distinctive Features of the Lankester Classification.—The distinctive feature of Lankester's classification is seen in his descriptions of the subclasses. To show the nature of the characters selected as definitive of the divisions recognized, the chief characters of the subclass (2), Gastropoda Anisopleura, will be quoted, and for any further details the reader is referred to the fully elaborated and illustrated article in the "Encyclopaedia Britannica" on Mollusca.

The definition includes the following characters, viz.:

"Gastropoda, in which, whilst the head and foot retain the bilateral symmetry of the archi-Mollusca, the visceral dome, including the mantle-flap dependent from it, and the region on which are placed the ctenidia, anus, generative and nephridial apertures, have been subjected to a rotation tending to bring the anus from its posterior median position, by a movement along the right side, forwards to a position above the right side of the animal's neck, or even to the middle line above the neck. . . . The shell is not a plate enclosed in a shell-sac, but the primitive shell-sac appears and disappears in the course of embryonic development, and a relatively large nautiloid shell (with rare exceptions) develops over the whole surface of the visceral hump and mantle skirt. . .

"The shell and visceral hump in the Anisopleura incline normally to the right side of the animal. . . . Atrophy of the representatives on one side of the body of paired organs is very usual." (p. 644.)

In these descriptions it will be noticed that characters chosen as distinctive are based upon comparison of the type under description with forms from which it is supposed to have been developed embryologically, or from which it is supposed to have descended by evolution.

The Gastropoda Anisopleura is conceived of as a Gastropod mollusk which has become modified in a particular way in the course of evolution.

CHAPTER XIV.

THE ACQUIREMENT OF CHARACTERS OF GENERIC, FAMILY, OR HIGHER RANK ILLUSTRATED BY A STUDY OF THE BRACHIOPODS.

In the foregoing chapters the history of organisms has been considered in its general principles.

We have noted how organisms are, in general, different for different periods of geologic time; how the peculiarities of structure and function, which have led to their classification into many different classes, orders, families, genera, and species, are intimately associated with differing conditions of environment.

The steps by which the individual organism acquires its morphological and physiological characteristics have been examined, and the course of this development for each individual has been found to be determined by the ancestry from which it sprang.

The principles of classification have been discussed, and from the investigation in this direction it has been learned that each organic individual develops in the course of its individual growth not only the specific, but the generic, family, ordinal, class, and branch-characters of its parents. These characters have various rank in the classification; those which are of higher taxonomic rank are found to be of more ancient, and those of lower rank of more recent, geological origin. Therefore we may conclude, as a general law, that the lower the taxonomic rank of the character the shorter has been its life-period, i.e., the period of time through which it has been repeated by ordinary generation.

The various opinions regarding the nature of species have been discussed. All naturalists find the employment of species necessary to their science, though the exact definition of the term and the exact determination of any concrete species are difficult to accomplish.

The examination has revealed the fact that the fundamental difference in opinion regarding species turns upon the belief as to the *mutability* or *immutability* of species.

The idea that species are mutable is intimately associated with the inquiry, What is the "origin of species"? In attempting to answer this question the deeper ones arise, i.e., What is evolved in evolution? and What is mutable?

The answer was that in any individual case all that is evolved is to be found in the variation exhibited in those characters by which it departs from the exact imitation of the characters of its ancestors, and that evolution consists in the acquirement of characters not possessed by the ancestors.

We examined the classifications of the Animal Kingdom particularly, and we found that, looked at analytically as composed of a vast number of different structures, or synthetically as a multitude of related organisms variously differentiated, and differentiated to various degrees along a few general lines of evolution, the Animal Kingdom is divisible into a number of definite groups, marked by definite organization, all the grander features of which were outlined in the Cambrian age, and the large majority of all the differentiations of even ordinal rank had been accomplished in the first quarter of the recorded history of organisms.

It is evident, therefore, that we must read the law of evolutional history in terms of the genera and species as they are distributed in families or in orders.

Generic and Specific Evolution Illustrated by the Brachiopoda. —In order to study the successive appearance of species and genera, it will be necessary to turn from the more general characters to the minuter marks distinguishing species from species, or at least genera from genera. For this purpose no better group of organisms can be selected than the Brachiopoda. In presenting the results of this analysis the paleontologist will miss that elaboration of the facts which would make the discussion of most practical use to him. The brief limits of this introductory treatise do not admit of this; and if the presentation of the facts shall stimulate some such readers to open up the immense field of investigation which is here suggested, the author's purpose in writing this book will be fully rewarded.

Brachiopods Thoroughly Differentiated in Early Paleozoic Time.—When we critically examine a group of organisms like the Brachiopods in their historical relations, we find a law of successive appearance in geologic time of new characters, but we are obliged to consider them minutely in order to understand what is the nature of the evolution. The more important characters were already present at the earliest period in which records are preserved.

Both of the orders of Brachiopods (Lyopomata and Arthropomata) appeared in the Cambrian, and they are represented by numerous individuals and genera; and, according to Waagen, there are three well-defined suborders of the Lyopomata, and all of these were expressed certainly as early as the base of the Silurian. If we take a later tabulation of the genera and classification of Brachiopods,* we find 11 families of the Lyopomata with 55 genera, and 19 families + 14 additional divisions recognized as of subfamily rank, with 220 recognized genera belonging to the suborder Arthropomata. Of the total 275 genera, recorded by Schuchert, 50 of the 55 Lyopomata and 139 of the 220 genera of the Arthropomata, or 189 of the total 275, i.e. about 68 per cent, appeared in Paleozoic time; and 17 genera of Lyopomata and 5 genera of Arthropomata are already known in rocks as ancient as the Cambrian system.

Many of these Extinct since Paleozoic Time.—These figures will give an idea of the great antiquity of the Brachiopods, and of the early elaboration of the differences which are expressed in the systematic classification into genera. Another fact can be expressed in mathematical form: not only were the Brachiopods greatly evolved in early geological time, but many of them have become extinct. Of the 189 genera of the Paleozoic time 7 lived on to Mesozoic time, and of these at least 2 genera still live.

Generic Life-periods of the Brachiopods.—Again, although

* Viz., that of Schuchert, in the *American Geologist*, vol. XI., March, 1893, p. 1?1, etc.

an ancient type of animals, and expressing great persistence in some lines of succession, they also present very clear indications of definite succession and limit in their generic life-periods, as may be expressed again numerically in the following way: To express this law we may select a group of related forms, grouped under three families by Schuchert, the *Terebratulidæ*, the *Dyscoliidæ*, and the *Terebratellidæ*, in which are known 66 genera. Of these genera 13 were initiated in the Paleozoic (1 Ordovician, 1 Silurian, 5 Devonian, 6 Carboniferous) according to present statistics, 38 in the Mesozoic (7 Triassic, 23 Jurassic, 8 Cretaceous), and 10 are known only in the Recent. Of these 66 genera 41, or about $\frac{3}{5}$, have a recorded continuance of only one era, 11 are recorded from two contiguous eras, 2 from three, 4 from four, 4 from five consecutive eras.

Climax of Generic Evolution at a Definite Period.—If we go one step further, and analyze the range of the genera of a single subfamily, we see the law of evolution expressed with greater clearness. Using Schuchert's list of genera, we find that the subfamily of *Dallininæ* (the first division of the family Terebratellidæ) contains 22 known genera: all of these are known not earlier than the first period of the Mesozoic. Of them, 3 genera are first seen in the Triassic, 13 first in the Jurassic, 2 first in the Cretaceous, 1 first in the Tertiary; 3 are only known as recent species. In this case it is perfectly evident that the group is a Mesozoic type, that it began to appear in numbers in the Triassic, that its greatest expansion was in the Jurassic, that as a subfamily it is not now extinct.

We draw from these facts the conclusion that there was constant evolution going on, that all along geological time old types were dying out and new ones were being initiated or introduced. It is by studying the characters expressed by these successive genera, and noting their relation to each other in the order of their succession, that we catch a glimpse of the actual facts of evolution as they have taken place in the past.

In order graphically to express the grand facts of the evolutional history of the various types of Brachiopoda the following diagram of the evolution-curves of the various divisions of Brachiopods, in terms of the initiation of new genera, is constructed (Fig. 58).

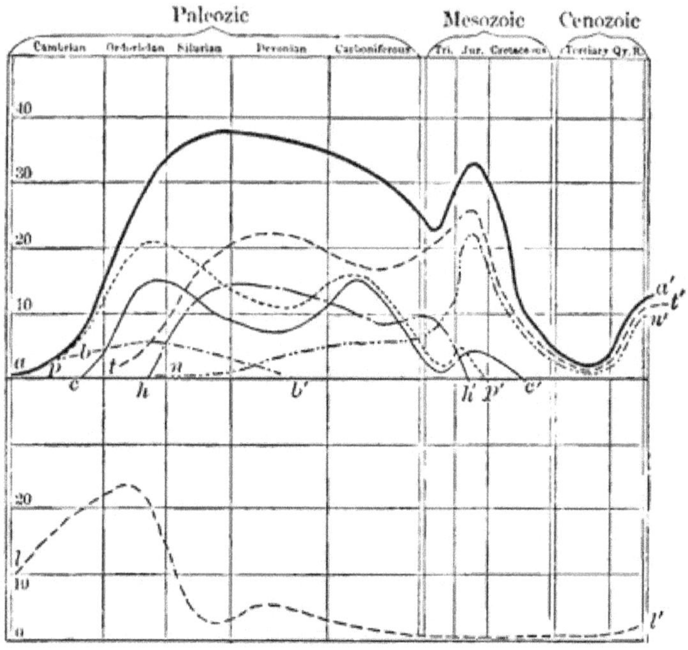

FIG. 58.—Evolution curves of the Brachiopods. The spaces from left to right represent the successive geological eras from the Cambrian to the Quaternary-Recent. The curves in the upper part of the diagram are those of the Arthropomata, the lower curve that of the subclass Lyopomata. Starting from the horizontal base-line, elevation above this line expresses the rate of evolution in terms of the number of new genera initiated in each division during the era, the cross-lines representing 10, 20, 30, etc., new genera, respectively. The curves connect the points so indicated for each group: *a–a'* = Arthropomata; *p–p'* = Protremata; *b–b'* = Trullacea; *c–c'* = Thecacea; *t–t'* = Telotremata; *h–h'* = Helicopegmata; *n–n'* = Ancylobrachia; *l–l'* = Lyopomata.

These facts expressed in numerical form are as follows:

TABLE OF THE NEW GENERA INITIATED IN EACH GEOLOGICAL AGE, GROUPED UNDER SUBCLASS, ORDERS, AND SUBORDERS (COMPILED FROM SCHUCHERT'S LISTS).

			C	O	S	D	Cr	T	J	K	Ty	Q	R
Arthropomata,	220	Gen.	5	25	37	34	34	23	33	12	2	4	11
Protremata,	81	"	5	21	18	11	17	1	6	2	0	0	0
Trullacea,	19	"	4	6	5	4	0	0	0	0	0	0	0
Thecacea,	62	"	1	15	13	7	17	1	6	2	0	0	0
Telotremata,	139	"	0	4	19	23	17	22	27	10	2	4	11
Rostracea,	20	"	?	2	3	4	2	4	2	1	0	1	1
Helicopegmata,	52	"	0	1	15	14	9	11	2	0	0	0	0
Ancylobrachia,	67	"	0	1	1	5	6	7	23	9	2	3	10

Its Interpretation.—The two subclasses (if we call the Brachiopods a class) Lyopomata and Arthropomata are thus shown in their strong historical contrast, the former (*l–l'*, Fig. 58) culminating its generic evolution in the Ordovician, while the Arthropomata (*a–a'*) culminates in the Silurian, but continues to differentiate until the middle of the Mesozoic; and the several distinct lines of differentiation are expressed by the curves for the several suborders, which are recognized in the classification of Schuchert.*

This classification recognizes the Brachiopoda as a class, and Arthropomata and Lyopomata as subclasses. The Arthropomata (*a–a'*) are divided into two orders, (1) the Protremata (*p–p'*) (Beecher), including two suborders, viz., Trullacea (*b–b'*), and Thecacea (*c–c'*); and a second order, Telotremata (*t–t'*), with the suborders Rostracea, Helicopegmata (*h–h'*), and Ancylobrachia, (*n–n'*).

From the irregularities of the curves made by these subordinal groupings of genera the indications are that the Thecacea (*c–c'*) is compounded of three distinct groups, having separate courses of evolution culminating in the Ordivician, in the Carboniferous, and in the Jurassic, respectively; the Trullacea (*b–b'*), the Helicopegmata (*h–h'*), and the Ancylobrachia (*n–n'*) are apparently natural groups; at least the present evidence expressed in the structural classification corresponds fairly well with the classification based upon the rate of differentiation of genera within each group.

The Trullacea are the earliest forms of articulate Brachiopoda known and their development was earliest of the families—too early for the exhibition of good evidence of actual differentiation; but the Helicopegmata (*h–h'*) show the beginning of differentiation as late as the beginning of the Silurian era, and as their history has also an ending at about the middle of Mesozoic time, a study of their history should throw some light upon the problems before us. The Ancylobrachia (*n–n'*) beginning its differentiation about the same time, and continuing to increase, reaching its culmination in the Mesozoic, presents descendants of the old genera and also some new genera even in Recent time. This group of genera may be studied in detail on account of the fuller

* A Classification of the Brachiopoda, by Charles Schuchert, *The Am. Geologist*, vol. XI. p. 141, March 1893.

records, the greater time-duration covered by its differentiation, and because living forms have been carefully examined and their structure and course of embryological development are well known.

Majority of Characters of Living Brachiopods traceable to Cambrian Ancestors.—From this tabulation of the range of the Brachiopoda it is evident that a great majority of the characters which any individual Brachiopod exhibits (as a specimen of the *Terebratulina septentrionalis*, now living in large numbers off the coast of Maine) (Fig. 59; see also Figs. 60 and 61) are of very ancient date, and can be accounted for by descent without modification through direct ancestors running back to the early Cambrian time.

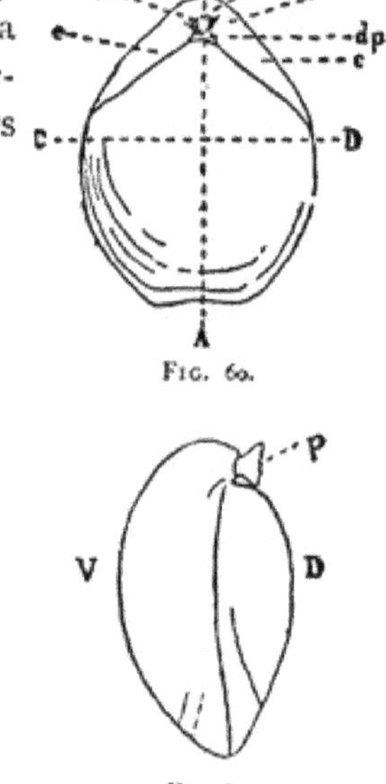

Fig. 60.

Fig. 61.

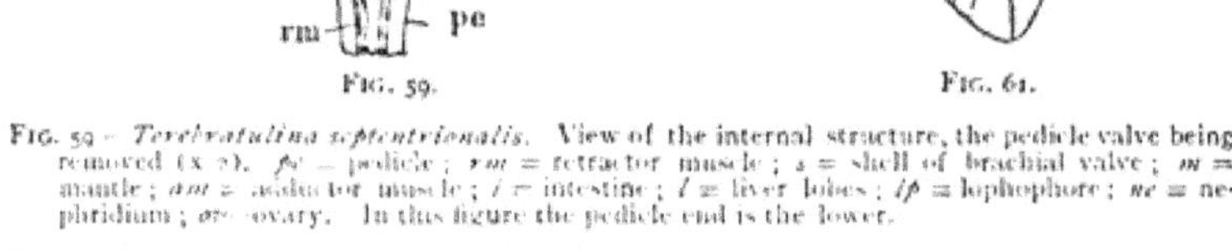

Fig. 59.

Fig. 59.—*Terebratulina septentrionalis.* View of the internal structure, the pedicle valve being removed (× 2). *pe* = pedicle; *rm* = retractor muscle; *s* = shell of brachial valve; *m* = mantle; *am* = adductor muscle; *i* = intestine; *l* = liver lobes; *lp* = lophophore; *ne* = nephridium; *ov* = ovary. In this figure the pedicle end is the lower.

Figs. 60, 61. Shell of a Terebratula. *AB* = antero-posterior axis; *CD* = horizontal axis; *V* = ventral or pedicle valve; *D* = dorsal or brachial valve; *p* = pedicle; *f* = foramen; *c* = cardinal slope; *a* = umbo; *u* = umbonal slope; *dp* = deltidial plates.

These characters may be enumerated in the following manner. The earliest Brachiopod possessed all the characters essential to each of the following taxonomic divisions, viz.:

A. *Organism.*—All the characters which it presents, distinguishing it from matter in an inorganic state, were differentiated before the Cambrian.

B. *Animal.*—All the characters distinguishing it from plants.

C. *Molluscoidea.*—The special characters of this branch were fully differentiated in the Cambrian era. These are (to follow Claus and Sedgwick): animals attached, as distinguished from moving or perambulatory organisms; the development of bilateral symmetry; the absence of metameric division—they are unsegmented; the differentiation of ciliated oral appendages; enclosure in a calcareous shell, with differentiation of organs into the various physiological systems of the Metazoa, viz., digestive, motor, neural, excretory, and those of reproduction.

D. Class *Brachiopoda* (and not Polyzoa).—This distinction includes the characters of two spirally rolled buccal arms; the development of bivalve, equilateral, dorsal, and ventral shells; the development of several ganglia connected by a pharyngeal ring. There must be included here also all the characters which are necessary to carrying on the functions of the different parts, mentioned under groups C and D.

E. *Lyopomata* and *Arthropomata.*—All that distinguishes these two orders was fully evolved, certainly, before the Cambrian era was far advanced, for we find several distinct families of the one and five of the other already in the Cambrian rocks. These differences are seen by comparing specimens of *Terebratulina* with a *Lingula*—both recent genera. The differentiation includes, in respect of intestine, a long and open intestine, with anal as well as oral orifice, and short, with postero-ventral end closed; in the shells the distinction between free-sliding valves and hinged, articulated valves, and the associated modification of muscular apparatus to move them laterally upon each other in the first case, and to open and shut them with a hinge in the other.

Perpetuation and Repetition of Characters a Common Law of Generation.—The more sharply distinguishable characters are mentioned above, but they include more than the ordinary observer would notice if handed a specimen and asked to describe what he saw—more, I say, but *not all* the characters that he would notice; for the ordinary observer will notice some of the specific characters more readily than he will the more essential characters of a strange object.

Before leaving the first era of our time-scale we have still more characters of family rank, others of subfamily rank, and enough elaboration of them to call for classification into fifty-five genera of Lyopomata and five genera of Arthropomata, and all these are found in the species now known from Cambrian rocks.

On the theory that the organisms now living are descended from ancestors in the past, the characters once having appeared in the ancestral line are most simply accounted for by supposing that they have been transmitted without change by the laws of ordinary generation. However the characters may have been originally produced, or came about in the first place, having once appeared in the Cambrian their continued reappearance in later stages of geological history calls for no other processes than those we see taking place on every hand, i.e., the successive reproduction of offspring by regular generation: no action of evolution is required. The preservation by continued generation of these ancient ancestral characters is no less remarkable than the slight modifications which have taken place in the course of geological ages.

Evolution Accounts for Divergence, not for Perpetuation or Transmission.—This familiar law of heredity will account for the continuance, as long as they appeared, of the families and genera of the Cambrian; the appearance of new families new genera, and new species requires on this theory the assumption of some other process. When we examine the length of recurrence of these Cambrian forms, of them we find only three genera of the Lyopomata and one genus of the Arthropomata are known from rocks above the Cambrian, and they are from the next succeeding system. Of the family characters of the Cambrian Brachiopoda, six Lyopomata, two Arthropomata families lived after the Cambrian: two of these lived on to the Carboniferous, two of them reached the Silurian, and only three reached the Ordovician. There were, however, four families and two genera that appeared in the second era, the Ordovician, which lived on to the present time, and it is not improbable that these types of differentiation may have taken place as early as the Cambrian.

Brachiopods Ancient Types and Early Differentiated.—From these facts we learn that the Brachiopods are very ancient animals; that at the first geological period they were very greatly differentiated in structure, and that, except in a very few cases, the forms that lived in later ages, though supposedly descended from the earliest types, suffered changes in their specific, generic, and in many cases family characters.

It is also evident that if we wish to study the history of Brachiopods we must read the evolution in terms of their specific, generic, and only in slight degree in any characters of as high rank as family, and not at all in characters of higher than family rank.

A glance at the range of the families and genera of the Lyopomata shows them not only to have been ancient, but to have reached their climax of evolution by the second geological period of time—the Ordovician. After the Ordovician no new families of Lyopomata are initiated, and the new genera fell from twenty-two new ones in the Ordovician to three in the Silurian, six in the Devonian, and after that seven new genera up to the living forms. This slight continuance of expansion may be driven much farther back by later discoveries.

Laws of Evolution Gathered from Study of the Early Families.—With such an early expansion of the suborder it is evident that the range of instructive history is limited to the earliest periods of geological time, and the few forms that still exist among the recent faunas are very slightly modified from the ancient types. In the case of the other suborder, Arthropomata, the evolution was continued to a later period. Family and subfamily differentiation was greatest in the first two geological eras, nine new families appearing in the Ordovician; but two or three new genera in each of the following eras, except in the Cretaceous and Tertiary, when the present information records only a single new subfamily in each.

Genera making their Initial Appearance in each Era.—The generic expansion kept up with greater force, as the number of genera making their initial appearance testifies. For the ten successive eras the initiations of new genera recorded up to the present time are as follows, viz.: 5, 25, 37, 34, 34, 24, 33, 11, 2, 4 for Quaternary and 11 for Recent. The greater number of recent genera not known in fossil state may be discounted by the vastly greater knowledge we have of recent organisms than of the faunas of any, even the most recent, extinct fossil faunas. The evolution kept up its differentiation of genera well into the Mesozoic time, when it began to lessen rapidly, and from the Jurassic to the Cretaceous dropped from 33 to 11 in the number of new genera appearing during the periods, and only two new ones appeared in the Tertiary.

Comparison of the Rate of Evolution of Generic, Family, and Ordinal Characters.—We may select this division of Brachiopods for more minute study of the historical laws expressed in the evolution of its successive forms. A study of the curve of results of this series of steps of evolution shows us at a glance that there are, at least, two nodes in the evolution, one culminating in the Silurian and one culminating in the Jurassic. Analysis of the structure of the forms reveals the fact that the evolution has taken place along several subordinate lines, which are expressed in taxonomy by division of the Arthropomata into two primary divisions, called by Beecher orders (*Protremata* and *Telotremata*), and these again into two groups of families, the *Trullacea* and *Thecacea* in the first order, and into the three groups *Rostracea*, *Helicopegmata* and *Ancylobrachia* of the second order, *Telotremata*.

Evolution Curves for the Several Families.—Each of these subdivisions was differentiated as early as the Ordovician, or second era, and their climaces are at somewhat different points in the time-scale.

The first group of families is the *Trullacea;* there were no new families of this type initiated after the Ordovician, and no new genera after the Devonian, and the whole group became extinct with the Paleozoic.

The second group is the *Thecacea.* Our curve of rate of

differentiation shows such irregularity that we are led to suspect within it three well-defined and separate courses of evolution, one of which culminated in the Silurian, one in the Carboniferous, and the third in the Jurassic. The presumption is that this group is not well arranged; the classification will need revision.

The third group is the *Rostracea*, and this is characterized by having a very long geological range; the chief family is the Rhynchonellidæ, which appears to extend from the Cambrian to the present, with its characteristic family-characters the same.

The fourth group is that of the *Helicopegmata* of Waagen. This includes the spire-bearing Brachiopods; the history of this group is well defined in families and in genera. The culminating point for both is in the Devonian, when the total number of forms is considered; but the greatest evolution of families is in the Silurian, new genera continuing to appear up to the end in the Jurassic.

The fifth group includes the *Ancylobrachia*. Although the first family of the group appeared in the Ordovician, the evolution of this type was very slow, but continuous to the very end in recent times. From its first appearance each succeeding era has seen the addition of a new family. The curve for generic differentiation is also emphatic, but it shows the evolution of this type of Brachiopods to have been late in geological time. Instead of being in the Paleozoic, the culmination of generic differentiation was in the Jurassic, when twenty-three new genera made their first appearance. The first four family groups of the Arthropomata had their culmination in the Paleozoic, and the fifth had its culmination near the middle of the Mesozoic.

Conclusions from Study of Generic Evolution Curves of the Brachiopods.—The examination of these differentiation or evolution curves of the generic and family life-histories of the Arthropomata can leave no doubt in our minds on a few important points:

1. The geological record, although imperfect, and not at all exhaustive in its declarations, reveals the fact that some types of organisms lived in one geological era, others in

another era, and leaves us in no doubt as to the general order of succession of the various genera.

2. Although it is not improbable that in almost every case the genera and the families will be found to have been initiated somewhat earlier than they are now reported, and new families and new genera will undoubtedly be discovered, nevertheless, the outlines of the differentiation curves are so emphatic in most cases that we have no reason to doubt that we already have the fundamental outlines of the history of each particular group of organisms clearly before us.

3. We have here unmistakable evidence that every genus and family had a definite time of initiation, and that this time of initiation for each has definite relationship to the time of initiation of other genera and families.

4. Another conclusion may be drawn from an inspection of the curves: the family differentiation for each grouping of higher rank, suborder or order, had its evident initiation, culmination, and decrease; also the generic differentiation for each family had its point of initiation, its period of rapid activity and culmination, and its period of decline; and in many cases the actual cessation not only of expansion, but of all appearance of the genus, is expressed.

CHAPTER XV.

WHAT IS EVOLVED IN EVOLUTION?—INTRINSIC AND EXTRINSIC CHARACTERS.

Laws of Evolution indicated by History of Brachiopods.—We have now gained a sufficient knowledge of the characters of Brachiopods to enable us to consider the question, What is indicated regarding the laws of organic history by these facts?

It is evident, first, that the history exhibits evolution. Evolution of what? We have been considering the time relations of the genera in families of Brachiopods: is it evolution of genera? Tables have been given of the phylogenetic relations of the families of hinged Brachiopods: have we been considering the evolution of families?

Before taking up these points a few words may be said on the question, "What is evolved?" In general, we may say, the history of organisms reveals a progressive evolution of the *morphological characters* which distinguish the successive organisms. Classes, orders, families, and genera are not the things which are evolved. These are names for the divisions of the classification we make of the evolved organisms. The classification, when historically considered, expresses the evolution; but the things classified are the individual organisms, each of which has its characters distributed through all the whole range of categories of the classification. Therefore it is incorrect to speak of the evolution of one or other of the categories—as a species or a genus: it is this or that character of the individual that was acquired by evolution, as contrasted with other characters acquired by natural generation from its parents.

Magellania Flavescens Examined as an Illustration.—In illustration of this proposition we may take, for instance, a *Magel-*

lania flavescens, obtained from the seas about Australia; we examine its shell; we find that it is a bivalve, equilateral shell, the two valves articulating, and the one larger than the other, and exhibiting a perforation through the beak for the protrusion of a stem-like peduncle for its attachment. All these characters are peculiar to the class Brachiopoda. They distinguish this individual organism from the organisms of every other class in the Animal Kingdom (Figs. 59, 60, 61, 62).

Evolution of the Class Characters.—Whence does the Magellania derive these characters? We at once say by descent from the parent Magellania from which it sprang. How did it attain the characters? By ontogenetic growth from an egg which expressed none of them. The law of heredity explains the appearance of the particular characters in this individual organism, and the law of ontogenetic growth explains the formation in the individual of these characters. But how did they come to be at all? or, to put the idea in another form, Why is it not a clam-shell? Heredity explains why it is like its ancestral predecessors; but what explains the fact that it is unlike organisms of all other classes? In answering this question we are led backwards, and find in the Tertiary Period forms presenting the same characters; and because there is thus traced a succession of forms with the same characters, we assume that descent will account for the succession. Still further back, in the Cretaceous, in the Triassic, in the Devonian, in the Silurian, and even in the lowest beds of the fossil-bearing series, the Lower Cambrian, we find fossils possessing the essential class characters of our living Magellania. There at the first stage of appearance of Brachiopods the difference is obvious between the Orthisina and the species of any other class than Brachiopods. We can go no further for facts. We have to confess that we have no knowledge of the origin of the class characters of Brachiopods; we only know that they were evolved as far back as Cambrian time, and that they have ever since been transmitted by ordinary generation.

Evolution of the Ordinal Characters.—In the same way we notice on the hinge margin the production of two processes each side of a triangular fissure which we call teeth and delthyrium. These we call ordinal characters; the Magellania is of the order Clistenterata, or hinged Brachiopods.

But these characters have a continuous succession back to the Lower Cambrian. Again, we notice on the smaller valve two plates, called teeth sockets, producing with the outer part of the hinge margin a groove or socket into which the teeth fit, and at the base of them a pair of calcified processes, called crura; but these too are traceable back to the Lower Cambrian (Fig. 62).

Calcified Loops which are Subordinal Characters were Evolved between the Cambrian and Silurian Eras.—The Magellania differs from some hinged Brachiopods in having, in addition to the crura, calcified bands of a peculiar form looped back upon themselves, which are technically called *loops*. These are characters of a part of the hinged Brachiopods, and they are called subordinal characters, separating the suborder Ancylobrachia from all other suborders of Brachiopods. But these loops cannot be traced backward further than the base of the Silurian; they are not known in the Ordovician or Cambrian. Regarding the characters of the specimen of as high as class and ordinal rank, we have no evidence regarding their origin save the law of hereditary transmission by ordinary generation; but Magellania has loops which it could not have gotten by the law of heredity, i.e; considered as a law of the transmission of like characters from ancestry to progeny. If we assume that the law of hereditary descent will satisfactorily explain the reappearance on successive organisms of a character which has once been formed, then we have the explanation of the class and ordinal characters of such a specimen as far back as to the Lower Cambrian, but its subordinal characters can by this means be accounted for only back to the Silurian. In other words, we are led by this train of reasoning to the conclusion that this Magellania had ancestors which did not possess its subordinal characters, among which are the calcified loop of a particular shape.

Each Case of Evolution a Case of the Appearance in some Individual of a Character not possessed by its Ancestors.—In the same way we learn from embryology, or the ontogenetic growth of this individual Magellania, that in the course of its individual life it has developed from an embryo condition in which its mature characters were not exhibited. By analogy we infer that these other characters of the loop were evolved from ancestors in which they did not appear; but before asking how, we observe that since the Silurian time the loop has appeared on successive forms up to the present time, exhibiting no greater differences than the ordinal or class characters in the same line have exhibited. The first specimen which exhibited a loop was distinct from previous forms by that character, and this, with other characters, caused it to be classed in a distinct suborder from all other forms. We use the term *evolution* to express the idea of appearance of such a character at first. All the various families of which we speak, and all the various genera, whose history we mark by range of life-period in the geological scale, had thus a place in the scale when their first known representatives appeared; and whatever the characters may have been (in the present case it is calcified brachial loops), every case is a case of first known appearance of such character, that is, it did not appear before, and the evolution consists in its coming into appearance on some organism whose supposed ancestors did not exhibit the character.

Evolution of Fundamental Characters Relatively Rapid.—The facts, to be sure, may be considered as very imperfect, but if we lengthen our lines backward we but lengthen the period in which the character has been repeated by ordinary generation without modification sufficient to upset our classification. Or if we extend the evolution over a hundred or a thousand generations it merely reduces the amount of the increment for each stage of the evolution. Thus we see that so far as the evidence testifies, the evolution of those characters which mark the differences between separate classes, orders, suborders, and even some families of organisms, has taken place in a relatively short period of time; taking as measure either the rate of general progress in the differentiation of organisms, or the length of the life-period of each particular genus or family. This is in harmony with a law of evolution formulated by Hyatt as given in a subsequent chapter (Chap. xviii.).

This Rapid Evolution difficult to Account for by any Working of Natural Selection.—Thus far it is the evolution of morphological characters with which we have been dealing. Genera, or species, are often spoken of as being evolved. When language is used in this way we mean that there is an orderly succession of genera. This orderly succession of forms we can readily conceive; but a genus is a group of species which possesses certain common characters of higher than specific rank. It is one thing to speak of the succession of the different forms and another thing to speak of the attainment by offspring of characters not possessed by ancestors. We are accustomed to the explanation by Darwin that the method of this attainment of new characters is by the gradual accumulation of varietal characters which are considered as arising spontaneously. Taking the case before us, we can imagine the form of the loop of the Magellania as having been acquired before its calcification; but the difference between the presence of a calcified loop and its absence was brought about within a brief period (geologically considered), while the modification of the loop, as indicated in the several genera of the Terebratulidæ, can be conceived of as having been produced gradually in the geological sense. Thus, when we consider evolution as applying to the production of differences, great difficulty is found in accounting for the structural differences, which are the basis of our classification into groups of family and higher rank, by the slow processes required for the working of natural selection upon normal variations.

What is Evolved?—Hence, in reply to the question "What is evolved?" it is evident that morphological characters are evolved—not species, genera, or any kind of *groups* of organisms. There is *an evolution of the characters* of the individual, and because this evolution takes place in many individuals at the same time, we recognize the evolution by the appearance of the modification in the many individuals, and group them into new genera or families, on account of their differences from other forms.

How Does the Evolution Proceed?—How does the evolution proceed? Not by the *ab initio* construction of a new organism, or a pair of them in each specific case, from which all the other representatives of the genus spring by natural generation without change—which is the old creational theory of origin of species; *but by the individual assuming a different course or extent of ontogenetic growth from the course or extent of growth of its ancestors, including acceleration in the growth of a part, or of an organ, with increase or specialization of its function.*

Intrinsic and Extrinsic Development, and Intrinsic and Extrinsic Characters.—This brings us to the consideration of the twofold nature of the morphological and physiological characters possessed by organisms. There are two fundamentally different ways in which we recognize the characters as differing one from the other when looked at from the evolutional point of view. When we mark the course of development from the egg to the adult chick we observe that there is a gradual building up, first of tissues, then of definite organs made of those tissues, from simple uniform cells; or, going further back, from the original nucleated, unsegmented cell itself. This is a process of differentiation of parts, as has been already defined, and with specialization of functions. *But it is a process of the increase of parts and functions by division of labor*, and is an expression of one of the fundamental laws of the organism as a whole. This kind of growth we may call *intrinsic development:* intrinsic, because it has to do with *the expansion or development of the organism as a whole, and involves the internal adjustment of the organism itself, and not simply the modification of one of its parts.* There is another kind of elaboration of organs and functions which consists in *the multiplication of like parts performing like functions, and results in difference in the size, the proportion, or the number of the morphological parts.* This kind of growth we may call *extrinsic development*, because it appears to be definitely correlated with the nature and amount of the external supply of materials for growth, and with the outward demands upon the activity of the functions concerned.

The distinction thus established in the mode of origin of the characters furnishes the basis for the classification of the characters into *intrinsic* and *extrinsic characters.*

Example of an Intrinsic Character.—To take an illustration: the character which distinguishes the Spiriferidæ from the Terebratulidæ and the Rhynchonellidæ, called the brachidium, is fundamentally an intrinsic character, because in the fixation and rigidity of parts there is implied an adjustment of

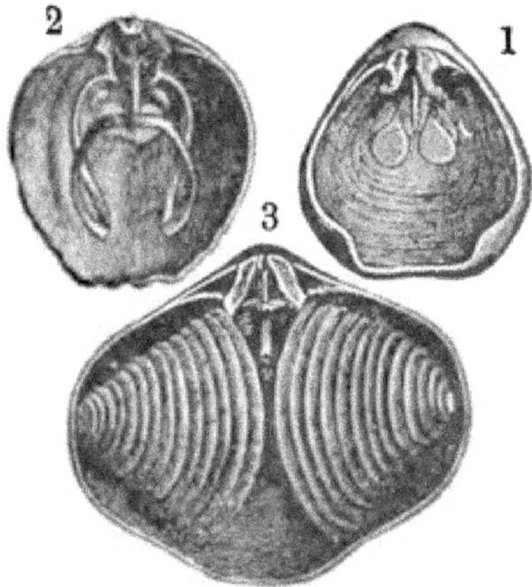

FIG. 62.—Brachial apparatus of (1) *Rhynchonella*, in which only the crura are developed; (2) *Magellania*, showing the crura with the looped bands of the brachidium; and (3) *Athyris*, with no loops but the brachidial bands extended in spiral coils.

the other parts of the organism to these conditions; and in the apprehension, the distribution, the deposition, and the supply of materials for constructing the apparatus, there is implied an adjustment of the whole organism to the work of constructing this new part. Even though the soft parts were essentially the same in the Orthis and the Spirifer, the modification in the Spirifer is a radical one, involving the whole organism, and not merely the particular part concerned.

Example of an Extrinsic Character.—On the other hand, the character distinguishing the spires of the Atrypidæ from those of the Spiriferidæ is the permanent turning of the point of the cone toward the centre of the valve in the Atrypa, and toward the upper outer angle of the shells in the Spirifer. This is a matter of adjustment which may involve a slight rearrangement of the relations of parts, but may involve no more; the difference in the shape of the shell itself may occasion such adjustment, as a tight shoe might distort the shape of the foot (Figs. 63, 64).

Characters Early and Rapidly Evolved were Chiefly Intrinsic Characters.—It will be observed that almost all of the characters, which have thus far been considered in tracing the differences distinguishing different classes, orders, families, and to some extent genera, are intrinsic and not extrinsic characters.

Application of the Terms Intrinsic and Extrinsic to the Elaboration of Machinery.—To illustrate the fundamental nature of this distinction we may call attention to a purely mechanical contrivance, the steam-engine, and the machinery run by it. The force here concerned is heat, which is transformed from burning wood into expansion of water into steam. The simple process is the transfer of *elevation of temperature* into *enlargement of the space* occupied by the steam. This expansion is in every direction. The engine is a device for concentrating the direction of expansion in one line, i.e., that of the axis of the piston-rod. So long as no greater elaboration of the mechanism is made in the engine, it is necessary to take the effect of the stroke upward only; the production of a hinge in the rod, and an attachment of the rod to a lever, make the walking-beam engine, which could, at the other end, work a pump; but the differentiation, which turned the link into a crank, causing continuous revolution of the wheel, was an intrinsic elaboration of machinery, involving a coadaptation of all the parts of the machine. Improvement in the way of elongation of the lever, or change of the relative size of the parts, in the modification of the wheel, in the shape and relative size of its parts, was purely extrinsic. Again, for the transfer of motion a belt and flat wheel was modified into a wheel with cogs, or the reverse—I do not know which; this is expressive of intrinsic elaboration of the device, while the increase of cogs in number, or size, or shape, or change of relative motion by different number of cogs on the two approximating wheels, is of the nature of extrinsic modification.

Summary and Conclusion.—From this illustration it becomes evident why it is rational to expect a different rate in the process of organic evolution from within, or intrinsic evolution, from the rate of the evolution from without, or extrinsic evolution. Both are at work at the same time, and every organism has its specific, its generic, and family characters, and those of higher order. Varietal characters in the process of extrinsic evolution may become invariable, and be ranked as specific accordingly; but when a character becomes fixed it is no longer variable, and because one species differs from another, and one genus from another, it does not follow that a specific character has by degrees become of family or ordinal rank. On the contrary, the cessation of plasticity which results when the varietal character becomes transmitted without change, and thus characterizes the species, makes it logically impossible to account for the difference in rank of the characters of an organism by any evolutional process. Rank of characters of the organism, as expressed in their place in the classification, is inherent in their use; and the same laws which are engaged in the origin of specific characters must also account for the origin of ordinal characters. The specific character does not become of ordinal rank, but whenever an ordinal character arose it must have first appeared as a variety. Herein consists the great importance of the facts of variation.

The accumulation of varietal modifications of parts or their intensification, their growing larger or smaller, stronger or weaker, is a matter fundamentally of addition or subtraction in the component units of lower order. Given a tissue made up of cells and performing a given function, and the modification of its form is but an expression of increased growth at one place or diminished growth somewhere else. It is easy to imagine conditions of environment, use, and disuse, adaptation to existing conditions or the opposite, as resulting in the modification of the form of the organ.

It is not difficult to imagine the same kind of phenomena working a selective discrimination among the variable degrees of such adaptation, and resulting in the preservation of certain variations and the elimination of others in the struggle for existence. The theory of origination of species by natural selection applies to cases of extrinsic evolution; but it is difficult to imagine how natural selection can operate in the production of the differences in structure which must be already differentiated before their relative fitness or unfitness to the conditions of environment can be tested. It is reasonable to expect, therefore, that all modifications of organic structure, which imply strictly intrinsic differentiation of the co-ordinated structure and function of the organism, were evolved by processes vastly more rapid than those of the extrinsic modification of structures already present in the race.

We have seen how Brachiopods furnish us with the data with which to trace the laws of the historical evolution of the more important characters exhibited by any particular Brachiopod. These characters have fallen into natural divisions, or groups of various rank, which are scientifically recognized as class, ordinal, subordinal, etc., characters. We have seen how the characters which we call subordinal, when they are arranged in the order of their morphological affinities, present a series of forms whose elaboration is as complete by the beginning of the Upper Silurian as it was at any later time; therefore we drew the conclusion that so far as the subordinal characters and those of higher rank are concerned, the differentiation expressed by these characters took place in the lower half of the Paleozoic time. As far as the facts are in evidence, we find that the characters of this kind were rapidly introduced: rapidly in relation to the degree of differentiation indicated by the characters, and rapidly in comparison with the length of time they persist without apparent modification. As two ontogenetic forces are at work in the growth of the individual, to which respectively we apply the terms *heredity* and *variability*, so we recognize upon analysis of the facts of the phylogeny two kinds of evolution: (I) a progressive evolution which operates from within and is associated with pre-existing conditions; this is called *intrinsic evolution;* (II) another kind of evolution, observed to be more intimately co-ordinate with external conditions, which may be regarded as fundamentally a process of adjustment or adaptation of the organism to its external environment; and this is *extrinsic evolution.*

In the ontogenetic development of the individual there is a rapid elaboration of those typical features of the organism which express its class, ordinal, and subordinal characters, the whole framework and plan of structure being elaborated before the individual comes into contact with external environment, while it is out of reach, so to speak, of the contests which are called struggle for existence. It is conceived that there were in like manner in evolution intrinsic modifications of internal structure, requiring for their functional operation adjustments of the whole mechanism of the body, and that these operations were relatively rapid, because they were the expression of evolutional force working from within, and in the determination of which the local and immediate conditions of environment had little or no part. As, for instance, in the plant, the special modification of ordinary tissues to produce the flower, and its complication of floral parts, relatively to the life-history of the plant is rapid, and the opening of the flower may in some sense be said to be occasioned by heat, sunshine, or, in general, by external conditions; but in a much more important sense it is true that the production of the flower is intrinsic, and is determined by ancestral, preexisting conditions, and not by those present only at the time of flowering.

CHAPTER XVI.

THE MODIFICATION OF GENERIC CHARACTERS, OR GENERIC LIFE-HISTORY.

IN the last chapter the conclusion was reached that evolution, which is the acquirement by organisms in the course of individual growth of characters not previously appearing in their ancestors, may be distinguished as of two kinds: one intrinsic, and expressing steps of progress in the differentiation of function and organization of the organism as a whole, working from within outward; the second extrinsic in nature, and expressed in the modification or adjustment of characters already differentiated to local and immediate conditions of environment.

We observed that as the particular characters examined are of higher and higher rank in classification they are more intensely intrinsic in nature, not only now, but were so in the earliest organisms of which we have any knowledge. And still further, that these more essential characters were earlier evolved, and the evidence seems to prove beyond doubt that their evolution was by steps more rapid than would be inferred from the relatively slow progress in the succession of the lesser characters, generic and specific.

Having noted the general laws of evolution respecting the more important characters of each individual, we next turn to an examination of the laws of evolution of the less important generic characters.

In the generic characters there appears to have been a rapid attainment of the total limit of modification expressed anywhere in the family, with a long persistence of the more widely divergent characters. When we examine the specific and varietal characters we observe a much slower rate of modification in individual race-series, but even here a remarkable degree of expansion of the main features of the variable characters appears very early in the history of each genus.

As an illustration of the rapid appearance of the full quota of extrinsic modifications of a new intrinsic element of structure we may examine the history of the spiral brachial appendages in the suborder Helicopegmata.

Statistics of the Life-history of the Spire-bearing Brachiopods (Helicopegmata).—The earliest trace of the spire-bearing Brachiopods is in the Ordovician, in a single simple form, the genus Zygospira.

At the next faunal stage, the base of the Upper Silurian system, there were representatives of each of the families into which the known Helicopegmata are divided (Atrypidæ, Spiriferidæ, and Athyridæ); and of the twelve subfamilies into which the seventy-three recognized genera are distributed, nine are also known from as early a stage as the Upper Silurian (viz., Zygospirinæ, Dayinæ, Atrypinæ, Suessiinæ, Trigonotretinæ, Rhynchospirinæ, Hindellinæ, Athyrinæ, and Meristellinæ). Of the others, Uncitinæ, first appearing in the Devonian, has the same kind of brachidium as the subfamily Suessiinæ; and the loop of Diplospirinæ, appearing first in Kayseria of the Devonian and having several genera in the Triassic, is rather to be considered as an extreme differentiation of the Athyroid type; and so far as the brachidium is concerned, Koninckinina of the Mesozoic is also an extreme differentiation of the same Paleozoic type.*

The Rapid Appearance of the Different Modifications of the Brachidium.—For the present discussion it matters not whether the calcification of the spirally-terminated brachidium of the Helicopegmata is a modification of that seen in the loop of the Ancylobrachia, or whether it arose from a form in which there was no calcified support; for both of the suborders, so far as evidence is at hand to show, first appeared in the Ordovician.

One intrinsic character distinguishing these suborders from all the previously existing Brachiopods is found in the presence

* In this discussion I have followed Schuchert's "A Revised Classification of the Spire-bearing Brachiopoda," *Am. Geol.*, vol. XIII. p. 102, etc., Feb. 1894.

in the former of the *calcified* supports, the brachidium, and it is the sudden or rapid appearance of modifications of structure of this brachidium which is under discussion.

TABLE SHOWING THE TAXONOMIC RELATIONS OF THE HELICOPEGMATA.

- Branch: MOLLUSCOIDEA
  - Class: Polyzoa / BRACHIOPODA
    - Subclass: Lyopomata / ARTHROPOMATA
      - Order: Protremata / TELOTREMATA
        - Suborder:
          - Rostracea
          - HELICOPEGMATA
            - Fam.: ATRYPIDÆ / SPIRIFERIDÆ / ATHYRIDÆ
          - Ancylobrachia

The above table is given to show the method of selection of this particular group of Helicopegmata for study. All the differentiation represented by the characters distinguishing the particular class, subclass, order, and suborder must be supposed to have already arisen before family characters of this particular suborder could take place.

I have adopted Dr. Beecher's ordinal classification, and take the order Telotremata, which appears to be the most fully differentiated of the orders of Brachiopods. The distinctive characters are found in the degree of differentiation of the delthyrium, or pedicle opening, and its covering, and of the brachidium or arm support. ("Pedicle opening shared by both valves in nepionic stages, *usually confined to one valve in later stages*, and becoming more or less limited by *two deltidial plates* in ephibolic stages. Arms supported by *calcareous crura, spirals*, or *loops*.") The distinctive ordinal characters I have italicized in this definition.*

It is within this order that we find the forms with special calcified parts called *deltidial* plates, *crura*, and *brachidium*, either *loops* or *spirals*. The subordinal distinctions are based upon the degrees and mode of elaboration of the brachial supports.

Rostracea is a new ordinal name proposed by Shuchert for the family Rhynchonellidæ of Gray, somewhat emended. It includes the genera with *rostrate* shells, no spondylium, and the presence of *crura*.

The *Helicopegmata* is the group proposed by Waagen to include the genera with two, calcareous, simple or double, spirally enrolled brachial supports, which may or may not be attached to each other by a variously constructed band or "loop."

The third suborder is Gray's *Ancylobrachia*, slightly emended by

---

* Beecher, "Development of the Brachiopoda," Pt. 1. *Am. Jour. Sci.*, vol. XLI. p. 355, 1891.

Schuchert, characterized by the possession of a calcareous loop for the support of the brachia.*

Three Families of the Helicopegmata.—In the classification of the Helicopegmata into families Mr. Schuchert's simple classification into the Atrypidæ, Spiriferidæ, and Athyridæ, based upon the essential structure of the brachidium, is adopted. His definitions are:

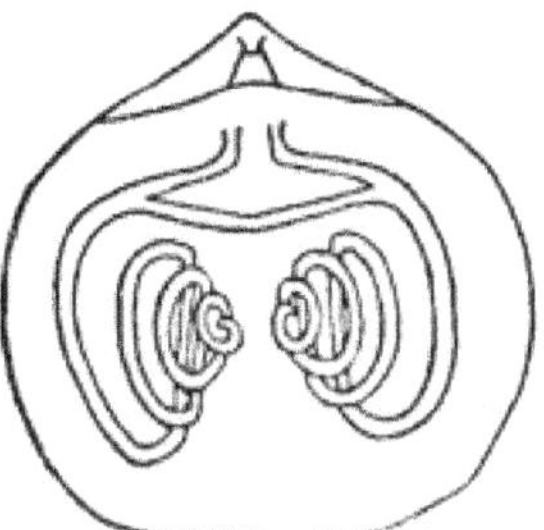

Fig. 63.—The brachidium of the Atrypidæ; *Zygospira modesta*, enlarged; view of interior from the side of brachial valve, which has been removed. (After Hall.)

1. In *Atrypidæ* the primary lamellæ are directly continuous with the crura, diverge widely, and have the spirals between them (Fig. 63).

2. In the *Spiriferidæ* the primary lamellæ are also directly continuous with the crura, but lie between the spirals, thus holding a position the reverse of that in the *Atrypidæ* (Fig. 64).

3. In the *Athyridæ* the primary lamellæ differ in direction from those in the other families in being more or less sharply recurved dorsally near their junction with the crura (Fig. 65).†

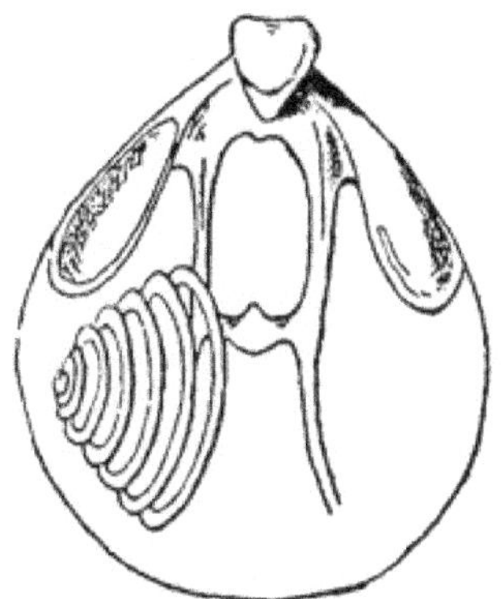

Fig. 64.

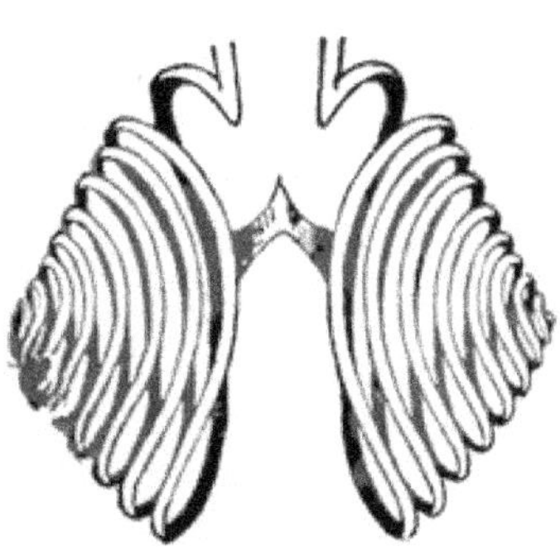

Fig. 65.

Fig. 64.—The brachidium of the Spiriferidæ, *Uncites gryphus* Defr.; interior of brachial valve, viewed from pedicle-valve side.

Fig. 65.—The brachidium of the Athyridæ, *Rhynchospira evax*, enlarged, and viewed from the pedicle-valve side. (After Hall.)

* Schuchert, "A Classification of the Brachiopoda," *Am. Geol.*, vol. XI. 141–167, 1893.

† Schuchert, "A Revised Classification of the Spire-bearing Brachiopoda," *Am. Geologist*, vol. XIII. p. 102, 1894.

Geological Range of the Families.—The following table of the geological range of the families, subfamilies, and genera will help to give a notion of the time-relations of the forms under discussion.

TABLE REPRESENTING THE GEOLOGICAL RANGE OF THE FAMILIES AND SUBFAMILIES OF THE HELICOPEGMATA, WITH THE NUMBER OF GENERA AT PRESENT RECORDED FOR EACH ERA.

	C	O	S	D	Cr	T	J	K	Ty	QR
HELICOPEGMATA.										
Families:										
Atrypidæ		2	5	4						
Spiriferidæ			5	11	8	4	3			
Athyridæ			11	14	12	20	1			
ATRYPIDÆ										
Subf.: Zygospirinæ		2	3	1						
Dayinæ			1							
Atrypinæ			1	3						
SPIRIFERIDÆ										
Subf.: Suessiinæ				1	2	2	2			
Uncitinæ				1	1					
Trigonotretinæ			5	6	5	2	1			
ATHYRIDÆ										
Subf.: Rhynchospirinæ			4	4	6	1				
Hindellinæ			4	3	1	1				
Athyrinæ			1	1	5	7				
Diplospirinæ				1	?	4				
Koninckininæ						6	1			
Meristellinæ			2	5	?	1				

Description of the Structure of the Brachidium.—The elements of the brachidium in the Helicopegmata are seen with considerable elaboration in the genus Athyris.

In the interior of the brachial valve are seen in the apical region, proceeding forward from the hinge-plate, two stiff processes called the *crura* (Fig. 79); attached to the crura, and in Athyris making a short twist toward the base of the crura, proceed two ribbon-like bands toward the wall of the shell, and thence along parallel to its inner surface toward the front: these are thus far called the *primary lamellæ*. At the front, and continuously with these lamellæ, the *spiral coil* begins by curving toward the opposite valve, thence upward parallel with its inner surface to near the crura, thence turning again toward the wall of the brachial valve, and in the case of Athyris proceeding onwards parallel but outside the primary lamella, the second ribbon of the spiral running a parallel course, but with each spiral diminishing the size of the coil, and finally stopping at the *apex* of the spiral cone, one of which is on each side of the median plane of the valve. The various volutions of the coils on each side are thus called primary, secondary, etc., lamellæ of the spiral coil of the brachidium.

Between and uniting the primary lamellæ of the two coils is developed a band, variously complicated in different genera, called the loop, saddle or jugum.

In Athyris the jugum has at the centre a process extending upward towards the space between the crura, which is called the *stem* of the jugum: this stem forks in the present case, and the two branches (Fig. 79, *b*) are called *arms* of the jugum (Fig. 79, *j*); they proceed on the outer side of the primary lamellæ almost in contact with them, forming *accessory lamellæ* (Fig. 79, *b*). In the genus Kayseria the accessory *lamellæ* are continued along the face of the lamellæ of the spirals to form on each side a secondary or accessory spiral coil.

Indirectly connected with the modifications of the brachidium is a calcified plate, arising from the interior walls of the brachial valve along the median line, to which the jugum or its processes are attached or come in contact; this is the *median septum*. A median septum may also be developed from the corresponding position in the interior of the pedicle valve.

Recent students of Brachiopods have found the structure of the brachidium of great value in classifying the species into generic groups; and we are indebted to the work of Glass, Whitfield, Bittner, Beecher, Clark, and others, that our knowledge, systematized in the hands of the veterans Davidson and Hall, is so full regarding these delicate parts of the Brachiopod structure.*

* For illustration and description of these characters of the Brachiopods the student is referred to "An Introduction to the Study of the Brachiopoda," by James Hall (published in the Reports of the State Geologist for 1891 and 1892; Albany, New York); to the elaborate final Report on the Brachiopoda, vol. VIII. of the Paleontology of New York, by the same author; to Dr. Oehlert's appendix

Significance of the Facts.—By turning back to the table representing the geological range of the several genera and families of the Helicopegmata (p. 280) it will be seen that the total life-range of all the representatives of the group extends over eleven periods of the time-scale (from the Neo-ordovician to the Jurassic). In the Neo-ordovician there appeared a few small representatives of one of the families, but in the next period (Eosilurian) all three of the families are represented. In other words, all of the family differentiation was attained in, we may say, the first decade of the life of the suborder, and there were in the Silurian 5 genera of Atrypidæ, 5 genera of Spiriferidæ, and 11 genera of Athyridæ.

All the essential extrinsic characters of the brachidium which ever appeared had arisen at the very outset or initial stage of the history of the group of organisms possessing the brachidium.

When we consider that in evolution the real increment in any case is seen in the acquirement of differences in the morphological characters of organisms, and it is not a new species or genus or order that is evolved, but it is the development by individuals of some part of their organization in a different form from that seen among their ancestors, the significance of this observation is apparent.

After this initial stage there are no representatives of the whole order Helicopegmata in which the relative position of the loop is not found to be of generic value in taxonomic classification, and there is no case in which the modification of this character surpasses the limits attained at this initial stage of evolution.

The Loop of the Ancylobrachia and the Brachidium of Helicopegmata.—This was in all probability near the time of divergence of the Ancylobrachia and Helicopegmata, and as has been suggested,* the fundamental difference between the calcified brachial supports of these important groups of

---

to Fischer's "Manuel de Conchyliologie" on "Brachiopodes;" to Zittel's "Handbuch der Palæontologie," vol. I., and to Davidson's classic treatise on the "British Fossil Brachiopoda."

* "On the Brachial Apparatus of Hinged Brachiopoda and on their Phylogeny," *Proc. Rochester Acad. Sci.*, vol. II. p. 113, etc., 1893.

Brachiopods (see Figs. 66–72) consists in the fact that the loop or jugum connecting the primary lamellæ in the Helicopegmata sets off from the sides of the lamellæ before they have begun to reverse their direction in forming the volution, and the continuation of these lamellæ is supplied with a calcified spiral support; while in the Ancylobrachia the connection does not take place till *after* the primary lamellæ have reversed their direction and are proceeding backward toward the crura. For them there is no calcified continuation of the lamellæ, but the brachial arms, although still preserving

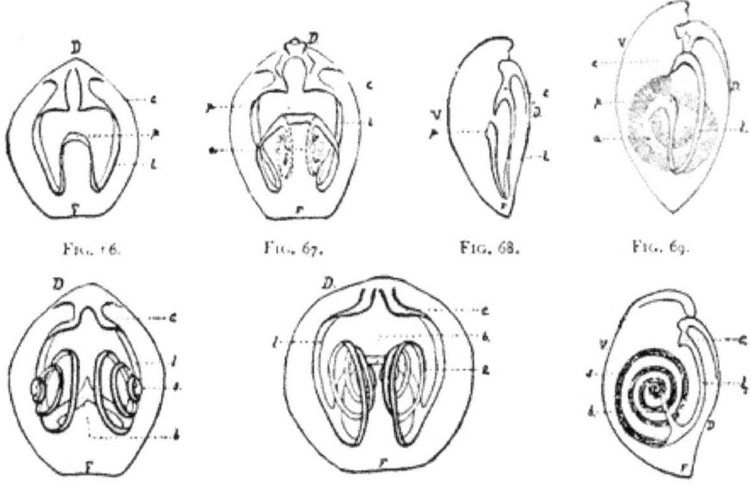

FIG. 66. FIG. 67. FIG. 68. FIG. 69.

FIG. 70. FIG. 71. FIG. 72.

FIGS. 66–72.—Diagrams expressing the relationship between the brachidial apparatus of Ancylobrachia and Helicopegmata. 66–69. The loop of the Ancylobrachia. *D* = brachial valve; *V* = pedicle valve; *c* = crura; *l* = primary lamella of the brachidium; *p* = the connecting bar of the loop corresponding to the jugum of the Helicopegmata (*b* in the lower diagrams); *a* = the fleshy spiral arms, not supported by calcified lamellæ. 70 = Brachidium of Zygospira, 71 of Anazyga, 72 of Dayia, seen from the side; the lettering the same as above, except *b* = jugum and *s* = spiral coils of the brachidium.

the spiral form, from the angle of the loop are entirely fleshy, and therefore not preserved in the fossil state.

If we examine in detail the kind and extent of modification exhibited in the various genera, in their relations of the time and order of appearance in the geological faunas, we gain a close view of the actual fact of evolution of new characters, as seen in the following particulars:

Relation of Jugum to the Primary Lamellæ.—1. The position of the jugum in relation to the point of outset of the primary lamellæ from the cruræ and the point of their turning back to form the first volution of the spiral, is perhaps one of the most fundamental differences, as it affects the whole mode of elaboration and position of parts of the brachidium. The extremes possible are for the jugum (1) to join the lamellæ immediately at their origin from the end of the cruræ, and (2) to be situated at the extreme front of the shell

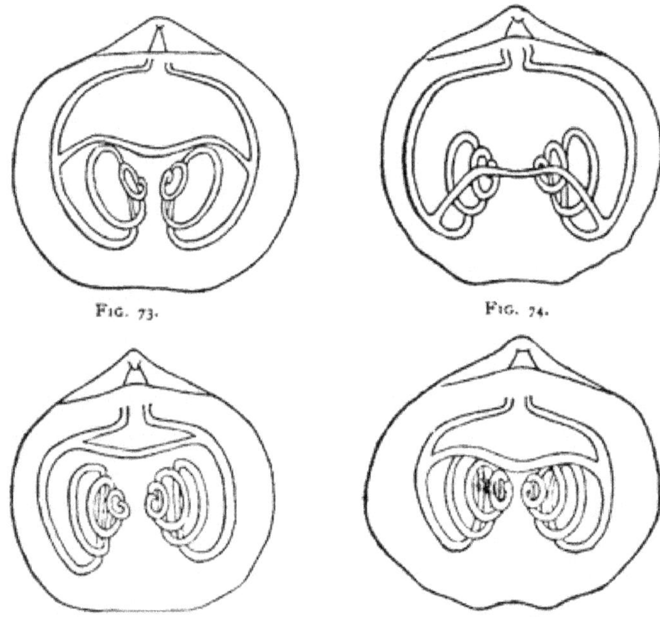

FIG. 73. FIG. 74. FIG. 75. FIG. 76.

FIGS. 73-76.—*Zygospira modesta.* (After Clarke.) Showing the variation in the position of the jugum.

joining the lamellæ where they begin to turn back to make the first volution of the shell (compare Figs. 75 and 74).

The rapidity with which the differentiation of structure in this particular took place is seen in a remarkable way by the examination of the earliest representatives of the Helicopegmata, as illustrated by the diagrams of the form of the brachidium of *Zygospira modesta* and of the closely allied form *Z. putilla* prepared by Mr. Clarke.*

The position of this jugum (or loop) is regarded by Hall and Clarke as of less than specific value. They say, "This is

* Pal. N. Y., vol. VIII. pt. 2, fasc. i. pp. 155 and 157.

not a specific character, but a matter of variation among individuals of a given species;" and remark further, "This mobility in the loop of Zygospira is without parallel among other genera." *

This case of the Zygospira loop is a striking example of rapid evolution. It has the appearance of being an insignificant feature, only a variation, because of the presence of all the intermediate variations at the initial stage.

Relation of the Primary Lamellæ to the Cruræ.—2. A second example is seen in the modification of the direction of the primary lamellæ after they set out from the end of the

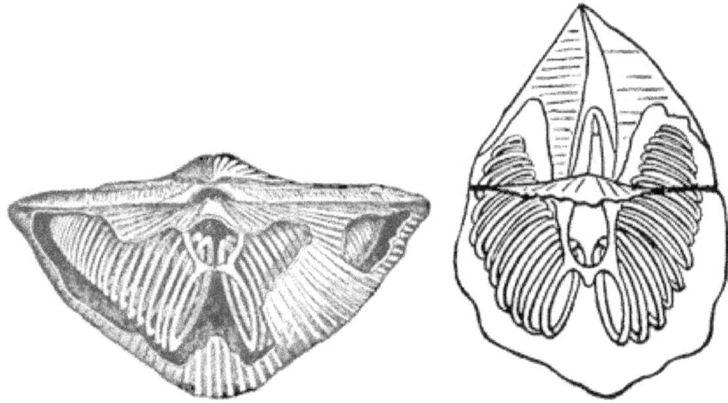

FIG. 77. FIG. 78.

FIG. 77.—A Spirifer, showing part of the brachial valve, the brachidium with the primary lamellæ, the jugum, and the spiral coils.

FIG. 78.—Cyrtina, the brachial valve removed, showing the brachidium with the spiral coils turning upwards into the produced umbonal part of the pedicle valve.

cruræ. There are two ways in which this direction differs: (*a*) The lamellæ may proceed directly toward the front of the shell away from the cruræ, as in the case of *Spirifer* and *Cyrtina* (Figs. 77, 78); or they may, immediately after their origin, take a sudden bend upon themselves, making a twist or double bend before proceeding along parallel to the inside surface of the shell, as in *Athyris* (see Fig. 79); observe also the brachidium in Figs. 64, 65 (p. 279). The latter is regarded as a characteristic of the family Athyridæ, and, although the family as a whole is the more differentiated and later to be dominant, there are several well-marked genera

* See l. c. p. 156.

showing the sudden reflection and twist in the origin of the primary lamellæ among the first species of the Eosilurian (viz., Dayia, Hindella, Merista, etc.), while the Atrypidæ and Spiriferidæ, in which the lamellæ are directly continuous with the cruræ, are fully expressed at the base of the Upper Silurian.

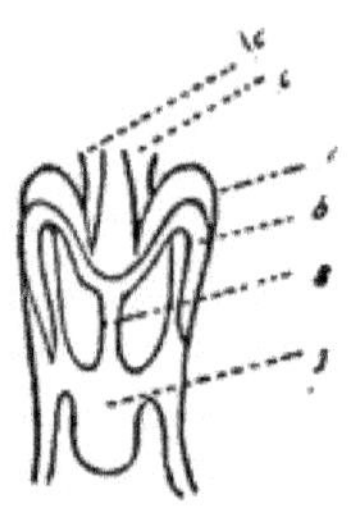

FIG. 70.—Athyris, showing part of the brachidium, the two cruræ, *cc*, succeeded by the upper twisted part of the primary lamellæ *l*, the flattened jugum, *j*, uniting the lamellæ of the two sides, and above it is the stem *s*, branching into two branches, *b*, which follow the direction and lie upon the upper part of the primary lamellæ.

The second particular in which difference is exhibited is seen in (*b*) the direction away from or else parallel to the plane of a median septum. In one extreme (see in *Zygospira*, Figs. 73–76) the lamellæ diverge at a right angle (or less) from the extremity of the cruræ toward the lateral borders of the shell, and curve outward and thence downward along this outer border to the front; and as they reflect in the course of the first volution, turn inward toward the centre. In this type the spirals have their apices directed more or less inward. Atrypa presents these characters, and Schuchert has adopted the characters of its brachidium as a mark of the family Atrypidæ (see p. 279).

In Spirifer the lamellæ proceed with almost no divergence in two nearly parallel lines, from the extremities of the cruræ directly toward the front along the inner surface of the brachial valve, and at the front curve directly toward the pedicle valve, and in making the first volution of the spiral return nearly to the starting-point at the end of the cruræ. The spiral thus formed has its apex directed outward toward the lateral border of the valve, and it is in this type of brachiopods that the great production of the lateral wing of the shell takes place, and the apices of the spires penetrate into the pointed extensions of the shells (see Figs. 77, 78). These two extreme types, however, first appear near together at the very base of the Upper Silurian.

The Number of Volutions of the Spiral.—3. Another divergence is in the number of volutions of the spiral. The earliest known Helicopegmata are generally of small size, and the volutions are not numerous: it is not improbable that the primitive form of spiral was with few volutions; but if this be the fact, the rapidity of their increase to the extreme, found in Atrypa and in some of the Spirifers (Fig. 77), was early reached in the basal fauna of the Upper Silurian, and it is observed that the embryonic forms have fewer coils to the spiral than the adults (Beecher). (Compare Protozyga (Hall) and Cyclospira with Atrypa reticularis or Spirifer.)

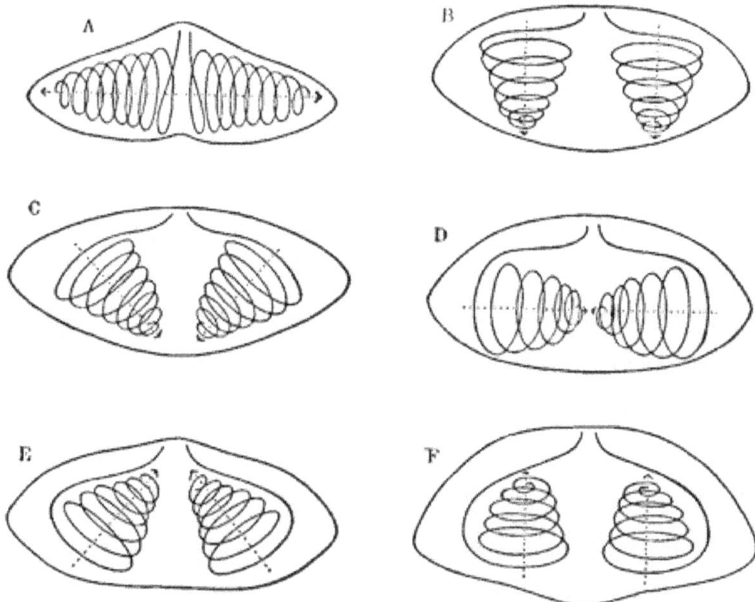

FIG. 80.—Diagram representing the various positions of the spiral coils in the brachidium of the Helicopegmata. The diagrams are drawn as transections viewed from the beak of the shell, the brachial valve being the upper and the pedicle the lower lines of each figure. *A*, the position with apices of the cones directed outward, as in Spirifer; *B*, apices directed toward the pedicle valve; *C*, apices directed toward the centre of the pedicle valve; *D*, apices nearly meeting on the median plane; *E*, apices directed obliquely inward toward centre of brachial valves; *F*, apices directed toward the pedicle valve with subparallel axes.

Direction of the Axes of the Spiral Cones.—4. Among the earlier representatives we have also every position of the spirals, so that the direction of the pointing of the axes and the apices of the cones reaches its full elaboration very early. In Zygospira the apices are directed obliquely toward the centre of the brachial valve (Fig. 80, *E*); in Atrypa and Atrypina toward the deepest part of the brachial valve (*F*), while in Spirifer and several other genera they are directed toward the outer margin of the two valves (*A*); in Cœlospira and Dayia outward toward the lateral slopes of the pedicle valve (the position is intermediate between *A* and *B*); in Catazyga toward the median plane just below the surface of the brachial valve (*C*); in Glassia toward the centre, and the apices nearly meet at the centre of the internal cavity (*D*); in Cyclospira they are coiled nearly parallel to the vertical axial plane, and the apices are slightly introverted.

Although in lines of species (which in their combination of characters show them to have close affinity and hence are grouped in generic groups) the direction of the axis of the spiral cone is pretty constant, we see that whatever importance may be attached to the different position of the spirals in relation to the other parts of the body, the differentiation of these features was quickly attained.

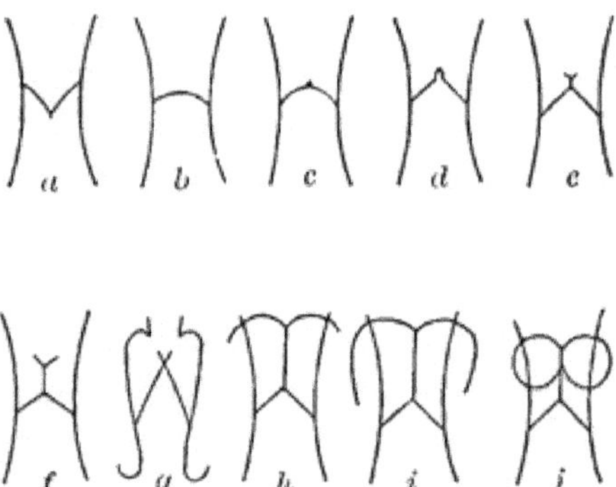

FIG. 81.—Diagrams of the various forms of the jugum in the Helicopegmata. *a* = Atrypina; *b* = Spirifer; *c* = Hindella; *d* = Hyattella; *e* = Retzia; *f* = Whitfieldia; *g* = Meristina; *h* = Athyris; *i* = Kayseria; *j* = Meristella.

The Form of the Loop.—5. The character presenting the greatest degree of divergence in the structure of the brachidium is the form of the loop or jugum. In the paper above referred to, Mr. Schuchert has suggested that the nature and complexity of the loop which joins the spirals are the more important characters for subfamily differentiation.

In Spirifer proper (Fig. 81, *b*) the loop is a simple band, about the size of the primary lamellæ, joining the two lamellæ together; in some cases in adults this was partly absorbed, leaving only two calcareous processes facing each other on the sides of the opposite lamellæ. In Zygospira the loop is simple, but arched or forming a double bow-like curve (Figs. 73–76). In Davia there is a simple process from the centre of the saddle running toward the base of the crura (see Fig. 81, *e*). There is added a bifurcated end in Whitfieldia (81, *f*). In Athyris (81, *h*) the ends of the branches are curved over to partly cover the primary lamellæ of the spirals. In Kayseria they are continued along parallel to the lamellæ of the spiral coil (81, *i*). This extension is only seen in the late Mesozoic forms. In Meristella (*j*) the branches of the process recurve and join together, forming on each side a loop, resembling the handles of a pair of scissors.

In this series of modifications the extreme degree of elaboration is met with among the Meristellinæ, and this subfamily was well represented among the Eosilurian faunas.

Characters of the Brachidium found to be good Distinctive Characters of Genera.—It has been acknowledged by all the more advanced students of Brachiopods, that the modifications of the brachidium are the most important characters to be found for determining the generic and higher affinities of these interesting forms, and great and most painstaking labor has been expended within the past ten or fifteen years in working out the structure of their delicate parts.

We may interpret this experience of systematists to mean that the various degrees of modification observed in these parts are found to be constant among species which by likeness in other characters are associated together into groups to form genera.

Plasticity a Characteristic of their Early Initial Stage.—We have already seen by analysis of the characters that almost without exception the plasticity of the characters, and the expression of the widest range of possible differentiation in each particular direction, were characteristics of the early stage in the history of the Helicopegmata. By the beginning of the Neosilurian the expansion of differentiation had reached its extreme in almost every particular.

Evolution of the Characters of the Brachidium Relatively Rapid.—When we consider that we have knowledge of only a few small types of this whole order earlier than the Eosilurian, and that the Helicopegmata lived on to the middle of the Mesozoic, and, third, that most species have a life-period of a third or half of the duration of the whole Silurian time, it is no exaggeration to say that the evolution of these modifications of the brachidium was, relatively to all laws of organic change in geology, extremely rapid.

Rate of Initiation of the Genera of Helicopegmata.—If now we reduce the facts of generic differentiation to graphic form, we find that the sudden or rapid differentiation is a fact, and is not due to imperfect evidence. Considering, as in previous cases, classification to be a mode of expressing degrees of difference, we may rely upon the mathematical relations of initiation of the groups of equal rank as an expression of the rate of initiation of new characters in general, or an approximate measure of the rate of geological evolution.

TABLE EXPRESSING THE RATE OF EXPANSION OF THE FAMILY, SUBFAMILY, AND GENERIC CHARACTERS OF THE HELICOPEGMATA.

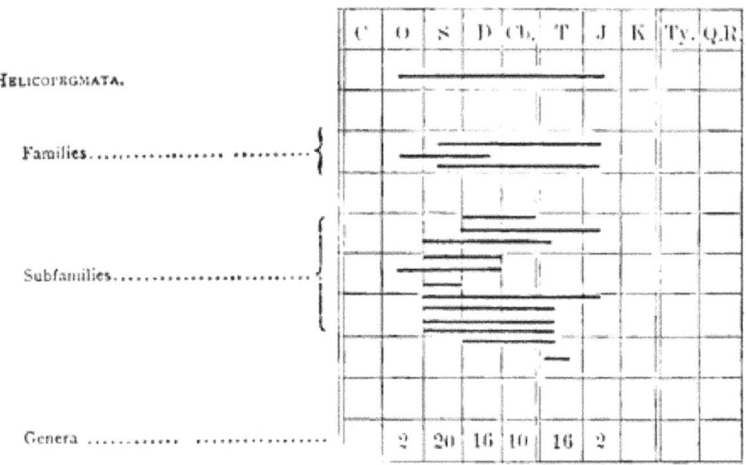

The Helicopegmata as a suborder is found to be represented in three family types of structure: one of these appeared first in the Ordovician, in a single subfamily, a single or possibly two genera, and but few species. At the opening of the next era, the Upper Silurian, the other two families appear, and seven out of the known twelve subfamilies were initiated.

If we consider the actual total number of generic types for the whole suborder, and some of the later of these genera are based upon very slight modification of characters, we find 76 in all. The rate of their initiation is: Ord. 2, Sil. 20, Dev. 16, Carb. 10, Trias. 16, Jur. 2; or by the time of the second stage in which any of the suborders are known one quarter of the total generic differentiation had taken place, and differentiation did not cease till six eras had been passed and the suborders became extinct.

Representing these facts in whatever way we may, they are positive in testifying to a rapid and early expression of the differences in structure which have served as the means of distinguishing different families, subfamilies, and genera; and a close inspection of the figures seems to indicate that in proportion to the higher taxonomic rank of the characters, the earlier or more rapid was their initiation.

General Law of Rate of Initiation of Generic Characters.—In general terms, the scientific fact here noted, irrespective of any theoretical explanation, is that, relative to the known geological range of species of the Helicopegmata, the grander differences in structure were very early to appear, and that the progress of differentiation after this early stage was largely in respect of varietal and specific characters and proportionally small in characters of higher rank.

The Life-period of Genera and the Initiation of a New Genus.—We have now examined some of the laws of genera as exhibited in the case of the Helicopegmata. The characters which are found to be of generic value, such as the particular structure of the calcareous framework supporting the brachial arms, have a definite history. Examining all the known Brachiopods, from the beginning of geologic time to the present, it is found that the structural characters peculiar to this suborder are confined to the time extending from the Lower Silurian to the Triassic or Jurassic era. As a particular example, for instance, the arrangement of the brachidium characteristic of the genus Meristella (see Fig. 81, *j*, with the complex loop forming two lateral rings, and the cone of the spirals pointing to the lateral margin of the shell, as in Fig. 80, *a*) begins in the Silurian, and is never seen after the Devonian. The genus is said to be characteristic of that period; and not only in America, but in Europe, in China, in South America, wherever Paleozoic rocks are known, Meristella is found characteristically in the Upper Silurian, running rarely a little below, but more frequently above, into the Devonian.

There comes a time in the history of organisms of a particular line of descent, when a certain definite arrangement of the parts of the organism becomes conspicuous, as this particular loop of the Meristella; the occurrence of individuals developing this peculiarity is limited below and above. This arrangement differs from that of the corresponding part in any other animals of the same time; and all the animals exhibiting this character may be considered as closely allied genetically, because in other characters they also show strong resemblance. This state of things is evidence of the *beginning* or *initiation of a new genus.*

If all the specimens known possessing this new character were examined and classified, they would be found to have minor differences of form, surface marking, etc., which furnish criteria for dividing them into several distinct species. Geologically, one of these species is the first to appear; it lives but a short time relatively, or it may continue to live during several periods. It is peculiar to one country, or it is the same throughout the world wherever the genus appears; but whether there be many or few species, the character which is called a generic character begins at some particular time: during a certain period it is frequently met with; after a time it ceases, and is never known to appear again. The particular combination of characters on some one organism constitutes its generic characters, and we may say that the genus so characterized has a certain definite *life-period.*

During the Life-period of the Genus its Characters Constant.—While other characters may vary, these generic characters do not change sufficiently to be noticed as of importance to the paleontologist. Not only the generic, but the family and the ordinal, characters, which are associated together under the generic name Meristella, are thus constant for all the specimens examined.

A Culminating Point or Acme in the Life-period of a Genus.—Again, we observe that the fossil specimens which present the characters (of Meristella for instance) are most abundant along the middle of this period; for the Meristellas it is about the Neosilurian; also in that period they are more frequently met with in distant parts of the world; and where they are most abundant the characters which serve to distinguish them into separate species are more numerous; and both before that epoch and afterward there are fewer and fewer, until we reach both ends, where the species are very rare.

Summary of the Geological Characteristics of a Genus.—To generalize the above observations, it may be said that the genus practically has a time of beginning and a time of ending. Practically, that is, according to the knowledge we possess, there was a geological time, represented by a particular horizon in the geological series of strata, when each genus began; there was a particular period, of shorter or longer extent, during which the genus was freely propagated, and abundant individuals flourished, leaving their remains in the strata, wherever the conditions were appropriate for their preservation. The genus had a period of decadence, or of growing old, the species became fewer and fewer, the individuals more rare, and finally the genus died out, and, so far as our knowledge goes, became extinct. These laws apply to Meristella, and in substance they apply to all genera we know of. The period from the initiation to the extinction of the genus is the life-period of that genus.

CHAPTER XVII.

THE PLASTICITY AND THE PERMANENCY OF CHARACTERS IN THE HISTORY OF ORGANISMS.

Races in Paleontology.—During the life-period of a genus constant changes are found to take place among the representatives of the genus as we follow them upward from stage to stage of their geological succession. The forms appearing at the first epoch, in the life-period of a genus, are generally found to be of different species from those occurring later; and in many genera there are enough specimens collected, and sufficient knowledge regarding them accumulated, to enable the paleontologist to recognize a series of forms regularly succeeding one the other, presenting slight modification from one stage to the next, but those of each stage showing closer resemblance to those immediately preceding them than to any other species of the same genus. The series of forms thus resembling each other may be called *races*, because of the very evident genetic relationship existing between the later and the earlier representatives of the series.

Phylogeny of the Race.—When we examine the details of form in such a series of succeeding forms or races of a genus, comparatively, it is often apparent that the changes undergone in respect to each character are progressive or of an accumulative nature, and thus they resemble the changes which the individual undergoes in ordinary growth. The technical name proposed by Haeckel for this morphological history of the race is *Phylogeny*, contrasting it with *Ontogeny* or the history of growth or development of the individual, from its relatively homogeneous condition in the ovum to the more or less differentiated adult organism.

Mutability and Phylogeny.—The Cuvierian school of naturalists believed in the immutability of species, and for them the principle of racial evolution or phylogeny was barred out. But Geoffroy St. Hilaire and Lamarck with their idea of mutability of species laid the way for a consistent theory of phylogenetic evolution, although in their time the knowledge of paleontology was not far enough advanced to furnish actual phylogenetic series of organisms. It was, however, not till Darwin had constructed a working hypothesis for the steps and manner by which new types of organisms can arise, that evolution became an accepted mode of explanation of the course of biological history.

The great advance which the present generation has witnessed in the interpretation of the science of organisms is the change in belief, which all naturalists have more or less thoroughly undergone, from the doctrine of immutability to that of mutability of species. Some theory of evolution and phylogenetic origin of species is the necessary outcome of this new doctrine. Darwin more than any other single man was the means of producing the change of conviction in regard to this point.

The Phylogenetic Theory of Evolution.—The phylogenetic theory of evolution is logically an expansion and application of the principle of organic growth, already recognized in the development of individual characters, to the evolution of specific and more fundamental differences. It is a recognition of an organic correlation between separate individuals. As growth takes place in the individual by the segmentation and separation of cells, with specialization of functions, first for different cells and finally for the complex structures called organs, the whole showing its organic unity by the mutual coöperation of all of the parts in the life of the whole, so the phylogenetic theory recognizes in the species, or the race of species, an organic unity of a higher sphere, in which the individuals play the part of mutually adjusted and coöperating parts in this greater organic whole.

The theory goes one step further, and includes the proposition, that as the principle *omne vivum ex ovo* is true in the life-history of individuals, so each species postulates a preexisting species. This is the philosophy of the theory, but it must be observed that the concrete facts illustrating these laws are always found together in the same organism. Each individual organism is the source and record of those facts which we separately interpret as evidence of cell-growth, individual growth, the differentiation of organs, and the phylogenetic evolution.

Thus there are series of organic forms succeeding each other in some regular order, known or unknown, which are bound together by organic, and in this case called particularly *genetic* relationship. The changes in form observed upon comparing the individuals at different points in the line of succession are accounted for by some law of evolution, and the origin of the different members of the series is said to be by generational descent, the later arising from the earlier. On account of the mutability of form in the process, species presenting different form, different function, and incapable of organic fertility are supposed to have arisen originally from a common parentage.

Mutability the Fundamental Law of Organisms; the Acquirement of Permanency Secondary.—This analysis brings us face to face with one of the chief inconsistencies in the prevalent conception of the nature of organisms. While the doctrine of mutability of species has generally taken the place of immutability, the proposition that *like produces like* in organic generation is still generally, and I suppose almost universally, accepted. It therefore becomes necessary to suppose that variation is *exceptional*, and that some reason for the *accumulation of variation* is necessary to account for the great divergencies seen in different species. Darwin's theory of natural selection is chiefly concerned in accounting for the accumulation, increase, and perpetuation of divergencies arising by natural variation.

If we extend the principle of mutability, and instead of regarding it as an accidental circumstance in the life-history of organisms, recognize it as the distinctive and fundamental characteristic of living beings, we escape this inconsistency.

In the physical and chemical world like causes do produce like effects; but in the organic world like produces like "with an increment," as Professor J. D. Dana put it. Mutability and variation are evidences of this increment. The increment is the great fact; the checking and limiting of it is secondary. The search has been for some cause of the variation; it is more probable that mutability is the normal law of organic action, and that permanency is the acquired law.

It is more probable that the use and tested adaptability of a variable part is the cause of checking the variability and of the transmission of the character with less or no variation, than that the variation is increased by this process. Adopting mutability as a fundamental law of all organic activity, and the construction of a theory of evolution becomes a simple matter. We have in that case to account for the acquirement of permanency of characters. This is found in the principle of ordinary generation, the instituting of habit, and the more and longer the species breed together the closer and more fixed will the characters become.

Early Plasticity Succeeded by Permanency expressed in Geological History.—Examination of the history of geological species suggests the truth of this hypothesis, for it is observed that many species, which by their abundance and good preservation in fossil state give us sufficient evidence in the case, exhibit greater plasticity in their characters at the early stage than in later stages of their history. A minute tracing of lines of succession of species shows greater plasticity at the beginning of the series than later, and this is expressed in the systematic description and tabulation of the facts by an increase in the number of the species.

In order to illustrate this law a special consideration will now be given to the facts regarding the laws of specific history as observed by the paleontologist.

Pritchard's Definition in which the Constancy of Transmission of Same Peculiarity is made the Criterion of Species.—Thus far we have been considering generic characters—that is, those characters which are constant for one or more species. The next question to consider is, What are the laws exhibited in the history of specific characters? There are various definitions of species which are more or less theoretical; but whatever our theory about the definition, the fact remains that all naturalists do recognize within slight limits of difference the reality of groups of organisms called by the name species.

In a previous page are given some of the definitions of species formulated by early naturalists. Alfred R. Wallace, who published as early as 1855 an article on the law which has regulated the introduction of new species (Darwin's "Origin of Species" was published in 1859), set forth some of the chief principles of the modern evolutionary conception of the history of organisms. Wallace made a careful study of species, and, perhaps as well if not better than any one else, understands the relationship between species and geographical distribution. In an article of his "On the Malayan Papilionidæ, or Swallow-tailed Butterflies, as Illustrative of the Theory of Natural Selection," published in 1864, is found the following definition of the word species:*

"In estimating these numbers [of the species of Papilionidæ] I have had the usual difficulty to encounter, of determining what to consider species and what varieties. The Malayan region, consisting of a large number of islands of generally great antiquity, possesses, compared to its actual area, a great number of distinct forms, often indeed distinguished by very slight characters, but in most cases so constant in large series of specimens, and so easily separable from each other, that I know not on what principle we can refuse to give them the name and rank of species. One of the best and most orthodox definitions is that of Pritchard, the great ethnologist, who says that '*separate origin and distinctness of race, evinced by a constant transmission of some characteristic peculiarity of organization*,' constitutes a species. Now leaving out the question of 'origin,' which we cannot determine, and taking only the proof of separate origin, '*the constant transmission of some characteristic peculiarity of organization*,' we have a definition which will compel us to neglect altogether the *amount* of difference between any two forms, and to consider only whether the differences that present themselves are *permanent*. The rule, therefore, I have endeavored to adopt is, that when the difference between two forms inhabiting separate areas seems quite constant, when it can be defined in words, and when it is not confined to a single peculiarity only, I have considered such forms to be species. When, however, the individuals of each locality vary among themselves, so as to cause the distinctions between the two forms to become inconsiderable and indefinite, or where the differences, though constant, are confined to one particular only, such as size, tint, or a single point of difference in marking or in outline, I class one of the forms as a variety of the other. I find as a general rule that the constancy of species is in inverse ratio to their range. . . . When a species exists over a wide area, it must have had, and probably still possesses, great powers of dispersion. . . . When, however, a species has a limited range, it indicates less active powers of dispersion, and the process of modification under changed

* "Contributions to the Theory of Natural Selection. A Series of Essays." p. 141. Macmillan & Co., 1870.

conditions is less interfered with. The species will therefore exist under one or more permanent forms, according as portions of it have been isolated at a more or less remote period."

Permanency of Characters in Living Forms Co-ordinate with Limitation in Distribution and Breeding.—From these quotations it will be seen that in the conception of an organic species the fundamental idea here emphasized is the reproduction of numerous individuals possessing likeness in all their morphological characters, except in those in which the offspring of a single brood may differ when compared together. This specific permanency involves absence of intermixing of the separate species, if in the same locality, or local separation of the species. In other words, co-ordinate with the likeness of form there is assumed to be limitation in breeding and limitation of local environment. This is the extent of the limitation which the study of living forms reveals.

Specific Variability Restricted with each Successive Generation in Fossil Forms.—When we examine geological species we find also a limitation in time of the repetition of like individuals. When species are studied historically, the law appears evident that the characters of specific value—those which serve to distinguish one species from another, according to the rules above formulated and generally practised—present a greater degree of range of variability at an early stage in the life-period of the genus than in the later stages of that period. To express this law in terms of the history of organisms, we say there are periods in the history of particular lines or races of organisms, of unusual variability or plasticity of some of the characters, and afterwards the history shows relatively long periods in which the characters expressing such plasticity are constant or present very slight divergence. Further, in this second period of slow modification, or persistence of form, the changes taking place in the phylogeny are slight, but they increase in a particular direction steadily and slowly with time.

Illustrations of the Acquirement of Permanency of Characters.—In order to illustrate these laws the following actual cases will be described in detail: the *Spirifers* at the base of the Silurian, as an illustration of extrinsic evolution; *Atrypa* *reticularis* and its allies, as an example of the permanency of the plastic condition; the bivalve shell Lamellibranch (*Ptychopteria*) of the Upper Devonian, illustrating the initiation of the species of a genus; and Mammals, in illustration of progressive evolution.

The History of the Spirifers.—When we attempt to discover the laws of phylogenetic succession we are obliged to consider specific and varietal characters.

As has been already shown, the length of the geologic time through which the characters of generic and higher rank are exhibited is, by the Brachiopods at least, measured by geologic periods: and there are series of Brachiopods extending through one or more geologic systems in which the generic characters expressed are alike, the various representatives from beginning to end exhibiting differences only in the lesser or specific characters.

For the study of the history of such specific characters the Spirifers may be taken as examples. The whole family Spiriferidæ begins, according to present knowledge, near the base of the Upper Silurian, and there are two known representatives in the Triassic. The genus *Spirifer* begins at the base of the Upper Silurian, and is well represented through the Silurian, Devonian, and Carboniferous. There are named 218 species in America. Hall in his "Genera of Brachiopods" recognizes over two hundred species. The species referred to this genus in the Mesozoic are probably of distinct generic rank; a large number besides are defined in other countries. Among the numerous species assigned to this genus there are great variations in a few particulars. In the whole genus there may be three hundred, or possibly four hundred, good species, or forms, presenting two or more describable characters, of which each is different from any other species. When we examine the whole genus, and note the characters which distinguish one species from others, and arrange the characters into classified groups, as they concern separate elements of the shell, they may be classified as modifications of a few elements of the form or structure of the shell.

The Permanent Characters of Generic or Higher Rank.—Examining the successive forms of Spirifers, we observe that there are long lines of individuals, each running through one or more geological periods, and repeating without noticeable change the precise morphologic characters of its ancestors down to the generic characters, and exhibiting differences only in the specific or less important elements. The specimens exhibiting this law we associate together as a genus and call them by the same generic name, expressive of the fact that they agree in all their morphologic elements, except such as distinguish different species of the genus. The characters of specific value vary during the life-history of the genus, but the generic characters remain unchanged; or, to apply a special designation to these two facts, the generic characters are *fixed*, and the specific characters are more or less *plastic*.

Characters which are Plastic at the First or Initial Stage of the Genus.—At the initial stage of the genus *Spirifer* the generic characters may be supposed to have become fixed. The still plastic characters are chiefly seen in a few definite morphologic elements. These are: (I) the *contour of the shell*, or in terms of growth, the relative rate of growth from the nucleus outward; this is seen in specimens with short hinge lines, in others with produced angles at the extremity of the hinge and in the intermediate forms; (II) the *vertical extent of hinge area*, ranging from low to high area; (III) the *delthyrium*, open to closed; (IV) the *surface*, evenly arched over, or producing a single median fold, or several folds, extending to the beak, or with intermediate development; (V) the *surface striations*, radiating and concentric, fine or coarse, continuous or interrupted, uniform or bifurcating as they develop toward the front.

The Fixation of Plastic Characters in a Generic Series.—It is the various degrees of modification of these characters that constitute the specific differences upon which the several species are defined. They are all plastic or variable at the first stage, and the individual species of the genus present certain limits of variation of each of the characters.

If we go a step farther and classify all the variable characters of the genus, we may discover in numerical terms the relation between retention of plasticity and the passage of time, or the effect of time in limiting the variability of the more or less plastic elements of the genus. The characters of Spirifers that are of chief generic and specific value are the following:

A. The form and arrangement of the spiral appendages.

B. The general proportions of the shell.

C. The delthyrium, deltidium, and fissure; their shape and development.

D. Hinge area, its length and height.

E. Surface markings; radiating striæ, fine and continuous or coarse and interrupted; including imbrication.

F. Medial fold and sinus.

G. Plications of surface—simple fold, or many and bifurcated folds.

H. Structure of shell—fibrous or punctate.

I. Spines, or setæ, or elevations, granular or otherwise.

K. Special development of septa, medial or deltidial.

Whatever evolution has taken place should be expressed in terms of some one or more of these characters, for these constitute the differences distinguishing the several known species.

A. *Spiral Appendages.**—So far as we know, these variations during the life-history of the subfamily or genus do not exceed slight adjustment of position and direction of the coils to the internal capacity of the shell, and variation in the number of the coils. Of both of these characters too few statistics are at hand to enable us to base upon them any law regarding the rate, or even direction, of evolution; but the modifications appear to be all easily explainable by the principle of extrinsic evolution, i.e., adaptation to external conditions in the process of ontogenesis.

B. *The General Proportions of the Shells.*—Taking an average of the extremes of form for the whole of the hinged Brachiopoda and constructing a medium form, the result would be an oval shell, with hinge line shorter than the greatest width, and the pedicle valve larger than the brachial, with low hinge area; a deltidium; no fold or sinus, further than a

* See, regarding this and other details, Paleontology of New York, vol. VIII., "An Introduction to the Study of Genera of Paleozoic Brachiopoda," pt. II., by James Hall, assisted by John M. Clarke, 1894.

tendency to lengthen the central part of the ventral and shorten the central part of brachial valve. Both valves would be convex, but slightly so. *Atrypa reticularis* is not far from such a medium form of an articulate Brachiopod. The Spirifers vary in the following directions in respect to these characters: The pedicle valve may be greatly developed about the beak, forming considerable contrast between the two valves. This variation is noted in species of the earliest stage, in individuals of most species when contrasted, and in different stages of growth of the same individual. The variation is most noticeable among species which are abundant

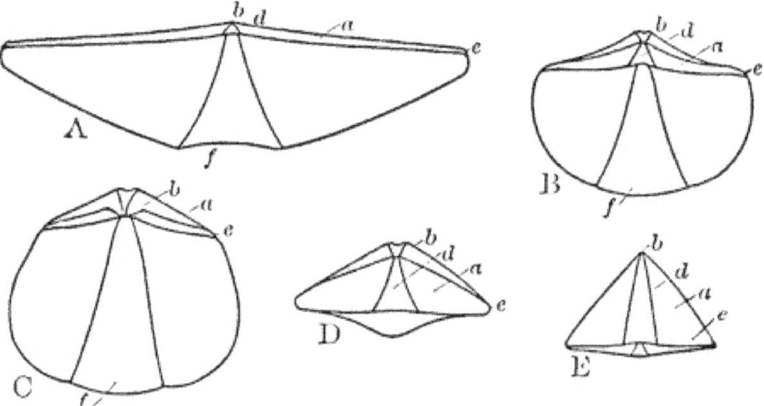

FIG. 82.—Variations in form of *Spirifer Verneuili*. (After Gosselet.) *A*, outline of the form *Cylindrici*, from the upper Frasnien; *B*, form *Hemicycli*, from the Frasnien; *C*, form *Ostiolati*, from the Famennien; *D*, cardinal view of form *Elongati*, Famennien; *E*, cardinal of extreme of *hemicycli* form, from the Frasnien; *f* = fold of the brachial valve; *b* = apex of the beak of pedicle valve; *d* = delthyrium; *a* = cardinal area; *e* = lateral extremity of the cardinal area.

and of wide range, and rare or local species are generally less variable in this particular (Fig. 82).* In size there is considerable variation, but most of the species of the Silurian are small for the genus, though in this respect perhaps the

* Fig. 82 illustrates some of the conspicuous differences in form assumed by the Spirifers. The variations are further interesting as occurring all on the same species, and appearing on specimens selected from the same geological province, from strata differing a little in age, but all from the upper half of the Devonian of northern Europe. Similar specimens have been seen in the corresponding rocks of New York State (see *Am. Jour. Sci.*, vol. XLIX, p. 473).

largest species of the Niagara or Silurian is not more than an average for the whole range of size. If the size is expressed on a scale of 10, 1 representing the smallest and 10 the largest, the range in the Silurian is about 1 to 4. There is a general tendency to increase in the size of species of the genus from their beginning to the Carboniferous. The Silurian species average about 3 on such a scale, the Devonian species 5, the Carboniferous 7, and the size of the Mesozoic species would average about 3.

The species which contain the larger individuals for their period are generally more abundant in numbers. There is evident adaptation of size and abundance to conditions of environment, for certain deposits contain abundant and large representatives of a particular species, while other deposits contain but few and small individuals. The character B, then, is evidently in its evolution purely extrinsic, the species adapts itself to environment, and in each race the adaptation is greater with advance of time up to the Carboniferous, where the whole race deteriorates, and in most species becomes extinct, only a few surviving, and those having some specially developed characters.

C. *The Delthyrium and Deltidium.*—The delthyrium is the opening through which the peduncle passes for the attachment of the shell, and its covering is the deltidium. In its early stage the young shell was always attached, and the deltherium was open. In some species there was very plainly a gradual closing of the fissure by a *pseudodeltidium*, a covering of shell growing over the fissure from beak downward. In others, there is this pseudodeltidium with a slight foramen permanently running through it (see Fig. 82, *A*, *B*, *D*, and *E*, *d*).

In others there is a permanent open fissure; at least, no calcified covering is present in the adult. The presumption is that there was variation in the length of time the individual was attached, some species becoming free at very early periods, others remaining attached throughout life. If we express this character mathematically, 1 referring to attainment of free state very early, 10 permanently attached, we find among the species of the lowest period of the Silurian (Niagara and corresponding formations), in each of the chief types, great fluctuation in this character; 1–10, perhaps, is not too much. In later periods there is variability, but each species is subject to less variation, so that mathematically the species might be said to have this character variable in separate cases, 1–3, 2–5, 3–7, 5–9, etc.; and there are certain lines of forms in which the general range of this variability continues the same from period to period.

As to the size of the fissure in proportion to the other parts of the shell, there is considerable variation, but it is probably co-ordinate with the development of the area, those with high area having narrow fissure, those with low area a broad fissure. The characters, therefore, of the delthyrium and its cover show, in respect of evolution, purely extrinsic modification, the characters reaching extreme range at first, and afterwards, in the various races, expressing modification by restriction of variation and adaptation to special or local conditions.

D. *Hinge Area.*—This may be very narrow and elongated, forming a long hinge-line, or it may be very high, forming a triangular and greatly developed area and ventral beak. I know of no species, or sets of forms, which express a greater range of modification of this feature than the two species called *Spirifer plicatella* and *Cyrtia exporrecta* of the Wenlock limestone. The specimens with high beak are generally called Cyrtia; the specimens with moderate or low beaks are Spirifer. This character ranges from 1–10 in the earliest stage. In other species (*S. crispus* and its associates) there is a less degree of modification of this character (Figs. 88–91). In later forms the range of modification for each species is generally confined within less limits. The extreme extent of the modification and the extreme forms themselves are generally met with where the species are most abundant, and the prevalence of one extreme or the other is expressed in the later end of a series, which from the close resemblance of the successive specimens constituting it may be considered to be a true genetic series or race. Here again we find evidence that whatever evolution takes place is extrinsic and results, theoretically, from adjustment to environment, selection in

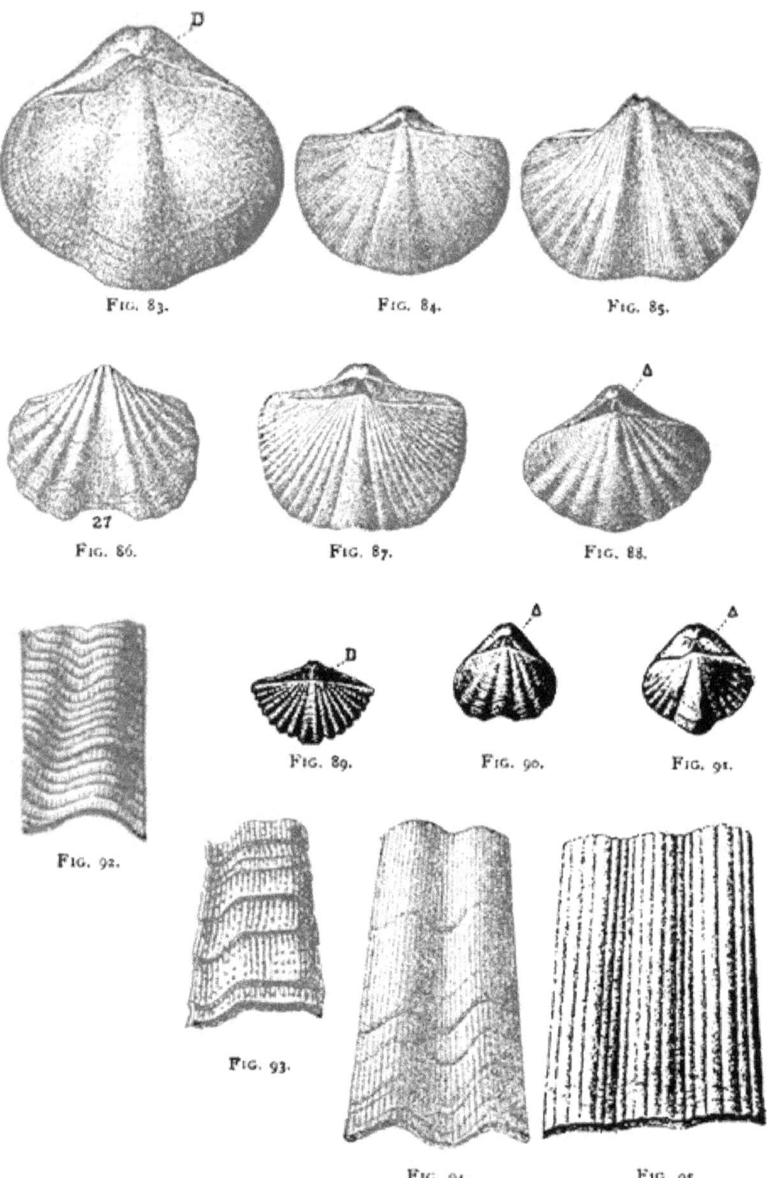

FIGS. 83–95.—Modifications of the surface features of the Spirifers, expressed in the species of the Niagara period. (After Hall.) 83. *Spirifer radiatus* Sow. 84, 85. *S. plicatella* Linn. 86. *S. eudora* Hall. 87. *S. Niagarensis* Hall. 88. *S. crispus* His. 89. *S. sulcatus* His. 90. *S. bicostatus* Hall. 91. *S. tenuistriatus* H. 92. Enlargement of the surface of *S. crispus*. 93. Surface, enlarged, of *S. sulcatus*. 94. Surface, greatly magnified, of *S. Niagarensis*. 95. Enlargement of surface of *S. eudora*.

breeding and limitation of range of variability by hereditary transmission.

E. *Surface Markings.*—The surface of Spirifers, when well preserved, are almost always covered with fine longitudinal or radiating lines, or these interrupted by concentric lines or irregular papillary elevations (see Figs. 83–95). Judging from the structure of living Brachiopods, these are associated with certain setose prolongations of the edge of the mantle, or bristles, and their appearance in the structure of the shell surface may be due to a growing around the individual bristles of the extreme edge of the shell, so that the striæ are of importance. In some (*Spirifer fimbriatus*, *lineatus*, etc.) the size is large enough to show the openings, which are quite complicated and resemble the opening of a double-barrelled gun. The modification of this feature is by increase or decrease in size of the striæ, by interruption regularly or irregularly. When interrupted regularly, it appears to be by a periodical stoppage of growth, and thickening of the shell lamellæ, forming on the surface imbricated structure, the striæ starting anew at each successive imbrication. The fact that they are surface striæ is also so accounted for, the deposit of shell filling up all the under side of the striæ. In *Sp. plicatella* (Figs. 84, 85) the whole surface is uniformly covered with these continuous radiating striæ. In *S. crispus* (Figs. 88, 92) the surface is interrupted by imbrications, and is covered by regular rows of the interrupted lines.

A comparison of series of successive species, which by their general combination of characters may be supposed to have been in direct line of genetic succession, shows a gradual diminishing in size of the striæ, and in case of the continuation and increase of imbrications there results an entire absence of the striæ—at least they fail to be discernible on specimens.

The particular size and form of these striæ seem to be a very delicate means of tracing the lines of hereditary succession, or what we may suppose to be such lines; for species which in other respects are very much alike can be easily distinguished by this character if the surface be well preserved. The modification of this character appears to be in two or three directions. In the series in which the striæ continue unbroken by imbrication there is an increase in their strength until, in Carboniferous times, the species of this series develop a spinous extension of the surface with minute tubes, extending outward from the shell: these tubes are seen in a few of the Devonian forms also. In the race with imbricated surface, where the imbrication is persistent and regular, the striate structure becomes entirely obliterated in the course of time (see figures of *S. crispus*, 90, 92). In others, where imbrication is irregular, in the Devonian and the Carboniferous eras there are species with roughened surface, irregular but granular (as *S. granuliferus* of the Hamilton), and this indicates a development of part, with obliteration of others, of these surface-reaching ends of the striæ. All the modification noticed in this respect is also extrinsic, and can be accounted for by processes of natural selection, slowly intensifying the character with repeated generation.

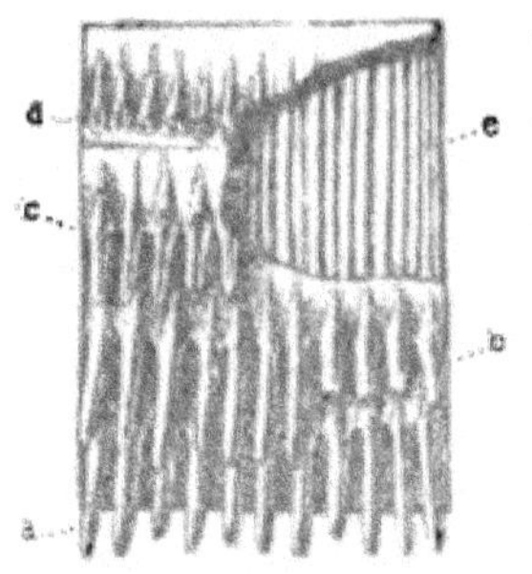

FIG. 96.—An enlargement of the surface of *Spirifer pseudolineatus* Hall. At *c* the test has been partially removed, exposing the tubular character of the spines below the surface of the shell; at *a* the spines are perfect; at *b*, broken away; at *e* they are represented as weathered, showing the tubular character of the spines; and at *d* they are broken close to the edge of the lamellæ, and show the opening of the tubes in the ends. The specimen is from the Keokuk limestone, Keokuk, Iowa. (After Hall.)

F and G. *Plication of Surface and Median Fold and Sinus.*—The next character to be noted is that of the *plication of the surface*; each species is pretty constant in the extent to which this modification reached, but in the early forms of the Niagara formation there is extreme range of variation, not only in the whole set, but in the species which are, in other respects, less variable.

Spirifer plicatella, variety *radiatus* (Fig. 83), is generally lacking in plications; but in Europe there are specimens (generally associated with the others) in which the plications are seen on the margin of the adults (see Figs. 84, 85); a few plications appearing on each side of the medial fold. In America the plicated form is called *S. Niagarensis* (Fig. 87), and is uniformly plicated to the beak. In the series *S. crispus* and *S. sulcatus* (Figs. 88–93) we find the same variability, specimens showing all grades of modification, from one or more to what might be represented by number six, on a scale of ten; and in *plicatella*, the variation is one to four. In some later forms the variation for each species is slight, rarely more than one or two tenths, using this means of designating the degree of plasticity. In the *Spirifer lævis*, found abundantly in the rocks at the foot of Fall Creek, Ithaca, N. Y., there are generally no plications, but occasionally a specimen is found with the margin for half an inch up corrugated by this modification. In this species the character is probably the remnant of a plasticity more strongly expressed in its ancestors.

The general development in number of plications is noted on some lines of species, especially those showing bifurcation of the plications during growth; and, as in the case of the median fold and sinus, this character is developed in the two directions of increase and decrease, in different races.

In one series increase, by dichotomy of the surface plications, beginning in adult forms and becoming more and more early in starting, affects first the centre of the shell, then the neighboring parts of the side until the whole surface is affected, but by slow degrees; so that, expressing the evolution in the same way as heretofore, the rate of development is approximately as follows: 1–3 in Lower Devonian, 2–4 in Upper Devonian, 3–7 in Lower Carboniferous, 6–10 in Upper Carboniferous.

This modification appears to be dependent upon, or expressive of, the rate of increase of the shell in either the radial or in the circumferential directions. If the circumference of the shell increases more rapidly than the growth in a radial direction the margin becomes too large for the shell at its normal distance from the beak, and it is necessarily puckered into folds to accommodate itself to its conditions; thus as it grows its surface becomes plicated into folds. When the growth in the radial direction keeps up with the increase in the circumferential direction the shell remains smooth, and no plications are developed. Thus the increase in the number of plications for a given shape of shell is evidently due to the acceleration or earlier starting of the differentially excessive growth in the circumferential direction. A general rule is, that the coarser plications are more prevalent among Silurian forms, while the forms with fine plications are more prevalent in the Carboniferous. Increase in the actual number of plications on a shell is, as a variation of the species, due to extension of the hinge-line and corresponding parts of the shell, and not to irregularity in the general size or number of the plications upon a given extent of surface.

H. *Structure of Shell.*—The shells of true *Spirifers* are fibrous in structure; the presence of punctation characterizes such closely allied genera as *Cyrtina*, *Syringothyris*, and *Spiriferina*. *Cyrtina* is present with the genus *Spirifer* in the Niagara, and continues about as long as that genus. They seem to be parallel genera, differing in the constant presence of this character in the genus *Cyrtina;* but this peculiarity of structure, the punctation of the shell, whatever it indicates, is more conspicuous among the later than among the early types of the family, and continues longer to be dominant. In its first or initial appearance, as a character, it seems to have been evolved intrinsically, among the distinctive differentiations of the family. The modification of structure, which distinguishes punctate from fibrous structure, appears associated with other modifications and to involve considerable internal adjustment. No evidence of the gradual appearance of the character has been discovered. In *Spiriferina* or *Cyrtina* the punctation is found wherever, among the earlier forms, the shells are well preserved. The punctate genera are sharply distinguished from the types with fibrous shell structure.

I. *Surface Spines, Granulation, etc.*—These are associated, more or less, with characters marked E. and affect the superficial layer of the shell (the periostrachum); their development is successive and accumulative, and is associated with particular series, and it appears to be a feature increasing with time, both as to size and strength of the characters. The characters develop quite in the extrinsic way in all the races in which they have been traced.

K. *Special Development of the Median Septum.*—This modification in different species of the Spirifers is extrinsic in its mode of evolution. One case has been traced with precision in a series of specimens of *Spirifer mesocostalis* Hall, which, in the Middle Devonian, shows in most specimens no trace of a median septum in the ventral valve, occasionally a variety appearing with a mere line representing the septum. At the base of the Upper Devonian (Ithaca group) frequent specimens with slight development of the septum are seen: a little higher, in the middle and upper part of the Upper Devonian, the septum is conspicuous and is strongly developed. In other species the development of septal parts appears to be varietal; the older shells, in general, expressing fuller calcification of parts which are supports or partitions between active organs of the animal. Among the Spiriferidæ there are several such lines of species, as the *Cyrtina* and the *Spiriferina;* and in fact the forms which are punctate are all more or less prone to develop calcified supports or partition plates.

Evolution of Extrinsic Specific Characters Comparatively Slow, although their Plasticity is Greater at the Initial Stage.—In all of these characters, which constitute the specific differentiæ of the species concerned, we observe a relative slowness of evolution which is quite consistent with the laws of natural selection, of gradual acceleration or retardation by hereditary means, and of the perpetuation of favorable characters by the dropping out of others; but at the same time we notice at the early stage of the life-history of the genus, or subfamily, a marked plasticity in respect of most of these characters which is in strong contrast with the fixity and persistence, without change, of the characters of higher rank which mark the family, and appear to have arisen at the same time.

Laws of Intrinsic and Extrinsic Evolution expressed in Variability and Permanency of Characters.—Among the first representatives of the family there are family characters which are repeated thereafter in numerous individuals for several periods of geologic time without noticeable change, and they did not appear before. There are also characters appearing on the first species which vary and show slight change all the way along thereafter, and are themselves less different from the characters of previous forms: relatively, one set of characters appears and thereafter a long line of successors follow with the same characters *not modified;* the other set are *plastic* at the first appearance, and only by degrees in the course of geologic time do they become fixed and permanent. It is this difference in the law of evolution of the characters, as traced in historical series, that has led to the distinction of the two modes of evolution, the one *intrinsic*, and the other *extrinsic*, as defined on a previous page. *Intrinsic evolution* is conceived of as normal expansion and differentiation of the organism itself from within, and is the expression, in some way, of an intrinsic tendency of the particular race of organisms. The other, *extrinsic evolution*, expresses the limitation and selection exerted upon the organism from without. Variability is thus the morphological expression of intrinsic evolution, and permanency or the transmission of characters without modification is the morphological expression of the effect of extrinsic forces.

Hall's Analysis of the Genus Spirifer and Classification of its Species.—The history of the evolution of the genus *Spirifer* may be seen from a somewhat different point of view by an examination of the classification of American Spirifers by James Hall, than whom we have no more critical observer of specific differences in fossils.* Professor Hall recognizes about two hundred species of Spirifers in the American Palæozoic rocks, none of which he considers worthy to be regarded as even of subgeneric rank in relation to the typical Spirifer stock. But there are certain groups of species naturally associated together in successive lines which may be regarded as genetically separate races, each line being characterized by an association of common characters and differing from the others by the relative development or elaboration of one or other of its characters.

Six such principal groups are recognized, called by Hall, I. Radiati; II. Lamellosi; Fimbriati; IV. Aperturati; V. Osteolati; VI. Glabrati.

Range of Species of Spirifer in American Formations.—In the following table the lists are arranged in such a way as to show for each particular race in each group the number of

* "An Introduction to the Study of the Genera of the Paleozoic Brachiopoda,"—Paly. N. Y., vol. VIII, part II, fascicle I, pp. 12-40, 1893.

species recorded from each successive formation in the North American rocks.

		Silurian			Devonian						Carboniferous						
		C	N	L	O	U	H	G	P	C	Ch	B	K	W	St	C	Cm
I. Radiati	Panciplicati......	—	—														
	Multiplicati......		—	—	—	—	—	—	—	—							
	Dupliciplicati....		—														
II. Lamellosi	Septati...........		—	—	—	—	—	—	—	—							
	Aseptati.					—	—	—	—	—	—						
	Submucronatus				—												
III. Fimbriati	Unici-spinei Crispus.		—	—	—												
	Unici-spinei Lævis...		—	—	—	—	—	—									
	Duplicispinei				—	—	—	—	—	—	—	—	—	—	—	—	—
IV. Aperturati	Disjunctus........				—	—	—	—	—	—							
	Hungerfordi.. ..		—	—	—	—	—	—	—	—							
	Striatus										—	—	—	—	—	—	—
	Texanus										—	—	—	—	—	—	—
	Imbrex										—	—	—	—			
	Suborbicularis...										—	—	—	—			
	Orestes..........				—	—	—	—	—	—	—	—	—	—	—	—	
	Divaricatus.. ...					—											
V. Ostiolati	Osteolati.........					—	—	—	—	—	—	—	—				
	Acuminatus....						—										
VI. Glabrati	Aseptati...					—	- -	- -	- -	- -	- -	- -	—	—	—	—	—

Each Type of Spirifer shows a Continuous Series of Species.—After making allowance for the gaps to be expected in our limited knowledge of fossil faunas, it will be seen, by a glance at the table, that each of these special types of Spirifer had a more or less continuous series of species, persisting for two or three or a dozen formations, represented, not by the same form, but by mutations of the earlier form which differ sufficiently from it to call for different specific names in the successive formations. Some of the species are reported for two contiguous or for several consecutive formations, but in these lists the life-period of most of the species is for the length of a single formation.

Each of the Chief Types Represented at the Initial Period of the Genus.—It will be noticed also that each of the four principal groups had representatives in the Niagara epoch, and only a single species of Spirifer is reported from a Lower horizon. The other two groups, Ostiolati and Glabrati, did not appear till the Devonian, but these both appeared together in the Upper Helderberg.

Three Epochs of Special Expansion with Slow and Gradual Change During the Rest of the History of the Genus.—Of the 20

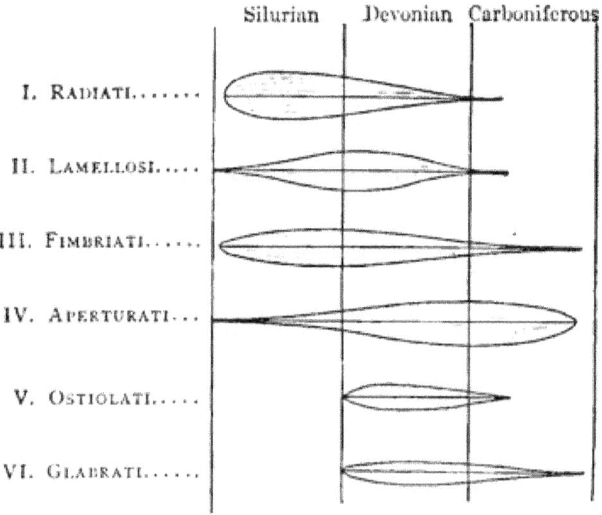

Fig. 97.—Table representing the expansion of the Spirifers in subgeneric groups, according to the classification noted by Hall, and elaborated on page 313.

races into which the known American species are subdivided, 7 are reported from the Niagara; 1 begins in the Lower Helderberg, 3, Oriskany, 4, Upper Helderberg, 1, Hamilton, and 4 at the base of the Carboniferous, i.e., over a third of the known races began at the first fauna in which the genus appears in North America (except the one species in the Clinton); 7 more began near the base of the Devonian, and 4 began at the opening of the Carboniferous. This special rapidity of appearance of new types at the three periods marking the beginnings of three geological systems points at the same time to the fact that these systems, which have been recognized as well-established natural divisions in the geological scale of formations throughout the northern hemisphere, were also distinguished at their beginnings by marked change in the life of the world. Not only do new types of genera and families appear, but even the specific types of a continuous race of species express the changes incident to the opening of a new period.

Whatever be the explanation, these facts make it evident that the divergence of a generic type into different subgeneric expressions was not by slow and accumulative process, but by relatively rapid expansions, followed by the continuance of the types with gradual but restricted modification until the race died out. The divergence of these types from each other was very early in the history of each race, and in the present case there was evidently a secondary divergence at the beginning of the Devonian, and a slight tertiary divergence, in the Aperturati group, at the beginning of the Carboniferous.

Characteristics of the Life-history of Atrypa reticularis.—*Atrypa reticularis* may be taken as an example of a species which exhibits scarcely any trace of what has been called extrinsic evolution, but has lived a long time, been very fertile, has been distributed all around the world, and has shown its adaptability to a great variety of environmental conditions, without suffering any appreciable morphological change. It began with the initiation of the genus, and lived throughout the life-period of the genus, which is almost equal to that of the family or suborder to which it belongs. The species has received a great many names, and been referred to many genera; but the more careful the study applied to it, the more clearly does it appear that under all proper discrimination of specific

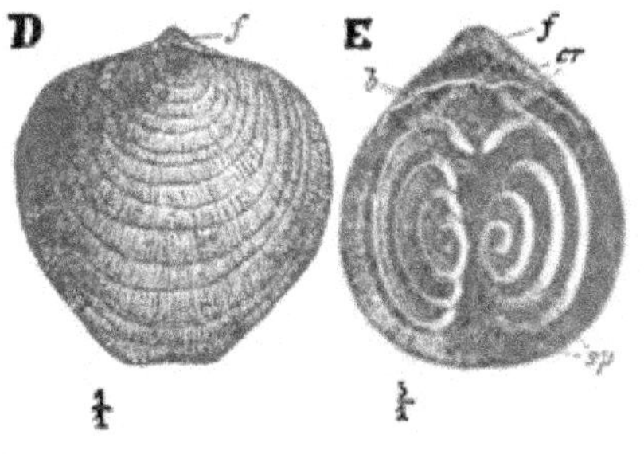

FIG. 98.—*A*, *B*, *Atrypa reticularis* Linn. *f* = foramen; *cr* = crura; *j* = jugum; *sp* = spiral coils of the brachidium. *A*, adult, natural size; *B*, young specimen, magnified one fifth. (From Steinmann and Döderlein.)

identity there is under consideration but one species, though it is constantly variable. The species exhibits constant plasticity of several of its characters, but never reaches that fixation into separate forms which has been interpreted as the result of the survival of the fittest by natural selection.

Considerable and Continuous Plasticity of the Species.—The width and form of the shell, the number of the striæ, and the concentric laminæ constitute some of the more conspicuous differentiæ of the various forms of the species; but, as Davidson says, "All these modifications can be traced in specimens from any locality."

In Murchison's "Siluria," (second edition,) is a remark regarding the species, so pertinent that it is worthy of quotation as it stands: "Among the Mollusca nearly all the species of *Atrypa*, *Orthis*, and *Spirifer* differ from those of the Silurian age" (speaking here of Devonian Brachiopods). "One shell, however, the *Atrypa reticularis*, must be mentioned as an exception to the prevalent rule of each great group being distinguished by peculiar forms; for this hardy species, with which the reader became so familiar in the Silurian rocks, lived on to the Devonian era, and is as common in the limestones and shales of Devonshire as in the older rocks. It even ranges to the farthest known geographical limits of the Devonian rocks of Armenia, the Caucasus, and China on the East, and to the Devonian deposits of America on the West."

Nature and Extent of the Variation.—The variations of this species interested the acute naturalist Edward Forbes, and he caused 117 specimens to be critically examined and the ribs of each to be counted, and also the number of concentric foliaceous expansions or fringes upon the surface. The number of ribs, counting those on old and young specimens, varied from ten to sixty, but there was found less divergence in respect to the development and frequency of the concentric fringes. Hisinger and Lindström, Davidson, Bronn, and McCoy, among the earlier paleontologists, agreed in considering the forms with fewer and larger plications, called *A. aspera* Schl. to be varieties of *A. reticularis*, but did not regard them as distinct species. Lindström observed "that the

Linnean form "*varies like all those species which possess an extended horizontal and vertical distribution.*" Barrande recognized two varieties of the species var. *Verneuiliana* and var. *Murchisoniana*. McCoy in "British Silurian Fossils," says "It varies, firstly, *in the convexity of the valves*, both as to degree, distance from the beak (at which it is greatest), and equality—some small varieties, and the young at all times, having the valves almost equally and evenly convex; secondly *in form*, some, and particularly the young and small varieties, being nearly orbicular; others being elongate, and nearly triangular from the width of the hinge-line and narrowness of the front; thirdly, in the number, thickness, and closeness of the ridges and the scales which cross them, both of which are often smaller and closer than in the typical variety;" and Lindström, speaking of the coarse-ribbed specimens in Gothland, says, "these variations are connected with the finely-ribbed varieties by every possible gradation and intermediate shape."

These opinions were written by naturalists looking upon species from the old point of view of immutability, but it will be noticed that the testimony is unmistakable as to *the great range of incessant variation* exhibited by the species.

Hall's Comment on the Variability of the Species.—James Hall, the veteran American paleontologist, in one of his latest and ripest publications,* speaking of the genus Atrypa, says: "Following closely the foregoing diagnosis will result in eliminating from this group the great majority of species passing under the name of *Atrypa*, and in retaining only those which conform to the well-known *A. reticularis*, primarily in the structure of the brachidium, and secondarily in the expression of the exterior. Such forms are comparatively few in number, and most authors have been disposed to regard them as representing unessential variations from the specific type of *A. reticularis*. There is, however, a multitude of designations which have been applied to contemporaneous variations or consecutive mutations of this specific type, some of them unnecessary, but many very useful both to the geologist and the systematist" (pp. 166–7).

* "Introduction to the Study of the Genera of the Paleozoic Brachiopoda" (1893).

The species, like the genus, ranges from near the base of the Upper Silurian to the Waverly or beginning of the Carboniferous age. "Almost coincident in time with the appearance of *Atrypa reticularis*, in its typical aspect, we find," writes Hall, "in the shales of the Niagara group shells

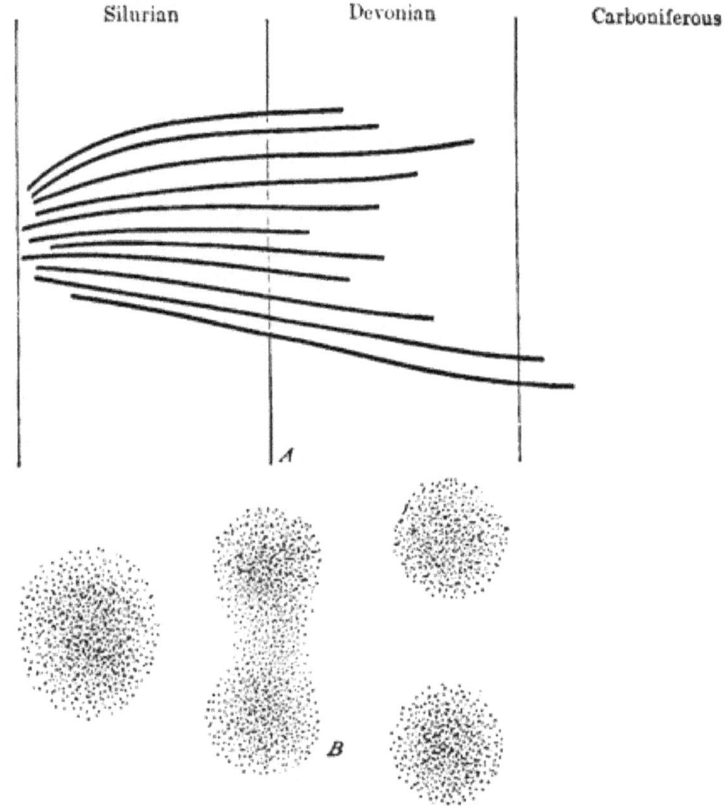

FIG. 91.—*A* is a graphic expression of the nature of the differentiation supposed to have taken place in the course of the history of the race, individuals of which are called *Atrypa reticularis*. The lines and their relative position and length represent the divergence in varietal modification and the continuance in generational repetition of like characters for the race. *B* represents the groupings of the individuals at three successive stages of its history; viz., at the beginning of the Silurian, near the beginning of the Devonian, and in the later part of the Devonian era. The rather distinct specific grouping seen in the latter case is observed to result from the dropping out of the intermediate forms as well as by the increasing dominance of the divergent forms.

which are persistently small, with few and coarse plications, more or less distinct median fold and sinus, and strong concentric lamellæ. These shells have been designated as *Atrypa rugosa* and *A. nodostriata* Hall" (p. 170); and these two types continued on to the close of the Devonian, living often together, but having an independent existence, and not reaching a completely specific differentiation till the close.

They are more properly claimed as varieties than as distinct species, this being chiefly due to the maintaining of variability and the failure of disappearance of intermediate forms linking the extreme and typical forms, which thus at the beginning of the life-period of the genus quite fully expressed their characteristics.

In the Closing Part of the Life-period of the Race the Extremes of Acceleration and Retardation Expressed.—In the last few pages the characters of Atrypa have been described, and it was pointed out that a certain part of the characters, those of the species *Atrypa reticularis* and closely allied species, have exhibited great persistence of variability. We observed that this species, or race as we may call it, began at the opening of the Silurian or possibly in the latter part of the Ordovician, was conspicuous in the Silurian and the Devonian, but appears to have become extinct at the close of the Devonian. At the close of the life-period of the genus the variability in respect of rate and extent of bifurcation of the surface plications presents a tendency in two predominating directions.

On the one hand, the bifurcation is rapid and extreme, and the whole surface of the adult appears covered with numerous fine plications: this would indicate rapid and continuous bifurcation during growth, or the character of bifurcating of the plications shows, in comparison with ancestors, *acceleration of development*.

On the other hand, there is a well-marked variety which becomes sharply distinct from the others in the Neodevonian and goes under another specific name, *Atrypa spinosa*, which shows the opposite tendency; the bifurcating has become almost *nil*. The adult shows no more plications than does the early stage of growth at the distance of one-fourth inch from the beak: this is an expression of *retardation* of this particular element in the growth of the shell.

Summary.—To define precisely those characters which are considered in the above analysis, the following summary may be given: In the geological series of forms described under the name *Atrypa reticularis* and its varieties, there are observed certain plications of the surface, of indefinite number, and increasing by bifurcation. The variability or plasticity is observed in respect to the rate and extent of the bifurcation, which in the early and middle part of the life-history is indefinite—i.e., there is in the species no fixation of the law of this bifurcation; but gradually there is acquired a tendency to permanency in the two directions of (*a*) extreme acceleration and of (*b*) extreme retardation of the rate of the bifurcation in the development of the individual, and the species which may be said to originate by this process, and to be characterized by the different extent of bifurcation attained, are thus gradually perfected (see Fig. 99). In a set of Iowa specimens examined by the author, a well-defined differentiation was noted; the two species are so nearly distinct that it is found, by arranging the forms in order of their resemblances and differences, that there are two well-defined groups, and the intermediate forms, although they almost touch, are so separate that careful study decides for every individual case on which side of the imaginary line it belongs. Thus *Atrypa reticularis* is an example of very slow evolution. The family characters appeared well defined with the earliest representatives of the suborder Helicopegmata; the generic differences were well elaborated at the first stage of the Eosilurian. This species was among the earliest representatives of the genus, and lived nearly as long as we have any trace of the genus. But the great variability or plasticity of certain characters is a peculiar characteristic of the early forms up to mid life of the genus, and might be called a specific character, and the fixation of this variability is very slowly assumed.

Conclusions Suggested by the Study of Atrypa Reticularis.—Natural selection is supposed to result in the fixing of variable characters, and the failure of natural selection to select would naturally result in a continuation of the variability. It is rational to conclude, therefore, that a species which continues to live without fixing its variable characters is particularly well adapted to live under a wide range of modified conditions. The wide geographical distribution of the species here under consideration confirms this conclusion. That a species does die out in course of time is illustrated by thousands of species which are represented abundantly in the rocks of some particular period, but thereafter are never seen again. Variability in ontogenesis is a necessity of living at all. The organ which in its minutest characters has ceased to change, has ceased to live; and if we extend this generalization to the law of phylogeny, we might expect to find, not a uniform, continuous evolution along all lines, but pulsations, so to speak, in the activity of phylogenetic evolution of organisms along each line. Taken as a whole, doubtless there is a gradual readjusting of parts; but each part is temporary, and is displaced by another. So long as great flexibility of any particular character, or set of characters, prevails, there will be rapid appearance of new forms; but after their initial appearance, the repeating of the characters by natural generation will tend to their fixation, and with the limitation of adjustability to environment there will result death upon the slightest maladjustment; thus, as the variability of the species becomes more and more narrow, the conditions under which it can thrive become more and more restricted, and the final result must be extinction.

Whenever the action of heredity becomes restricted—that is, when sterility limits the range of variation within which generation is possible—this condition of fertility must work toward the final extinction of the race. Thus, according to this theory, if a species be found breeding perfectly true, we can conceive it to have reached the end of its life-period, and likely soon to become extinct. The theory in this respect can be tested by the facts; and although statistics as to the actual fact on this particular point are wanting, it has been frequently noticed in fossil species which have been carefully observed by the author that it is a conspicuous law, that in respect to those characters which serve as distinctive marks of species, there is greater general variability in the early stages of the life of the genus than in the later stages. The following fact is an expression of the same law, viz.: the species occurring at the early stage of a genus are generally more difficult to separate, and there are more intermediate links among earlier than among later species of a genus. After examining and trying a number of hypotheses to account for these facts, the following definition seems to be fairly satisfactory: *The species in its specific characters shows a greater degree of variability or plasticity in the earlier than in the later stages of its history.*

Atrypa was an illustration of the remarkable continuance of the stage of plasticity, but we observe that the particular limitation of range of the plasticity became thereafter a specific characteristic of the race. The greatest and the least number of plications attained by any representative of the genus are probably met with within what has been called, in a broad sense, the species *Atrypa reticularis.* Another law of specific modification is seen in the gradual *narrowing of the limits of the plasticity*—one series concentrating about the forms with few plications, the other series concentrating about the forms with many;—the one expressing the law of *retardation* of growth for this character, the other the law of *acceleration* for the same character.

The Initiation of the Species of Ptychopteria.—Ptychopteria* is a remarkable instance of variability among the initial representatives of a genus. The case is as follows: A genus of Lamellibranchs, having some well-defined generic characters, is first seen in the upper sandstones of the Neodevonian in Western New York and Pennsylvania. A few years ago the genus Ptychopteria was defined and figured by the New York State Geologist,† and nearly a score of species were described from different localities and, possibly, different geological horizons. About the time of the publication of the species a block of sandstone, about a cubic foot in size, was found in Chautauqua County, fallen from a ledge of the Panama sandstone, containing many hundreds of specimens of shells of this genus. These were carefully collected, sorted, and classified according to the characters by which the several species defined by Hall had been distinguished. An analysis of the species already described showed the following

* The facts of the case were briefly alluded to in a paper "On Devonian Lamellibranchiates and Species making,"—referring to species which paleontologists make, and not to the origin of species. *Am. Jour. Sci.*, vol. XXXII, p. 196.

† Paleontology, New York, vol. v, "Lamellibranchiata."

to be the distinguishing differences: the chief of them were certain surface markings, the prominence and the angle formed by the shell along a line called the umbonal ridge, the angle formed by this umbonal ridge and the line of the cardinal margin, and the contour shape of the shells. A careful study of the characters exhibited by all the known species was made, and instead of finding the new specimens to represent a new species, they practically represented the whole genus. Every specific character which was described for the known species was expressed in a series of 32 specimens. One feature, of great importance in producing the shape of the shell, is the angle formed by the umbonal ridge and the hinge-line. This character varied regularly in the series from less than 30° to over 60°, and these were also the limits of difference in the described species. The geological horizon in which this set of specimens occurred was probably the lowest in which the genus has been seen. The specimens were slightly smaller in size than most of the species described from other regions, but the uniformity in size and their occurrence altogether in a single block of stone, well preserved as originally imbedded, are proofs that the specimens were very closely related genetically, and were not very far separated from a common ancestor. The variations may be assumed to have been pure variations, in the strict sense of the word, that is, of common origin and possessing common fertility.

This series seems to admit of only one explanation for the origin of the several species of the genus Ptychopteria—i.e., the fixation, by isolation or subjection to various conditions of environment, of the variable characters of the initial stage of the genus as it appeared in the Panama sandstone.

The Law of Progressive Evolution of Mammals as Formulated by Osborne.—The force of the evidence of Brachiopods may be weakened in the minds of some by the consideration of the very low rank of these organisms in the Animal Kingdom. But the same methods of minute analysis lead to like conclusions in the study of mammals, the highest type of organic structure. Professor H. F. Osborne, at the conclusion of his recent address, as Vice-president of the American Association of Science, on "The Rise of the Mammalia in North America,"* in which a minute study is made of the law of evolution as expressed in the teeth of mammals, says: "The evolution of a family like the Titanotheres presents an uninterrupted march in one direction. While apparently prosperous and attaining a great size, it was really passing into a great corral of inadaptation to the grasses which were introduced in the Middle Miocene. So with other families and lesser lines, extinction came in at the end of a term of development and high specialization. . . . A certain trend of development is taken leading to an adaptive or inadaptive final issue; but extinction or survival of the fittest seems to exert little influence *en route*. The changes *en route* lead us to believe either in predestination—a kind of internal perfecting tendency, or in kinetogenesis. For the trend of evolution is not the happy resultant of many trials, but is heralded in structures of the same form all the world over and in age after age, by similar minute changes advancing irresistibly from inutility to utility. It is an absolutely definite and lawful progression. The infinite number of contemporary developing, degenerating, and stationary characters preclude the possibility of fortuity. There is some law introducing and regulating each of these variations, as in the variations of individual growth." †

---

* *Am. Jour. Sci.*, vol. XLVI. pp. 379–392 and 448–466. † pp. 465, 466.

CHAPTER XVIII.

THE RATE OF MORPHOLOGICAL DIFFERENTIATION IN A GENETIC SERIES; ILLUSTRATED BY A STUDY OF THE CEPHALOPODS.

The Evidence Furnished by the Cephalopods.—Having used Brachiopods for what they are worth towards illustrating the laws of evolution, another group of organisms may be examined in the same way, to ascertain what they testify regarding the same points of history.

Cephalopoda present some general peculiarities contrasting them with the Brachiopoda. The Cephalopoda are constructed on a plan which is shared with two or three other large groups of organisms. The class Cephalopoda is, with Gastropoda and Lamellibranchiata, and, according to some authors, Pteropoda, only one of the classes of the branch Mollusca. We are able, therefore, to distinguish its class characters from those of closely allied classes. This we could not do satisfactorily with the Brachiopoda, which stands out sharply distinguished from all other classes of organisms from the earliest geological time. We find the first traces of the Cephalopoda above the first, or Cambrian, period, i.e., we have a well-defined fauna in which no Cephalopoda existed, so far as our records testify.

Lankester's Schematic Mollusk.—In attempting to introduce a beginner to a knowledge of the Cephalopod mollusk the method of Lankester, so admirably expressed in his article "Mollusca" in the Encyclopædia Britannica, and afterwards published with others under the title "Zoological Articles, etc," presents some excellent features.

Professor Lankester constructs a schematic mollusk as represented in Fig. 100.

This schematic mollusk possesses "in an unexaggerated form the various structural arrangements which are more or less specialized, exaggerated, or even suppressed in particular members of the group." It represents, as near as our knowledge will enable us to do, the actual mollusk ancestor from which the various living forms have sprung, and therefore does not represent any actually living species of mollusk. However, the accuracy of the schematic type is evident when

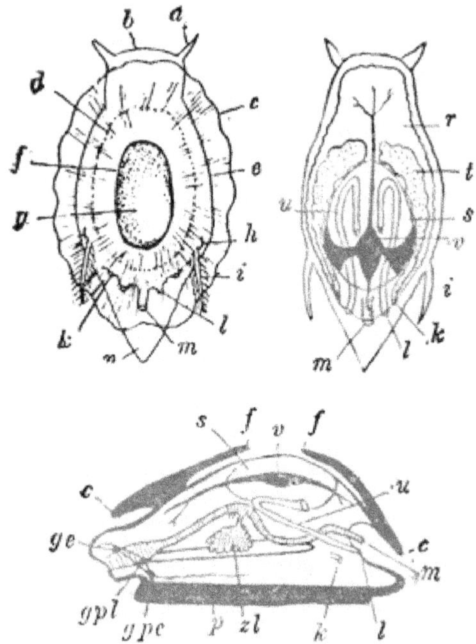

FIG. 100.—Diagrams showing the arrangement of the organs in an ideal Mollusk. (After Lankester.) *a*, tentacle; *b*, head; *c*, margin of mantle; *d*, margin of shell; *e*, edge of body; *f*, edge of shell depression; *g*, shell; *gc*, cerebral ganglion; *gpe*, pedal ganglion; *gpl*, plural ganglion; *h*, osphradium; *i*, ctenidium; *k*, reproductive pore; *l*, nephridial pore; *m*, anus; *n* and *p*, foot; *r*, cœlom; *s*, pericardium; *t*, testis; *u*, nephridium; *v*, ventricle of heart; *zl*, liver.

we attempt to compare with it a living specimen of some one species of mollusk.

It is an attempt to give form and definite relation to the terms of a systematic definition of the characters of the branch Mollusca. In his diagrams of a series of mollusks the same method is used to give formative expression to the characteristics of the several classes.

The value of this representation for our purpose is to show the extent of structural elaboration which the evolution of organisms had actually reached at the time when we first meet with a representative of the class Cephalopoda.

Supposed Characteristics of the Primitive Mollusk.—In this earliest mollusk bilateral symmetry was fully developed. The nervous system was expressed in bilateral pairs of ganglia and nerves. The organs of sense were in pairs: two eyes and two otocysts were present. The body form was normally symmetrical, its spiral coiling or one-sided development coming as a specialization of growth. The cephalic is sharply distinguished from the visceral part of the body. The shell is associated with the visceral part, and is not auxiliary to the functions of motion, but is protective in nature. The head—anterior part—is distinctly connected with motor functions; and organs of the motor and sense functions are separate and widely differentiated. Motion is elaborated into distinct organs for offence and for prehension.

The alimentary function is dominated by a single central canal, with an anterior mouth, about which are the accessory organs of excretion (nephridia or rudimentary kidneys), and there is a circulatory system, with a heart and a pair of auricles and one ventricle. Locomotion is a conspicuous function, and the presence of an enlargement of the mantle as a foot-organ is one of the most characteristic features of the mollusk.

The differentiation of this foot-organ is also one of the most fundamental of the characters distinguishing the classes of Mollusca, and the adaptation of the part to special modes of locomotion was developed at a very early stage, as indicated by the presence of distinct Gastropoda, Pteropoda, Cephalopoda, and Lamellibranchiata at as early as Ordovician time.

Differentiation of the Foot-organ in Mollusks.—This differentiation is represented in Lankester's diagram of a series of mollusks to show the form of the foot and its regions, and the relation of the visceral hump to the antero-posterior and dorso-ventral axes (Fig. 101). In these figures are seen the simple continuous flat foot of the Chiton (1), or isopleural Gastropod, which retains the bilateral structure of the primitive mollusk. In the Gastropoda anisopleura, or typical Gastropods, the specialization does not greatly affect the foot, which is still symmetrical and occupies similar relations to the rest of the body; but the twisting of the body coincident with the spiral shell which is developed as a cover, affects the proportionate size and vigor of the organs on the two sides, so that the organs are in fact not strictly symmetrical in the

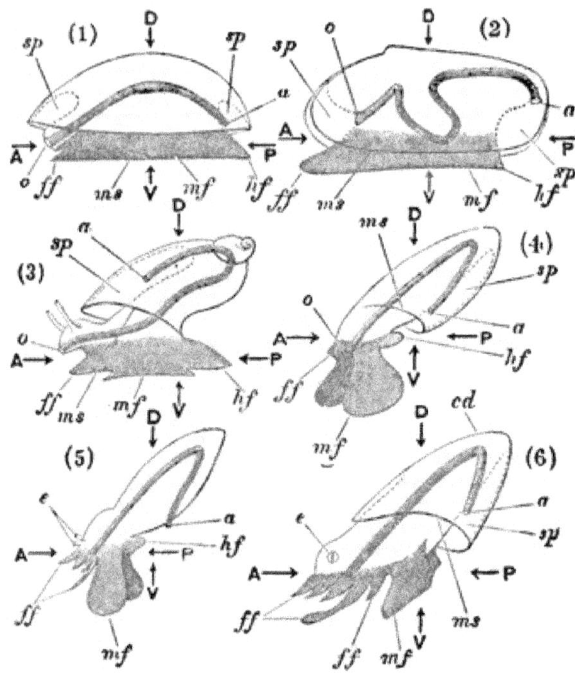

FIG. 101.—Diagrams of a series of Mollusks to show the form of the foot and its regions, and the relation of the visceral hump to the antero-posterior and dorso-ventral axes. (1) A Chiton. (2) A Lamellibranch. (3) An Anisopleurous Gastropod. (4) Thecosomatous Pteropod. (5) A Gymnosomatous Pteropod. (6) A Siphonopod (Cuttle). *A*, *P*, antero-posterior horizontal axis; *D*, *V*, dorso-ventral vertical axis at right angles to *A*, *P*; *o*, mouth; *a*, anus; *ms*, edge of the mantle-skirt or flap; *sp*, sub-pallial chamber or space; *ff*, fore-foot; *mf*, mid-foot; *hf*, hind-foot; *e*, cephalic eyes; *cd*, centro-dorsal point (in 6 only). (After Lankester.)

adult (3). The Pteropods have the foot modified for free swimming into two lateral flappers or wings (Fig. 101, (4)), and the Cephalopods proper have the right and left lobes, corresponding to the wing-expansion of the Pteropods, folded under to form a funnel-like tube or siphon, which accomplishes locomotion by forcing water outward and forward, the anterior part of the foot being differentiated into special grasping organs auxiliary to the mouth functions.

The specialization of the tubular mode of locomotion and the differentiation of the foot into a funnel and tentacles are characteristics of this highest type of mollusk; and its relation to the Pteropod wings is seen in the fact that in the Dibranchiate order the lateral lobes are fused together to form a closed tube—the siphon, while in the Tetrabranchiate order they are only brought close together, and not fused into a continuous tube. There is also the differentiation of distinct swimming flappers in some of the Dibranchiates, in addition to the siphon, which is specialized as an organ for distribution of ink into the water, and by darkening the water compensates for its slow rate of escape by locomotion from any cause of danger.

The Structure of the Cephalopods.—Although the purpose of this volume does not include the detailed description of organisms, a better understanding of the remarks that follow may be reached by a brief review of the essential structural elements of the Cephalopods. For this purpose the following translation of extracts from Zittel's description will be useful, and for further details the reader is referred to his excellent Handbook of Paleontology.*

FIRST ORDER, TETRABRANCHIATA.—Cephalopods with shell; furnished with four branching gills, or branchiæ, funnel formed by union of two lobes of foot, but not permanently united; no ink sac or pouch. In the place of arms, numerous tentacles, slender, elongated and without suckers or hooks; shell chambered.

The Animal.—All that we know of the organization of the Tetrabranchiata is based upon the genus Nautilus, the only one of the order now living, the shell of which is seen in most museums; but the animal is very rare, and has been seen alive in only a few instances. The animal occupies the last chamber of the shell, with the ventral side turned outward (the coiling of the shell thus being toward the dorsal side); the body is short and thick; the head separated from the trunk by a slight constriction. In place of arms, about ninety contractile filiform tentacles inserted in muscular sheaths surround the mouth; they are grouped in several bundles, and in an order a little different in the male and female. The tentacles situated on the dorsal side are soldered together to form a thick muscular lobe which can close the opening of the shell when the animal has withdrawn into the last chamber. The funnel is a

* "Handbuch der Palæontologie, I. Abtheilung: Palæozoologie," von Karl A. Zittel, vol. II, 1881–1885, pp. 332, etc.

very thick, enrolled muscular fold, of which the external borders are interlaced one with the other; the tentacles and funnel, as the innervation shows, correspond to the foot of the Gastropod. At the base of the lateral ocular tentacles is found on each side a large eye, with short peduncle; in the midst of the crown of tentacles is situated the buccal (mouth) cavity surrounded by thick walls, with a fleshy tongue, the root of which is composed of many series of plates and hooks. The jaws, of extraordinary strength, recall in form the beak of a parrot. The large branchiæ, or gills, are found in two pairs at the base of the funnel: they penetrate freely into the respiratory cavity; between these open the anal orifice and a little further back, the organs of generation.

The respiratory cavity and head are covered by a thin lobe or mantle, especially developed on the ventral side, and secreting the shell of the outer chamber.

The animal is attached to the shell by a powerful muscle of oval form, placed below the eyes, and inserted on the internal wall of the chamber

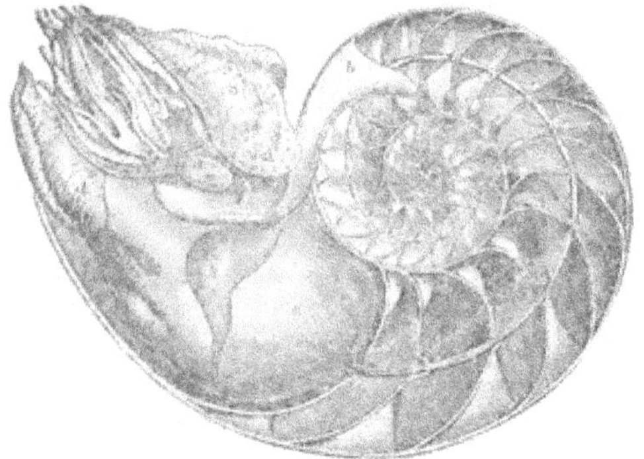

FIG. 102.—*Nautilus pompilius.* (After Owen.) *a* = mantle; *b* = dorsal aspect of mantle; *c* = hood; *d* = funnel; *e* = nidamental gland; *h* = shell muscle; *o* = eye.

of habitation, where it leaves slight impressions. From the rounded posterior extremity of the animal proceeds a membranous hollow cord, furnished with blood-vessels, the Siphon, which passes by a rounded opening through the last partition-wall into the chambered part of the shell, and continues thus in an uninterrupted manner to the initial chamber.

The Shell.—By the internal chambering or partitioning of the shell (Fig. 102), which is characteristic of them, the shells of the Tetrabranchiata are distinguished from all the shells of Mollusca hitherto considered. The last, distinguished by its greater capacity, serves as the chamber of habitation for the animal; all the rest of the shell is divided into chambers by transverse partitions, called Septa, which succeed each other at regular intervals. The chambers are filled with air (gas), and united together by the Siphon. The exterior form of the shell presents extraordinary variation; in general, it may be considered as a straight conical tube, augmenting little by little in thickness, which continues to incurve, sometimes in a straight line, and often in a curved line. There are, consequently,

FIG. 103.—*Orthoceras timidum* Barr.

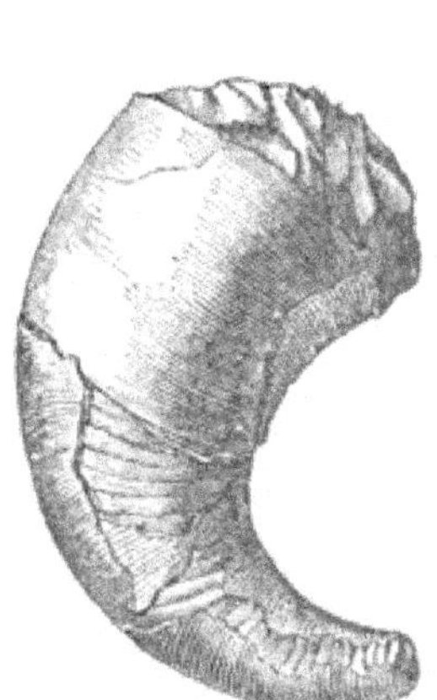

FIG. 104.—*Cyrtoceras Murchisoni* Barr.

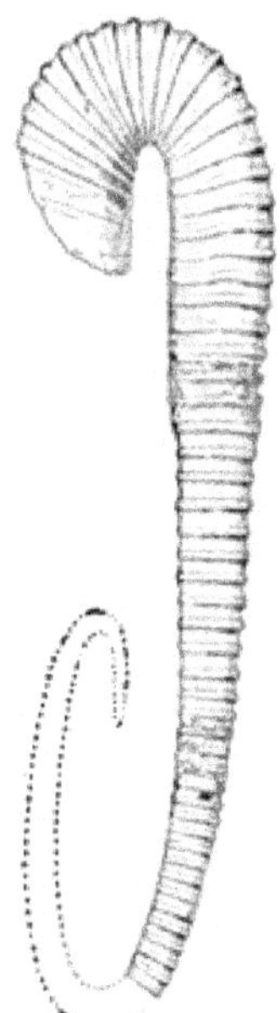

FIG. 105.—*Hamites rotundus* Sow.

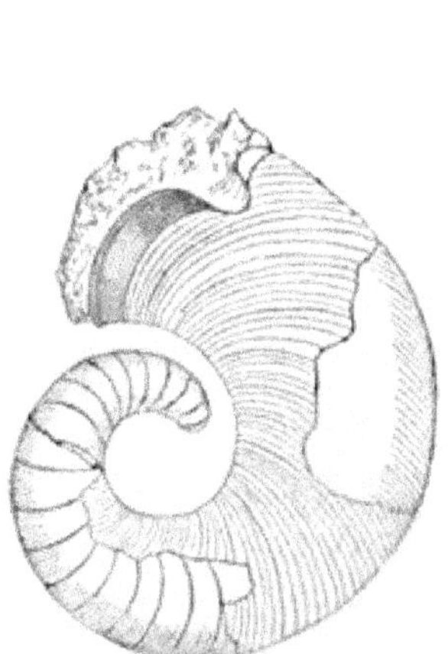

FIG. 106.—*Gyroceras alatum* Barr.

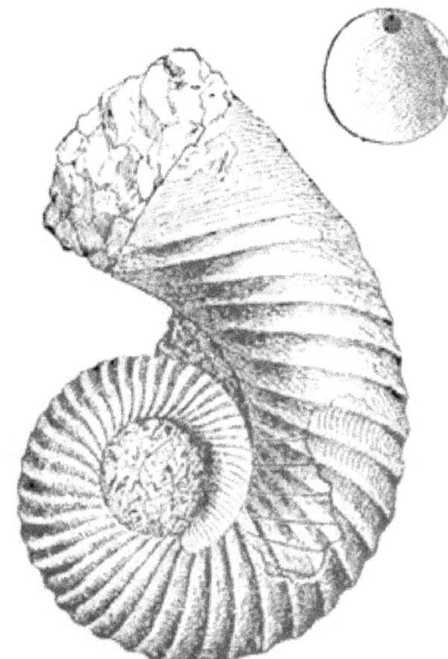

FIG. 107.—*Trochoceras optatum* Barr.

shells of straight, staff-like form (Orthoceras, Fig. 103; Baculites), slightly curved (Cyrtoceras, Fig. 104), hooked (Hamites, Fig. 105), spirally enrolled (Gyroceras, Fig. 106), or coiled in manner of a snail shell (Trochoceras, Fig. 107). If the turns of the spirally enrolled tube are in the same plane, and touch each other, the shell is disk-formed (Clymenia, Trocholites, Nautilus, Ammonites); if they turn in form of a screw, the shell is helicoidal (Cochloceras, Turrilites, Fig. 108). It is not rare that the last coil is elongated in straight line, and detached from the rest of the anterior by enrolled spiral (Lituites); sometimes it is curved still more slowly in the form of a hook (Ancyloceras, Macroscaphites). In many of the shells spirally coiled in the same plane the last turn encloses the previous turns either entirely or in part. If this envelopment goes so far that the preceding turns are entirely concealed, and that only the last one remains visible, the shell is called involute. If the older coils are still visible in the centre, there is then an umbilicus, and according to the degree of involution the shell is said to have narrow or broad umbilicus. In the evolute, or open spiral, the turns do not touch each other so that one can see between them. By their ornamentation also the shells of Tetrabranchiata show considerable diversity: on the one hand there are forms of which the surface is covered only by fine striæ of growth, and on the other are forms presenting a rich ornamentation of the surface. The surface markings are smooth lines, punctate, granulate, and more or less prominent lines, foliaceous excrescences, rings, protuberances, simple or bifurcate ribs, tubercles, or spires, isolated or arranged in series. The ornaments which follow the general direction of the longitudinal axis of the whorl go under the name of longitudinal or spiral sculpture, while those which are arranged obliquely, or at right angles to these, are called transverse or radiating ornamentations.

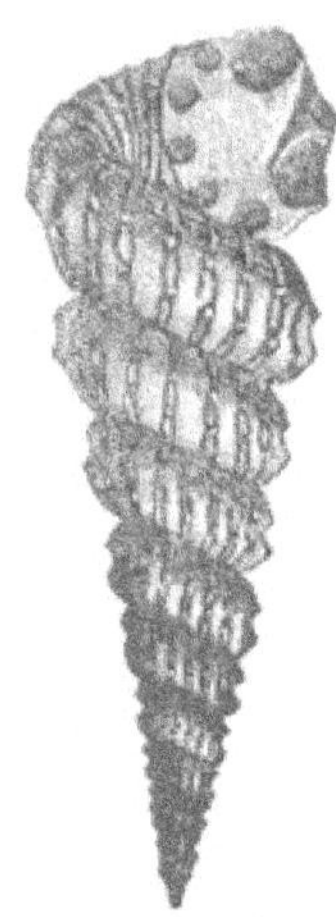

FIG. 108.—*Turrilites catenatus* d'Orb.

The position of the animal of the Nautilus (see Fig. 102) offers the only good evidence by which to orient the shell of the Tetrabranchiates As it turns the ventral side of the animal outwards, R. Owen has designated the external or arched part of the shell, the ventral side, and the opposite internal part, the dorsal side. All the ancient authors, who occupied themselves exclusively with the shells, called, in the spirally enrolled forms, the external side of the shell back, and the internal side the ventral side of the shell. According to Barrande, the external arched part of the spirally twisted fossil forms does not always correspond to the ventral side of the animal; the convex ventral side of the shell is distinguished, particularly in the Nautilus, by a depression of the buccal border. It is admitted, therefore, that always where such a sinus exists in the buccal border it indicates the position of the siphon, and consequently the ventral side of the animal. According to Barrande, the sinus is found frequently in fossil Nautilids, sometimes upon the external arched side, sometimes upon the concave inner side. There are thus, evidently, exogastric and endogastric shells. In the majority of the fossil shells of Cephalopoda, and particularly in the Ammonites, data are wanting for deducing the organization of the animal; in that case the terms internal side and external side are used, which prejudge nothing. A vertical line running from the external to the internal side gives the height; a second line, perpendicular to the preceding, gives the breadth, or thickness of the turn.

In the involute shells, the growth, as was first recognized by Reinecke, and later verified by Leop. von Buch, takes place according to a definite law. Moseley and Naumann show that the law of growth corresponds to a logarithmic spiral; consequently, the height and breadth of all the turns are in the same proportion; the quotient of the height of two successive turns gives the rate of growth of the mouth in height; the quotient of the corresponding breadths gives the rate of increase in breadth; the quotient of the diameter of the entire shell by the height of the last turn expresses the rate of growth of the discoid (Scheibenzunahme). The calculations of Moseley and Naumann were afterwards confirmed by G. Sandberger and Grabau.

The constitution of the internal partitions (septa) which limit the different air-chambers is of considerable importance. Their number varies extraordinarily in the different genera and the different species, but it is quite constant in one and the same species; they are at increasing intervals from each other, according to law, proportionate to the growth of the shell, and it is only the last two partitions (septa) which precede the final chamber, which are at a somewhat less distance apart. Probably all the chambers have successively served as dwelling-chambers, and it is only after a new partition was formed that it was transformed into an air-chamber, which was no longer in communication, except by the Siphon, with the last chamber. The mud and the sand were not able, generally, to penetrate into the interior of the fossil shells when they were buried intact, except in the last chamber, or by the siphonal opening into the last air-chamber only. This is the reason why the chambers are very often not filled with rock, but are coated or filled with crystals of calcite, of quartz, of pyrite, of celestite, of barite, etc., which have been precipitated from the infiltrated chemical solutions.

The line of attachment of the partition to the internal wall of the shell is called the *Suture.* (See Fig. 27, p. 106, and Figs. 112–118, p. 346.) It is not exteriorly visible unless the shell is removed or dissolved. It is seen more distinctly on the fossil moulds, in which the shell is wanting. In the Nautilus, and in many of the shells of fossil Tetrabranchiates, the septa attach themselves to the internal surface of the shell by a slightly arched sutural line. Moreover, very often the line of the suture, on account of the undulating curvature and a fluting of the septum, acquires a high degree of complication resembling the branching of moss. There are all degrees of variation from lines the most simple to those most complex. Besides, as the lines have essentially the same sinuosity for all the specimens of one species, and on the contrary show differences quite striking in different species and separate genera, they furnish thus one of the most important systematic characters. In the Nautilidæ the lines of the sutures are generally simple (Fig. 106); in the Goniatites and Clymenias (Fig. 112) the undulating and slashed suture forms prominent *saddles* before and curved sinuses behind, called *lobes.* A later differentiation is met with in the Ceratites, etc.,

(Fig. 114), the lobes being denticulated by secondary notches. In the Ammonites (Fig. 115) the saddes also, as well as the lobes, are denticulated in the most varied manner, notched, cut, or ramified, in form of branches, or foliated. The curvature of the suture line, as well as the formation of the saddles and lobes, takes place symmetrically in such a manner that a median line in the direction of the height divides the turns into two equal parts. The exterior lobe is called the external or *siphonal lobe*, when the siphon is on the exterior side. For Leop. von Buch it is the dorsal lobe, because he called this the back of the shell, but for recent authors, who consider the external side to be the ventral part, it is the ventral lobe. The opposite unpaired lobe is the *internal lobe* (or, according to opinions, antisiphonal lobe, or dorsal, formerly ventral lobe). Between the two are found the lateral lobes and the lateral saddles, situated on the body of the whorls, and the lobes and saddles concealed between the line of contact of the contiguous whorls and the internal lobe: among the former, the saddle which is found on the side of the external lobe is the external saddle, the two following are the first and second lateral saddles; all the others, up to the line of junction of the two whorls, are the auxiliary saddles; near the internal lobe is found, generally, an internal saddle, which is distinguished by its size from the other concealed internal auxiliary saddles. For the lobes, the first lateral lobe is that which is between the external saddle and the first lateral saddle; the following one is the second lateral lobe; all the others are called auxiliary lobes.

The beautiful researches of Hyatt and Branco have shown that the complicated lines of the suture of the Ammonites do not attain their normal form until the animal has developed a greater or less number of the chambers. The first sutures of all the Ammonites are always as simple as those of the Nautilidæ, Clymenias, or Goniatites (Figs. 112, 116); it is only little by little that the undulating lines become marked by secondary notches, and the complication of the line of the suture proceeds always from the exterior to the interior side. The complication of the suture line—which augments with age, so that the young sutures, more simple in Ammonites, resemble those of the geologically more ancient Goniatites and Nautilidæ, —shows, probably, that this differentiation indicates at the same time a perfection of the organism. It is truly difficult to discover wherein this consists. It is possible that the strongly ramified borders of the septa serve to increase the solidity (firmness) of the shells; for, in general, the shells of Nautilidæ, provided with simple suture lines, are considerably thicker than the shells of Ammonites—ordinarily as thin as paper. If one breaks cautiously, little by little, the enrolled shell of a Tetrabranchiate, there are distinguished the first whorls, and finally also the initial chamber of the whole coil. In the fossil evolute, or baculiform, shells this first chamber is, ordinarily, abbreviated or broken, and it is extremely rare that it is preserved.

According to Barrande, Hyatt, and Branco, there are two kinds of initial chambers in the Tetrabranchiates which can be distinguished by fundamental characters. In the Nautilus, and many of the paleozoic genera, the initial chamber is in the form of a truncated cone, slightly arched or straight, enlarged in front; upon the posterior convex wall, which terminates the truncated cone, is observed a depressed cicatrix, linear (Nautilus), circular (Cyrtoceras), elliptical (Trochoceras, Phragmocus), or sometimes cruciform.

The initial chamber of Clymenia, the Goniatites, and the Ammonites is formed in an entirely different manner. In all these this spirally enrolled chamber has a vesiculous, spherical, or ovoid form, generally a little depressed and transversely striated; no scar or impression has been met with, and the siphon begins at the anterior wall. It is not probable that the initial chambers of the form of a truncated cone of the Nautilidæ are homologous with the spherical enrolled initial chambers of the Ammonitidæ; on the contrary, the presence of a cicatrix makes it probable that this impression represents either the point of attachment, or the opening of communication, closed at a later stage, of a frail vesicle, perhaps membranous, which corresponds with the initial chamber of the Ammonites. According to this view, proposed by Hyatt, the initial chamber of the Nautilidæ should be equivalent to the second chamber of the Goniatites and the Ammonites.

The Siphon is a tubular prolongation of the skin of the posterior part of the body; it traverses all the chambers, and in Nautilus begins under the form of a closed tube covered with nacre, in the initial chamber, or truncated cone, where it touches the internal posterior wall at the same place, where exteriorly is seen the cicatrix. In the Ammonites and the Goniatites the siphon begins with a spherical swelling situated immediately behind the anterior wall of the initial vesicle (nucleus), consequently perforating only the first septum, without penetrating more deeply into the chamber. According to Hyatt, the part of the siphon penetrating into the embryonal chamber was, in general, only a depression of the first partition. Munier-Chalmas has observed in the Ammonites a particular prolongation of the siphon in the initial chamber which ought to have replaced the true siphon in the embryonic stage; this prosiphon, as he calls it, is attached to the siphon, which begins in a reflected cul-de-sac, and is very variable in form. It forms sometimes an enlarged membrane, sometimes a cylindrical tube; the prosiphon does not communicate with the siphon, properly speaking.

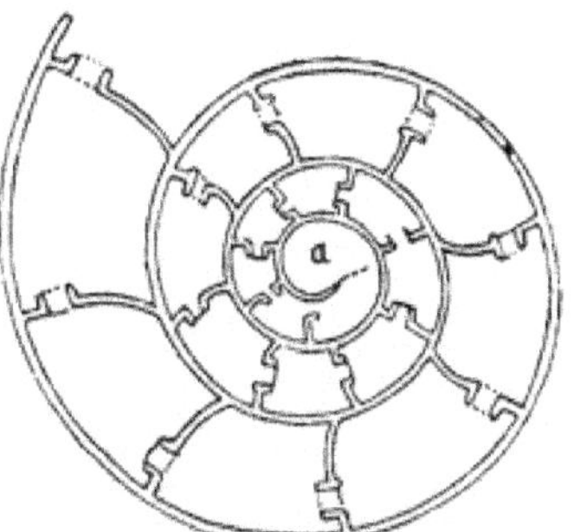

FIG. 109.—Median section of a young Ammonite (*Tropites*) showing the siphonal collar and its change of direction on passing from the initial chamber *a*, where it is retrosiphonate or turned backwards, to the third whorl, where it is prosiphonate or turned forwards. (After Zittel.)

In the recent Nautilus the siphon is a rather solid membranous tube covered exteriorly by a thin coating of brown color, earthy, formed of fine calcareous grains. In the Ammonites (see Fig. 102) this exterior calcareous envelope seems to take on a more substantial consistency, so that the siphon is enclosed in a delicate calcareous tube. It is necessary not to confuse this envelope of the siphon itself with that which is called the siphonal collar, which is met with always where the siphon penetrates the septum.

The siphonal collar (Fig. 109) is a reflection or production of the septum of greater or less length, directed, generally, in Nautilus, backward, and in Ammonites forward, and possesses the same structure with the septum. Ordinarily, the siphonal collar has only short length, and forms in front and behind the septum a sheath in the form of a band or collar about the siphon; but, sometimes, they pass from one septum to the other and form there a close continuous tube, or they have the form of an open funnel, slightly contracted behind, and prolonged to the next following septum, or even go beyond it, thus implanting themselves one within another (telescoping, Endoceras). The siphon is found in the median plane of the shell, and it is only exceptionally that it deviates a little from this plane. In this plane its position vacillates from the external side to the internal side in the different genera and the different species.

In the Ammonitidæ it is constantly on the external side of the shell. In the Nautilidæ its position does not remain constant in one and the same genus: it may be external, internal, central, or intermediary.

Numerical Rate of Differentiation expressed in Terms of the Initiation of New Genera.—A study of the statistics of classification in relation to time will exhibit in this, as it has in previous cases, the grand features of the historical differentiation of the cephalopods.

First, we may consider what are the conclusions to be drawn from the succession of new genera as to the rate and order of the differentiations of the class Cephalopoda.

The classification itself is expressive of differentiation, as has been already observed. The division of the class into two orders is expressive of a very marked differentiation in structure. The genus is a group of organisms with the same ordinal and family structure, but exhibiting some particular characters, such as shape, relative size of parts, or special development of some part, which are the same for several different species; hence we recognize the number of genera to be a numerical expression of the amount of differentiation attained in the family at any particular period of time, and the greater the number of genera in a particular family, at a particular time, the greater is the amount of differentiation expressed in the family-history at that period, and the number of genera beginning or living in each period becomes a rough indication of the rate of expansion or evolution of the race under consideration. The total number of genera in the order Tetrabranchiata is 123 (Zittel). Two grand subdivisions of subordinal rank are made, including, respectively, Nautiloidea 29 genera, and Ammonoidea 94 genera. 28 genera of the 29 Nautiloidea had appeared in the Silurian.

One genus, Aturia, is considered to be a distinct new genus of the Tertiary; 16 genera were already well exhibited in the Lower Silurian, or Ordovician. Only 8 genera lived into the Devonian, only 5 to the Carboniferous, and but 2 (Orthoceras and Nautilus, the perfectly straight form and the tightly coiled form) survived from Paleozoic into Mesozoic time.

The other suborder, Ammonoidea, has 94 genera; of these, one genus is known as early as the Silurian (Goniatites), one new genus (Clymenia) was added in the Devonian, and in the latter part of the Carboniferous 5 more genera were initiated. Of the rest, all appeared in the Mesozoic, 41 genera beginning in the Triassic, 28 new genera starting in the Jurassic, and 18 new ones appearing, for the first time, in the Cretaceous. Not a single genus of the whole suborder survived the Cretaceous period. Thus the Nautiloidea are peculiarly Paleozoic in range, although there is still living the genus Nautilus, and the Ammonoidea are peculiarly Mesozoic, and every genus of this suborder is now extinct.

The other order, Dibranchiata, is less capable of showing its history: the hard parts were of inferior character and less in proportion to the fleshy parts, and upon the death of the animal were much more likely to be destroyed; 33 genera are known, and all are Mesozoic, or more recent. There were 3 genera in the Jurassic, 15, Triassic, 8, Cretaceous, 10, Tertiary and 3 now living.

Second. The lesson, regarding the evolution of the ordinal and subordinal characters and their generic expansion, which we derive from these statistics is as follows:

Rate of Differentiation of the Suborder Nautiloidea.—The Nautiloids (Orthoceras, Nautilus, and their kindred genera) first appeared in the Ordovician. Before the close of the Silurian this type had reached its fullest expansion, and began in a very marked manner to drop out of the race; 5 genera did not survive from Ordovician into Silurian, and of the 22 Silurian genera only 8 survived into the Devonian. Of this type the two genera to live the longest were Orthoceras, the simplest expression of the type, and Nautilus, probably the most differentiated; and the latter continued to live up to present time. At least, of the structures preserved to tell us the story these two are the extremes—one, Orthoceras, a simple slender cone, straight, and with regular septa dividing it into chambers, and with a central siphuncle; the other, Nautilus, a closely coiled disciform shell, with siphuncle also central,

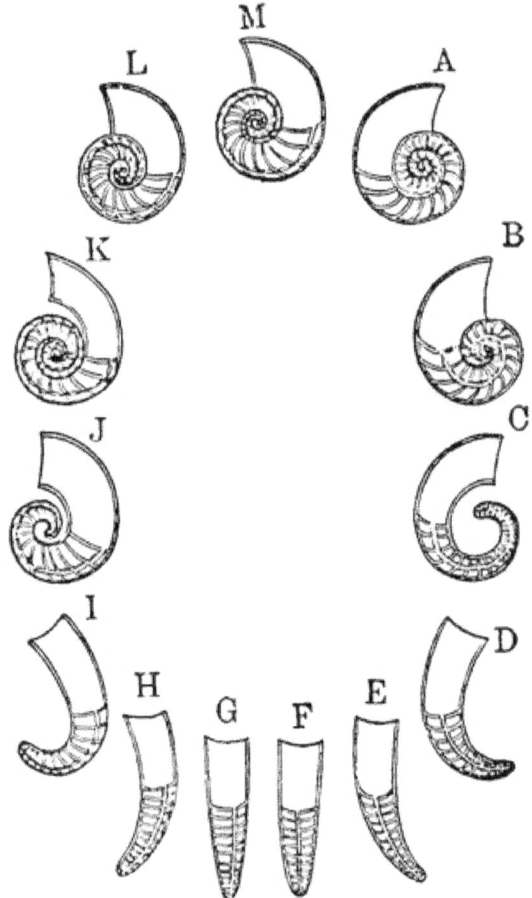

FIG. 110.—Theoretic sections through the middle of the shells to show the variations in the curvature and coiling of Paleozoic Cephalopod shells. *A*, Clymenia; *B*, Nautilus; *C*, Nautiloceras; *D*, *E*, Aploceras; *F*, Orthoceras; *G*, Melia; *H*, *I*, Cyrtoceras; *J*, Gyroceras; *K*, Ophidioceras; *L*, Cryptoceras; *M*, Goniatites. (After Gaudry.)

outer chamber large, and whorls with ventral side out. The two features which best express in these shells the amount or degree of differentiation are, the amount and direction of the curvatures of the shells and the position of the siphuncles.

The characters which serve most readily to distinguish the Ammonoids from the Nautiloids are the sutures. In the Nautiloids, above described, the suture is always straight, or but slightly curved. In all the Ammonoids the suture is more or less lobed or notched.

Mode of Curvature of the Nautiloid Shell.—Third. Before considering the Ammonoids, we may notice the law of variation expressed by this curvature of the shell. In the Nautiloids there are four types of form expressed in the direction of growth of the cone:

1. The shell is straight, or nearly so (see Orthoceras, Fig. 110, *F*).

2. The shell is simply arched (see Cyrtoceras, *H*, *I*, *D*).

3. The shell is discoidal, rolled in a spiral in single plane (*J*).

(*a*) This spiral may be loose (*C* or *J*, Gyroceran);

(*b*) or close-coiled, (Goniatites, *M*);

(*c*) or involute (Nautilus, *B*).

4. The shell may be spirally coiled in a screw plane, or helicoidal. (See Fig. 108, p. 332).

When we separate out for special consideration the mode and amount of curvature of the shell, we are first struck with the evidence of progressive modification from the straight to the close-coiled forms; but when the relation of these modifications to the time of their first appearance is noted we learn that forms of the different types of modification occur at the earliest period (the Ordovician) in which the suborder appears, and if we were to seek a series representing gradual modification from one extreme to the other, we could find them well expressed at this initial stage of the subordinal history.

The Rate of Initiation of the Orthoceratidæ.—For instance, take the species of America alone, and of the straight or slightly bent form, Orthoceratidæ, there are recorded by Miller 5 recognized genera and 170 species in the first stage of this family, Ordovician; in the Silurian there were 3 genera and 81 species; in the Devonian 3 genera with 145 species; in the Carboniferous, 2 genera with 43 species.

Rate of Initiation of the Cyrtoceratidæ.—The same general law is seen in the (2) arched forms included in the family

Cyrtoceratidæ. In America 2 genera (Cyrtoceras and Phragmoceras) began in the Ordovician and continue throughout the Paleozoic, and Miller records of them 72 species in the Ordovician era, and 42 Silurian, 20 Devonian, and 8 Carboniferous species.

Rate of Initiation of the Nautilidæ.—Take the third type (3), the discoidally spiral form Nautilus, and its various generic allies. The Nautilidæ has in America 5 well-marked genera. 4 genera, including 35 species, are Ordovician; 4 genera, including 17 species, Upper Silurian; 2 genera, including 35 species, Devonian; 2 genera, including 62 species, Carboniferous. In this case the apparently different law expressed in the number of genera and their decrease, and in the number of species and their increase, is due to the combination in the family of two sets of genera, the one set of which have their maximum representation of species early in the Paleozoic; the other increases in the number of its species as we ascend. *Lituites*, for instance, has 15 species in Ordovician, 7 in Silurian, and then became extinct. On the other hand, *Nautilus* has 13 species recorded for the Ordovician, 4, Silurian, 15, Devonian, 59, Carboniferous; and the genus continues on to the present time.

History of Trochoceras by Species.—The helicoidal type (4), including, for America, the one genus Trochoceras of the family Trochoceratidæ, is specifically represented as follows: Ordovician 1, Silurian 7, Devonian 10; and then it ceases.

General Law of Evolution of Shell Cuvature in the Nautiloidea.—Thus, to generalize, we find that this grand feature of the Nautiloidea, the form assumed by the shell in its growth, expresses the fulness of its differentiation among the representatives of the first or initial period of the existence of the race. All the several types of form run along together and continue nearly, or quite, to the close of the Paleozoic, and there, with the exception of two genera, become suddenly extinct.

Rate of Initiation of New Species in the American Region.—Fourth. As if to emphasize the law above expressed regarding the initiation of new genera, the statistics of the initiation of species in the American rocks point in the same direction.

When we observe the number of the different species of a genus, recorded in the rocks of each period in which the genus occurs, we find that the greater number of species of the genus, as well as the greater number of genera of the family, are recorded from the initial geological period, which in this case is the Ordovician; and the genera and the species gradually *decrease* in number for each following period until the close of the Paleozoic, with the exception of the genera *Nautilus* and *Trochoceras*, whose expansion appears to be later and its life-period longer. Even in these cases, however, the law is relatively the same.

This, again, is expressive of the general law before stated, that *the chief expansion of any type of organisms takes place at a relatively early period in its life-history.* This law was observed in the case of the Brachiopods, and is observed here among the Cephalopods. There are some modifications or exceptions to it, which the facts regarding other groups suggest; but the general law is sufficiently well attested to be defined in these general terms.

Hyatt's Formulation of the Law of Rapid Expansion of Differentiation at the Point of Origin of a New Type of Organism.— Hyatt has given expression to this law in an article on "Genera of Fossil Cephalopods." * The generalization is based upon a very exhaustive study of the Cephalopods. He had access to the collections in the Agassiz Museum of Natural History, which was the most complete in this country; and he also visited all the museums in this country and in Europe where Cephalopods are found, and made particular examination of every species he could learn of throughout the scientific world. Speaking of the Nautiloidea and Ammonoidea both, he wrote: "These groups originate suddenly and spread out with great rapidity, and in some cases, as in the Arietidæ of the Lower Lias, are traceable to an origin in one well-defined species, which occurs in close proximity to the whole group in the lowest bed of the same formation. These facts, and the acknowledged sudden appearance of large numbers of all the distinct types of In-

* Published in 1883, in the *Proceedings of the Boston Society of Natural History.*

vertebrata in the Paleozoic, and of the greater number of all existing and fossil types before the expiration of Paleozoic time, speak strongly for the quicker evolution of forms in the Paleozoic, and indicate a general law of evolution. This, we think, can be formulated as follows: *Types are evolved more quickly and exhibit greater structural differences between genetic groups of the same stock while still near the point of origin, than they do subsequently. The variations, or differences, may take place quickly in the fundamental structural characteristics, and even the embryos may become different when in the earliest period, but, subsequently, only more superficial structures become subject to great variations.*" *

Summary.—If we ask, In what particulars does the structure of Cephalopods illustrate this law? we may answer in brief, that we notice it first in the class characters of the Cephalopoda. In the description of the class we found the Cephalopods most closely allied to the Pteropods. This is conspicuously observed in the difference in structure of the locomotor apparatus of the foot. In the Pteropod there are two lateral flaps used like wings, or paddles, for locomotion. The Cephalopods are modified to form a siphonal funnel which accomplishes locomotion by forcing water violently out and forward from this funnel; other structural peculiarities are associated with this modification.

The Pteropods are abundant in the Cambrian faunas, and appear to have attained a relative dominance never afterward held, but in this first fauna there were no Cephalopods. The Cephalopods of the next (Ordovician) period were extremely abundant, and the Tetrabranchiata type was expressed by 17 of its 29 genera at the initial Ordovician stage (including here the Upper Tremadoc, whose fauna seems more appropriately associated with Ordovician than with Cambrian faunas).

It is seen, secondly, when the Ammonoid type of the Cephalopods made its appearance in the Goniatites. The Goniatites came out in full force in the Devonian, with a few species in beds doubtfully referred to the Upper Silurian but called Lower Devonian by Kayser. The most characteristic

* See "Phylogeny of an Acquired Characteristic," by Alpheus Hyatt, *Proc. Phil. Soc.*, vol. XXXII., No. 143, p. 371.

difference in the hard shell is seen in the curved and lobed suture of the Goniatites as contrasted with the simple suture of the Nautiloids.

The law is again seen in force in the evolution of the Ammonites, beginning in the Sicily and India Permian beds; by the early part of the Trias this new type had expressed a wonderful expansion. Out of the 92 genera described and recognized by Zittel, 45 occur in the Triassic, representing 9 out of the 13 known families.

Again, in the Jurassic the great differentiation of type expressed in the Dibranchiates took place, not in a single form, but both decapod and octopod modifications appear together.

Thus we find this distinguishing character of the Dibranchiate (the consolidation of the siphonal tube, after the tube with disunited edges had existed from Ordovician time throughout the Paleozoic) making its first appearance at the beginning of the Mesozoic, but thereafter continuing on in successive and various forms until the present time.

In each of these cases, of the initiation of new types of the Cephalopod mode of organization, there was a rapid evolution of the chief modifications of the new type near the period of its first initiation among the geological faunas of the world.

CHAPTER XIX.

PROGRESSIVE MODIFICATION OF AN EXTRINSIC CHARACTER; ILLUSTRATED BY THE EVOLUTION OF THE SUTURE LINES OF AMMONOIDS.

The Ammonoids Illustrate the Law of Acquirement of Differences by Gradual Modification.—The Ammonoids illustrate another of the laws of evolution in a particularly satisfactory manner.

When we examine the representatives of the same genus, or family, or order, at the beginning and at the close of its life-period, it is very common to find the two representatives differing in one or more characters, which may be described as differing in the degree or extent of their development. The number of parts has increased; some part which is small in one is large in the other; some structure which is simple in the earlier is complex in the later; or parts which are indefinite in form, or similar in the beginning are definite and particular in form and structure at the close.

It is rare, however, to be able to collect examples to show the various stages by which the one was elaborated by degrees of modification into the other. The famous case of the development of the specialized horse foot out of a five-toed ancestor is familiar to all, with the beautiful theory of the way by which the modification came about. This is a case of relative rather than of direct evolution, since the prominence of the one toe and line of connecting bones is produced by the aborting and withdrawal from use, and finally from development, of the normal number of parts which were present at the beginning of the series. The Ammonoids, as we shall see, illustrate the case of actual increase in complexity, gradually and continuously; the order of succession in the steps of progress being clearly and regularly expressed by the actual appearance of each form at the particular geological stage in which it should appear according to the law of genetic evolution of the characters of the race.

Description of the Characters of the Ammonoids.—In order to place before the reader a concise description of the characters of Ammonoids, the definitions of Zittel may again be followed, furnishing as they do the precise characters needed for an understanding of the problem under discussion.

Zittel's definition of the characters of the Ammonoidea is as follows:

SECOND SUBORDER: AMMONOIDEA. Shell generally enrolled or spiral, discoidal, more rarely spirally coiled, evolute, arched or straight; opening simple or furnished with lateral and ventral prolongations. Suture-line undulating, notched or with slashed or dentate lobes and saddles; siphuncle cylindrical, always marginal, without internal deposit; initial chamber spherical or ovoid, frequently an aptychus or anaptychus.

In the description of the fundamental characteristics of the sutures and their development we follow Zittel's synopsis.

The embryonal chamber (nucleus, ovisac) of the Ammonoids has a spherical or transversely ovoid shape (Fig. 112, *a*); it is smooth, separated by a contraction from the rest of the shell, and always enrolled spirally about an imaginary axis. Its anterior aspect is, in consequence, essentially different from its lateral profile, its sides having a projection in form of an umbilicus. The embryonal chamber, of which the height varies from 0.3 to 0.7 mm., is limited in front by the primary septum. The constitution of the first suture gives, according to the beautiful researches of Branco, excellent basis for classification. In the most ancient Ammonoids it forms a straight line, more or less simple, and then resembles the first suture of Nautiloids; Branco calls these forms the *Asellati* (Fig. 111, *A*)

In a second group the first sutural line proceeds forward to form an arch towards the exterior, and forms a large simple ventral saddle, *Latisellati* (Fig. 111, *B*).

The third group is distinguished by the relatively narrow ventral saddle, on each side of which is developed a lateral lobe and generally also a small lateral saddle, *Angustisellati* (Fig. 111, *C*).

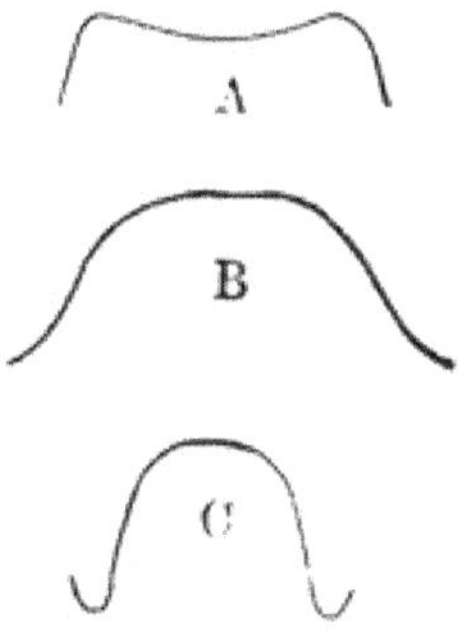

FIG. 111.—Ventral views of the edges of the embryonal chamber, representing *A* the primary septum of an asellate Goniatite, *B* of a latisellate Ammonite (*Arcestes*), and *C* of an angustisellate Ammonite (*Phylloceras*). (After Branco.)

While the first suture of all Ammonoids is comparatively simple, more or less considerable complication is produced by the later development of the shell. Only a few of the more ancient types possess a sutural line altogether simple, like that of the Nautilids. Almost always, even in Paleozoic forms, the suture attained at least the Goniatite stadium, that is, an undulating or notched suture formed of simple lobes and saddles (Fig. 112).

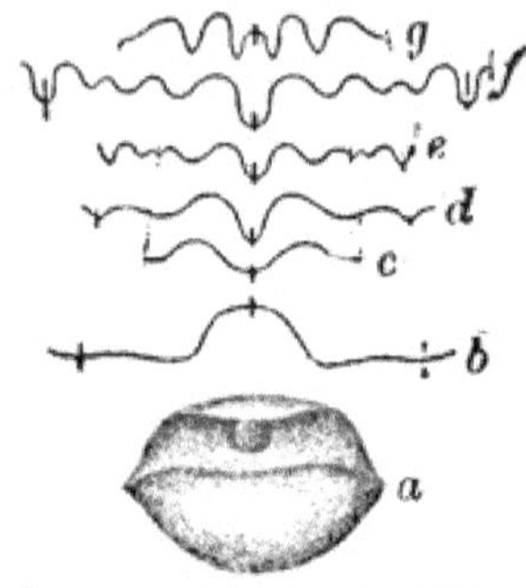

FIG. 112.—Development of the suture of *Goniatites diadema* Goldf. (After Branco.)

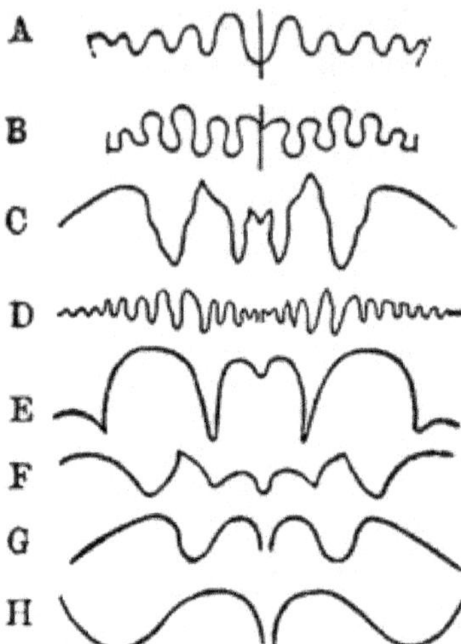

FIG. 113.—Sutures of the various tribes of Goniatites. (According to the Sandbergers.) *A* = Linguati, *G. tuberculosis costatus*; *B* = Lanceolati, *G. Becheri*; *C* = Genufracti, *G. sphericus*; *D* = Serrati, *G. sagittarius*; *E* = Crenati, *G. intumescens*; *F* = Acutolaterales, *G. terebratus*; *G* = Magnosellares, *G. retrorsus*; *H* = Nautilini, *G. subnautilinus*.

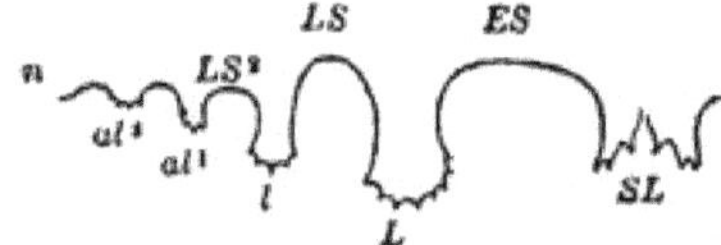

FIG. 114.—Suture of *Ceratites nodosus*.

A later complication is observed in the Ceratite stadium, in which the saddles remain intact while, on the contrary, the lobes are notched by slight denticulations. The more elaborate differentiation is reached in the Ammonite stadium, in which the lobes and saddles are gashed by secondary notches in the most variable manner.

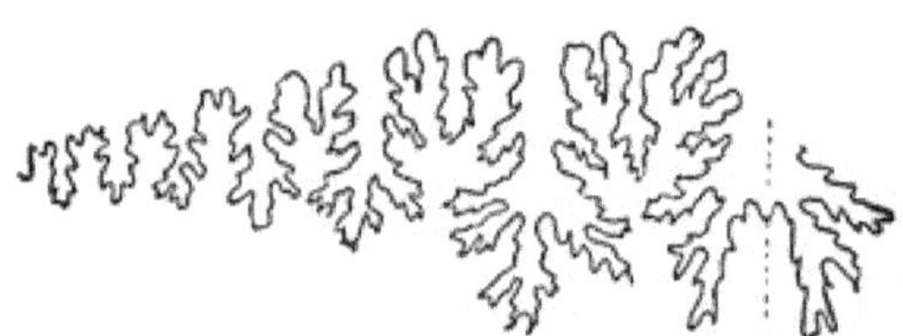

FIG. 115.—Suture of an Ammonite, *Desmoceras latidorsatum*. (After Zittel.)

As the Goniatites appeared, in general, before the Ceratites, and these in part before the true Ammonites, it is believed that these three genera may be considered to be the three principal stadia of development of the Ammonoids. This view is further confirmed by the fact that the suture line of all Ammonites in the course of the first whorl passes through the Goniatite stadium (Fig. 116, *II* to *X*). According to the researches of Hyatt and Branco, however, the Ceratite stadium is, in general, passed over and the Goniatite stadium passes directly into the Ammonite stadium. The development of the sutural line by folding of the septum advances from without inwards; on the contrary, the new lobes and the new saddles are intercalated, almost always, at the lateral suture of the whorl, and rarely on the external ridge.

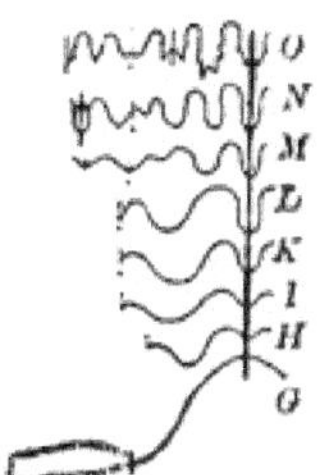

Fig. 116.—Development of the suture of an Ammonite (*Tropites subbullatus*). G = 1st suture. H = 2d, I = 3d, L = 7th, MN = sutures of second whorl, O = definitive suture. (After Branco.)

The second suture is distinguished from the first in almost all Ammonoids by the development of an external ventral lobe, more or less deep, simple or bifid, which gives rise to two external saddles caused by the dichotomy of the original simple saddle. It is rare that it is confined to these three elements; generally, there is added besides a lateral lobe and a lateral saddle. In the more simple forms the suture has by that time acquired its definite shape, and all the later chambers present the same design at the point of their attachment. There generally occurs, however, a multiplication of the lobes and of the saddles, and the external lobe takes part in it by one small median saddle becoming bifid.

Such is the characteristic development of the suture in the Goniatites, the Clymenias, and a small number of the Triassic Ammonites. In the Ceratites and the true Ammonites there takes place exactly the same differentiation at the outset as in the Goniatites; but later, when the shell has reached the size of 3 mm. in diameter, begins the secondary slashing of the lobes and of the saddles of the exterior and of the interior. (See O of Fig. 116.)

At the size of 4 mm. the Ammonites are generally in possession of these characteristic suture lines, which from that time on remain constant, or at least suffer very slight change. In the determination of the several species it is necessary to compare the suture lines of only the mature forms. The external lobe does not tend to become bifid in the Goniatites and Ammonites, the most ancient geologically, as in a stadium of relatively tardy growth. In the relatively young Angustisellati the division into

Fig. 117.—Suture of *Pinacoceras Metternichi*. (After Zittel.)

two lobes is distinctly accomplished. In a single form, or even in series of forms, or in the most closely related species, the geologically younger representatives generally possess the more differentiated suture lines; on the contrary, however, it is not possible to deduce the geological age of an Ammonite from the structure of the suture line alone. In the Trias there are forms (Pinacoceras, Fig. 117) which present lobes so finely slashed and so complicated that one can scarcely observe similar ones in the most recent formations; on the other hand, there are known Ammonites (Buchicera) from the Middle and Upper Cretaceous, the sutures of which represent the Ceratite stadium (Fig. 118; also compare with Fig. 114) by retrocession, if they be not quite the same genera. In all typical Ammonites there is developed, besides the external ventral lobe, which, in the forms with an external siphuncle, is called often also siphonal lobe, two main lobes on the side—the first and second lateral lobe. Besides the external lobe, there are two large external saddles; and besides the lateral lobes, the two primary lateral saddles. The external is almost always profoundly slashed into two points by the development of a secondary median saddle, while the internal lobe (dorsal lobe) opposite ordinarily remains entire. The external saddles are also able to be divided sometimes by deep secondary indentations. In some genera (Pinacoceras) the differentiation of the external part of the external saddle goes so far that there are intercalated between it and the external lobe a greater or less number of supernumerary saddles and lobes. All the saddles and all the lobes from the second lateral saddle to the internal contact suture of the whorl are called external; those which are within the contact sutures up to the inner saddle receive the name internal auxiliary lobes and saddles.

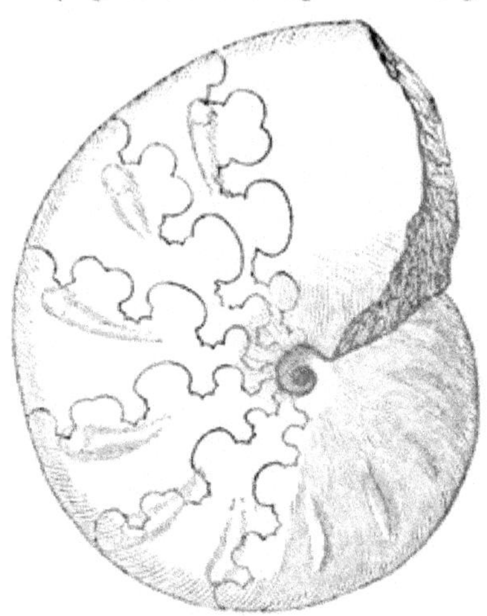

FIG. 118.—*Tissotia Fourneli* Bayle. Cenomanian, Algeria. (After Bayle.)

The variability in the number and size of the lobes is, generally, in relation with the form of the shell. If the whorls are circular, one observes, ordinarily, only a few lobes, and in that case they are of nearly equal dimensions (Lytoceras); upon a wide ventral side the external lobe and the external saddle acquire considerable dimensions; the more flat the sides are and the thinner the ventral part, the larger the size of the lateral lobes and lateral saddles, and the more numerous the auxiliary lobes.

Two Divisions of the Retrosiphonatæ: Goniatites and Clymenias.—In following the course of evolution of this group, as indicated by the modifications of the suture-line, we begin with the first division of the Ammonoidea—the *Retrosiphonatæ* of Fischer. The two groups are the Goniatites and the Clymenias. The fundamental and constant difference is found in the relative position of the siphuncle. In the Goniatites the siphuncle is external and in the Clymenias always internal.

The Goniatitinæ, of Hyatt's classification, begin in the

Silurian and are dominant in the Devonian, and the undisputed Goniatitinæ are not continuous beyond the Carboniferous. Sagiceras and like forms are Triassic, and are intermediate between this and the true Ammonite type. The Goniatitidæ (v. Buch, emend. Zittel) contain about 300 species, all of which are Paleozoic.

Quick Evolution of the Clymeniidæ.—Of the Clymeniidæ, about 30 species are known—all from the Upper Devonian. When, however, the character of the suture is made the chief means of classification, we find a considerable range of modification in the Clymeniidæ, and of the other characters: the shape of body whorls, rounded, angular, tuberculated, etc., and amount of involution of whorls, all indicate great modification, so that authors have classified even this special little group of forms into many genera. Hyatt proposes 3 families, with 9 genera in all, based upon the minute studies of Gümbel. Hyatt remarks, regarding the Clymeniidæ:

> "This extraordinary series shows the phenomena of quick evolution in three series of forms. Cyrtoclymenidæ, with a series beginning with an Arcestes-like form, and passing through discoidal and compressed to quadragonal forms; Cymaclymenidæ, a similar parallel series, but with more complex sutures; and Gonioclymenidæ, also a similar series, but with more involute forms than the last, and the sutures becoming Ammonitic, with median ventral lobes and saddles, divided by a pair of marginal lobes." *

When we compare this series of suture-lines with those of a single Goniatite, at different stages of individual growth (Fig. 112), the evolution may be expressed as a case of rapid acceleration, with some variation added.

Classification of the Goniatites.—The attempt to classify the Goniatites by their sutures has resulted in various systems, in each of which the particular form of the mature suture-line has been the criterion of classification.

Beyrich proposed six groups, which he called (1) Nautilini, (2) Simplices, (3) Æquales, (4) Irregulares, (5) Primordiales, (6) Carbonarii.

Sandberger made a more minute analysis, based upon the form of the lobes and saddles making up the suture. His nomenclature is: (1) Linguati, (2) Lanceolati (= Æquales in part of Beyrich), (3) Genufracti (= Carbonarii Beyr.), (4)

* See "Genera, Foss. Ceph.," p. 313.

Serrati (= Irregulares Beyr.), (5) Crenati (= Primordiales Beyr.), (6) Acutolaterales, (7) Magnosellares (= Simplices Beyr.), (8) Nautilini (= Nautilini Beyr.). (See Fig. 113.)

Hyatt distributed the Goniatites into several families, including in each the several groups based on sutural characters as follows: (1) Nautilinidæ (Nautilini Beyr.), (2) Primordialidæ (Primordiales Beyr., and Crenati Sandb.), (3) Magnosellaridæ (Magnosellares Sandb., Acutolaterales Sandb., Simplices Beyr., *p.p.*, and Æquales Beyr.), (4) Glyphioceratidæ (Carbonarii Beyr., Simplices Beyr., *p.p.*, Genufracti Sandb., Indivisi Bronn), (5) Prolecanitidæ (Lanceolati, Linguati, Serrati Sandb., Irregulares Beyr.). (Hyatt included here the genera Medlicottia, Sageceras, and Lobites, referred to the Ammonites by Zittel.)

Differences in the Sutures of the Ammonoidea explained as Various Degrees of Crimping of the Edge of the Diaphragms.—The sutures may be considered as simply the edges of the diaphragm which is built by the animal across the conical shell in which it lives, to constitute air-chambers of the vacated part as the animal grows in size. A simple explanation is suggested by the mechanical principle that the natural result of attempting to force a diaphragm into a tube too small for it would be the crimping of the edges of the diaphragm. With this clue applied to the interpretation of the sutures, we discover that all the various sutures may be defined in terms of difference in degree of complexity of the crimping of the edge of the septum.

Classification of the Types of Sutures.—Gathering statistics of all the known forms, and studying their embryological development as well as their actual differences, we find the following facts to be true regarding the modifications of the suture-lines which result from the crimping or fluting of the outer margin of the septum where it is attached to the wall of the chamber of the shell:

A. *The Nautilian Type of Suture.*—In the Nautilidæ the suture is simple, either straight or slightly curved, but never folded, i.e., in its complete circumference not exceeding a single oscillation of curvature (see Fig. 102). This is the Nautilian or simple type of suture.

B. *The Goniatitic Type of Suture.*—In the Goniatites we find the suture lobed, forming rounded or bluntly angular curvatures; these curvatures in the simplest stage of the protoconch are arched forward at the siphonal side (Fig. 112, *a, b*). In the growth of the individual, as well as in the different genera or subgenera of Goniatitidæ, the lobation never exceeds the repetition of these forward and backward curvatures of the suture. The multiplication of the curvatures is accomplished by the infolding of the node of each curve (Fig. 112, *d, e, f, g*, and Fig. 113).

This constitutes the Goniatitic type of suture, and consists, with all its complexity and variation, of a system of curvatures forward and backward; the forward curvatures (upward in the figure) are called *saddles*, the backward curves (downward in the figure) are the *lobes*.

The various modifications of this type of suture are produced by different degrees of division of the lobes and saddles in different parts of the circumference of the whorl. This kind of bending of the suture may be called lobation of the suture, and may be defined as the type of suture formed by the primary crimping of its edges.

C. *The Ceratitic, Helictitic, and Medlicottian Types of Suture.*—The primary lobes and saddles may be again crimped so that the lobes are cut by a series of lesser lobes, the saddles are dentate by secondary slits, or the sides of the curves connecting the lobes and saddles are secondarily lobed; this modification constitutes a secondary system of lobation of the suture; and there are three stages of this mode of *crimping* of the edge of the septum.

FIG. 119.—Suture of *Medlicottia primas*. (After Zittel.)

1. The Ceratitic type, in which only the lobes (*L. l. al_1, al_2* of Fig. 114) or the backward curves of the septum edge are secondarily crimped.

2. The Helictitic type, in which the saddles (see *ES*, *LS*, Fig. 114) is alone secondarily crimped.

3. The Medlicottian type, in which the sides of the saddles and lobes, or lines connecting them, are dentate or secondarily crimped (Fig. 119).

To distinguish these three from the former type they may be classified as the *crenulated* or secondarily crimped type.

D. *The Ammonitic Type of Suture.*—There is a still higher complication of this system of sutures. The secondary curvatures may be themselves tertiarily crimped or notched, forming a tertiary system of lobation of the suture; this gives us the Ammonitic type of suture, and the suture is called *foliate* to various degrees of elaboration in different genera.

E. *The Pinacoceran Type of Suture.*—A further extreme of differentiation is attained in the crimping of the edge of the septum of Pinacoceras of the Trias (Keuper), of which twenty-seven species are reported, in them the tertiary lobes are again dentate or crimped, forming the quaternary system of lobation. This is the highest stage of elaboration recorded for the suture line of the Ammonoids (Fig. 117).

Relation of Order of Succession of Initiation to Order of Ontogenetic Development and Evolutional History.—The natural law of sequence of these various types of lobation of the suture is that given above: (1) Nautilian, (2) Goniatitic, (3) Ceratitic, (4) Ammonitic, (5) Pinacoceran,—that is, the order of succession is (1st) the simple, (2d) the lobed, (3d) the crenulate or secondarily lobed, (4th) the foliate or tertiarily lobed, (5th) the quaternarily lobed form of Pinacoceras,—and is so far an arrangement of a series of related characters in normal progressive order.

The question naturally forces itself upon us, What relation has this normal order of sequence of the characters to ontogenetic development and to phylogenetic evolution?

Order of the Ontogenetic Growth of these Characters.—1. First, in ontogenetic growth (as illustrated in Fig. 116) we find this order to be the order of sequence in the development of the shell of an individual. The first, or protoconch, stage has a Nautilian or simple suture, or what is the primitive form of that suture (Fig. 116, *G*, and Fig. 111, *A*); the second stage (Fig. 116, *H*) shows the formation of a siphonal lobe by the indenting of the primary siphonal saddle. The Goniatitic modifications are seen in the sutures *K*, *L*, *M*, *O* of Fig. 116, and suture *O* expresses the combination of the Ceratitic and Medlicottian types of crenate suture; but it is the secondary lobation clearly, although in this particular specimen it has not its simplest expression.

This is the general law of ontogenetic growth as developed by the authors who have specially examined these facts; but in Ammonites, as Zittel says, the Ceratitic stage is wanting or passed over. This we may interpret to be due to the fact that the Ceratitic type of suture alone is not expressive of a stage of evolution; but the true fact expressed by Ceratites, so far as its relations to differentiation of suture line are concerned, is its crenate or secondary lobation. This secondary lobation may take place in the lobes, on the sides, or on the saddles, and is a stage which, in the individual growth, is quickly passed over; the order of sequence is preserved by the secondary lobation always preceding the tertiary lobation. The particular part of the curved surface which first suffers the secondary crimping appears to be the lobe, as is seen in Trobites.

Chronological Succession of the Characters.—2. When we look at the chronological relations of this differentiation, we find that the time of first appearance or initiation of the several types of suture lines corresponds with the normal state of differentiation of the character. That is, the Nautilian suture line is the first to appear, in the Ordovician. This continues to be the only one until the close of the Silurian, when the Goniatitic suture line appears. These two are the only types existing, so far as known, until we reach a late Carboniferous stage—the Permo-carboniferous, or Permian—when the third, the Ceratitic and Medlicottian types appear, seen in the genera Sageceras, Medlicottia, and Xenodiscus. But in this same geological period, in the Salt range group of India, is found first appearing the form of suture characteristic of the fourth or Ammonitic stage, in the two genera Cyclolobus and Arcestes. Thus, before the close of the Paleozoic faunas, as now defined, there is seen developed each, except the extreme Pinacoceran, stage of this character. Immediately after, in the Trias, the Ammonitic and Ceratitic types are both well developed and represented by many genera. The historical order of initiation of the several types of sutural modification is thoroughly consistent with the order which an analysis of the nature of the modifications themselves suggests to be the natural order of sequence.

When we examine the order of sequence of the stages of dominance of the several types of suture the former conclusions are also confirmed. The Nautilian, the Goniatitic, the Ceratitic and its modifications, and the Ammonitic and its modifications, became dominant in the normal order. And the appearance of the extreme Pinacoceran type in the Trias, with its failure ever to become dominant, is in keeping with the general principle that it is rarely the case that extreme modifications of a type are either longest to live or the best adapted to struggle with competing types of organization.

Rate of Elaboration of the Various Types of Suture.—3. When we look at still another relation of this series of facts, and ask, What was the relative rate of expansion of this character in comparison with the life-period of the race expressing the modification ? we learn that regarding the character as originating in the straight Orthoceran form, the first stage of sutural modification was reached when the first Goniatite appeared; this was near the base of the Devonian. The Ceratitic and Ammonitic stages had both appeared before the close of the Paleozoic, and by the early Trias the Pinacoceran had appeared; hence the extreme expansion of this character had taken place between the base of Devonian and base of Trias, but the life-period of this particular race of organisms reached its close rather suddenly at the end of the Cretaceous; and we may infer that the extreme limit of modification of this particular character had been attained before the race expressing it had half finished its course.

Rapidity of Modification of each Type soon after it was Initiated.—4. When we consider the degree and rapidity of development in each of these types of suture-lines, we observe that after the character had once appeared it was expressed in numerous species and genera, and it expressed a tendency to expand in a definite direction in all the lines which assumed it, but its rate of development in the different lines was not uniform.

The rapidity of development of this character may have been determined, more or less, by environment, but the facts seem to preclude the possibility of the determination of the nature of the differentiation, or of the order of the sequence of its expansion, by environment. We see here an exhibition of evolution proceeding in a definite and continuous line of expansion. It consists in a differential expansion in a definite direction and in a definite manner, by slow stages of progress from generation to generation; and it is as distinctly a predetermined law of evolution for the race as increase of size and development of organs is a predetermined law for the individual living organism at its birth. Environment checks or accelerates it just the same as temperature or climate affects the vigor of growth of the tree; but the law of expansion from Nautilian to Goniatitic, and then to Ammonitic suture is the only one which the race can follow out; and the expression of this law is as sure to follow in case the generations succeed each other, as the tree is sure to bear its appropriate fruit in case it lives and grows.

Summary of the Laws of Evolution of the Suture-Lines of the Ammonoidea.—The following may be given as a summary of these interesting laws recognized in the history of the suture-lines of the Cephalopod shells. The various suture-lines of the chambered Cephalopod shells can be distinguished by the differences in degree of complexity of the crimping of the edge of the septum, viz.:

(*a*) In the *Orthoceran* and *Nautilian* type the edge of the septum is straight, or the curving is not enough to produce more than a single oscillation of the suture-line during its complete circumference.

(*b*) The *Goniatite* septum presents a lobed suture, but the edges of all the lobes and saddles are simple.

(*c*) In the third type the lobes and saddles are variously crenulated. In the *Ceratite* the crenulation affects the base of the lobes, in *Helictites* the top of the saddles is crenulated, and in *Medlicottia* the lobes, the saddles, and the connecting parts of the suture are crenulated.

(*d*) In the typical *Ammonite* there is a tertiary crimping of the suture-line, i.e., each of the archings of the line corresponding to the crenulations of *Medlicottia* is again crenulated, forming a complexly foliate suture.

(*e*) In the adult forms of *Pinacoceras* there is a still further elaboration of the crimping, the tertiary archings of the Ammonite are again crenulated, forming a quaternary stage of corrugation.

The series presents a gradual elaboration of the crimping of the edge of the septum, forming a suture line, 1st, *simple* 2d, *primarily lobed*, 3d, secondarily corrugated (the *crenulated type*), 4th, tertiarily corrugated (the *foliate type*), and 5th, with the *quaternary corrugations* of Pinacoceras.

In their historical bearings it may be said of this series that:

1. It is the order in which the various types made their first appearance in the geological series.

2. It is the order in which the several types became dominant.

3. It is the order of elaboration in the ontogenetic growth of the individual.

4. It is the normal order of mechanical relation borne by the several types to each other; each type is a mechanical elaboration of the next preceding type.

The convolutions of the suture are crimpings of the edge of a more or less flat disk,—the septum,—and these convolutions are the simplest mode of adjustment of the edge of such a disk, whose circumference increases more rapidly than its radius.

Considering only the differences in the sutures, it would be correct to state that if we assume that the one is derived by modification from the other, it would be mechanically impossible for the Ammonite's septum and suture to be formed without passing through the stages represented by the Nautilus, Goniatites, and Ceratites. In other words, this exhaustive analysis of this one element of structure of cephalopod shells shows us that the actual history of these organisms has been exactly that which a serial classification on the basis of differences of this part would suggest, and that no other classification or order of succession could take place by natural descent.

Evolution of the Suture results in the improvement of the Structure of the Shell.—When we look at the complex foliated septum of the Ammonite in relation to its use, we are struck with the economical use of materials for greatest strength with least weight. The principle of using thin plates of corrugated material in place of solid supports in engineering and building is well understood by man, and from this point of view it appears evident that the result of the evolution of the cephalopod septum has been the improvement of the device concerned.

In conclusion, the analysis of the structure of the Cephalopoda, based upon a comparison of the different modifications of their structure and upon the historical study of the fossil remains of this class of animals, shows very clearly that there is an intimate co-ordination between (*a*) the morphological differentiation of the characters, and (*b*) the historical sequence of initiation and of dominance in numbers of the individuals exhibiting them. Thus we notice, upon examination of the characters of the two great divisions Tetrabranchiata and Dibranchiata, that the group which appeared later, and after the first had flourished and the great majority of its families and genera had become extinct, was the one in which is found the greater amount of differentiation of each of the characters by which the two groups are distinguished.

It is also to be observed that, among the characters, including all that is known of the group, by which the grand divisions of the Tetrabranchiata are discriminated those which were less differentiated morphologically were first to appear. In the case of the modification of the sutures, about which the facts have been minutely studied, the types follow each other in regular successive order from the less differentiated to the more highly differentiated; and the same order is observed in the numerical dominance of the several types. We notice also that this order of increasing differentiation, which may be traced in the case of the suture of the Ammonoidea, is the natural order of evolution when viewed from the points of view of (*a*) mechanical differentiation, that is, the greatest amount of effective use for the least expense of energy or material; (*b*) from the point of view of ontogenetic growth, that is, the natural order by which the structure is produced in the normal growth of an individual organism; and (*c*) from the point of view of historical sequence.

But this is not a case of the survival of the fittest,—it is the evolution of the fittest,—and, from this point of view, too, it is not the fittest that survives; for of these ancient forms it is the Nautilus, and not the Ammonite, that survives; but of the order of initiation there is no mistake—the Ammonite does not appear before the Nautiloid; and the sequence Goniatite, Ceratite, Ammonite is not reversed, but is the order which the structure would suggest. The general law of survival of the fittest is exhibited in the general dominance of one type over another, but a structure once developed may persist entirely beyond the period of its relative importance or relative stage of perfection, as is wonderfully exhibited in the Lingulas of the modern sea, which are traceable back to the Cambrian period through a line of ancestry that was very highly modified in many parallel lines, of which only Lingula survives.

CHAPTER XX.

THE LAWS OF EVOLUTION EMPHASIZED BY THE STUDY OF THE GEOLOGICAL HISTORY OF ORGANISMS.

Testimony of Vertebrates.—The vertebrates might be used with great force to illustrate the general laws of evolution. No better example than the vertebrates could be selected to illustrate the fundamental law of the gradual increase, in differentiation and in rank, of the great classes of a branch in the order of their successive appearance and dominance in the geological formations.

In the lowest system of stratified rocks, the Cambrian, no trace of vertebrates has yet been found. In the Ordovician and Silurian only the lowest type of fishes, and they very rare, have been seen. Fishes were abundant in the Devonian. The Lower Carboniferous shows the first amphibians; and large-sized and extinct types of amphibians prevailed in the Carboniferous era. In this era also a few traces of true reptiles have been found. In the Triassic the great Dinosaurian reptiles were abundant on the land. In the Jurassic the shallower seas swarmed with the Enaliosaurs or sea-lizards, and in the lower Jurassic (Lias) the flying reptiles infested the air and culminated the reptilian domination of the Mesozoic time.

While reptiles were the masters of sea, land, and air, the lower types of mammals—the marsupials, and probably monotremes—began to appear in feeble representatives as early as the Triassic, and in the Cretaceous birds, too, make their appearance: though true birds in structure, they compete with the flying reptiles in their use of reptilian teeth for offence and defence.

Remarkable and Extreme Evolution of the Mammals in the Eocene.—As we examine the earlier beds of the Tertiary rocks we observe for the first time the dominance of mammals; and perhaps no more remarkable fact is established in the history of organisms than the sudden expansion of the placental mammals in the Eocene.

Over fifty genera, representing the chief ordinal types of the placental mammals, are already reported from the lowest Eocene, none having been discovered in the underlying Cretaceous. In Europe alone Zittel reports for the fauna of the Upper Eocene about 110 genera and about 200 species. To show the richness of this fauna, in spite of the imperfection of the records, he cites the facts that "Our present European land mammalian fauna contains 54 genera with about 150 species, and of these 60 per cent belong to the microfauna, consisting of the smaller forms of Rodents, Insectivora, Bats, and Carnivora, for which the conditions of preservation in earlier epochs were very unfavorable" ("The Geological Development, Descent, and Distribution of the Mammalia," by Karl A. von Zittel, *Geol. Mag.*, Dec., III., vol. X., Sept., Oct., Nov., 1893).

If we glance at the whole group of mammals, we find the actually known forms included in three subclasses: the (I) Prototheria, with the order Monotremata; (II) Metatheria, represented by the order Marsupialia; and (III) Eutheria, or the Placentalia.

There can be no doubt as to the higher rank of the Placentalia over the marsupial and monotreme types. No certain traces of the Placentalia are known to occur below the Eocene. Stegadon, a genus of the Tillodontia, is thought to have appeared possibly in earlier beds.

Of these mammals, ten orders, fossil and recent, are recognized. Two of these are marine—Sirenia and Cetacea. The Edentata is a South American order, and has its representatives in the earliest known South American mammalian fauna (Vera Cruz fauna of Patagonia), which is probably equivalent to the northern Eocene.

If we omit the above three orders, of the remaining seven orders of land mammalia five are represented in the older Eocene of Europe—the Ungulates, with 5 suborders; the Rodents, the Insectivores, the (Carnivora) Creodonta, the Prosimiæ—the forerunners of, if not true, Primates.

True Carnivores appeared in the newer Eocene, Cheiroptera in the middle Eocene, and true Primates in the older Miocene.

In these orders of placental mammals 56 genera appeared for the first time in the older Eocene, and there were successively added to them, in the middle Eocene 40 new genera, in the newer Eocene 105 new genera, Oligocene 5, older Miocene 49, newer Miocene 34, Pliocene 27; or previous to the opening of the Pleistocene 260 genera, distributed among the seven land orders of mammals, of which the first traces were obtained from the older Eocene beds of North America and Europe. The Australian, South American, and African types are not here included; and it must be remembered also that new discoveries are constantly adding to these statistics, and in general they augment the earlier more than the later totals.

Again, the fact that (? Prototheria and) Metatheria were already well developed in genera in the Mesozoic does not lessen the significance of the remarkable expansion of the mammals in the older Eocene period; nor does the imperfection of knowledge lessen the testimony to the relatively sudden expansion which the evidence now in hand indicates. The approach to recent time, and the increasingly better representation of the land faunas among the preserved remains, does not invalidate the truth of the general proposition, that all the grand features of structural modification, expressed in the subclass of placental mammals, made their appearance in distinct genera with great rapidity at the first stage of appearance of the Placentalia.

The prominent differences, expressed in the limbs, teeth, form, and habits, in the hoofed animals, the odd and even toes, the gnawing rodents, the flesh-eating Carnivores (Creodonts), the insect-eaters, the flying bats, and the climbing monkeys, were all seen among the members of the first fauna of the new type of placental mammals, in the Eocene period.

Synthetic Types Illustrated by the Vertebrates of the Mesozoic.—No better illustration of the principle of the "synthetic" or "comprehensive" character of early types of organization is to be found than that presented by the Dinosaurian reptiles and the reptilian birds of the Mesozoic. Here we find biped reptiles, three-toed and with avian pelvic structure; flying reptiles, with beaks instead of teeth; birds with teeth, and birds with long vertebrated tails.

So many points of combination of features have been seen in the Mesozoic fauna, which are now only found separated in the two great classes Aves and Reptilia, that zoölogists have been forced to provide an intermediate group to include these ancient types, or to expand and combine the two classes into the one superclass Sauropsida of Huxley.

Specialization of Five Fingers in Reptiles and its Relation to Later Specializations.—The principle of synthesis, or combination, in an early type, of the characteristics of two or more sepa-

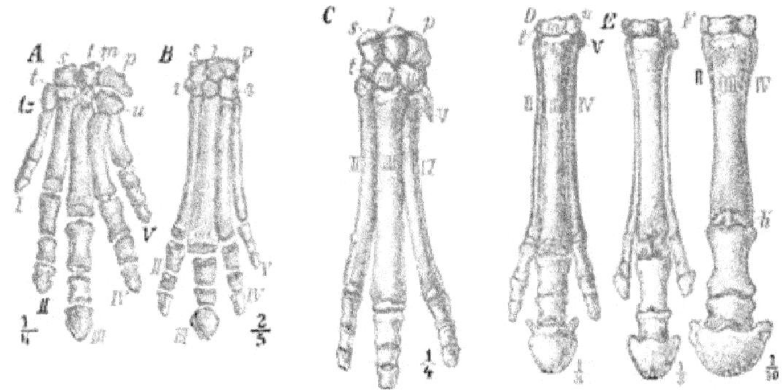

FIG. 120.—Left forefoot of *A*, *Phenacodus primævus* Cope, Eocene; *B*, *Hyracotherium venticolum* Cope, Eocene; *C*, *Palæotherium medium* Cuv., Oligocene; *D*, *Anchitherium aurelianense* Blainv., Miocene; *E*, *Hippotherium gracile* Kaup, Pliocene; *F*, *Equus caballus* L., Recent. *l* = lunar; *m* = magnum; *p* = cuneiform; *s* = scaphoid; *t* = trapezoid; *tz* = trapezium; *u* = unciform; I–V = 1st to 5th finger or metacarpal bones; *mc* = metacarpal. (Steinmann and Döderlein.)

rate types of a later stage, is seen in the case of the Permian reptile *Mesosaurus tumidus* Cope, in which five tarsals are present, rather than four—the normal number of later reptiles. Such a fact shows, according to Cope, that five is the primitive number of tarsals, and that four is a specialization—just as we find in general in the evolution of paws, feet, and hands the full number of parts was provided before the specialized reduced number was evolved. The fewer number of fingers or of bones, entering into the mechanism of the foot or hand, is the result of selection and specialization of parts rather than the direct production of any new function or part. The Eocene *Phenacodus primævus* Cope illustrates this principle in the evolution of the forefoot of mammals, as shown in the figure on the opposite page.

Finger-bones and Teeth as Tests of Degree of Differentiation.—In tracing the history of mammals we find the principle of five fingers already developed before mammals began. Hence the wonderful modifications noted by Owen, Kowalevsky, Ryder, Marsh, Cope, and others, in the arrangement of the bones of the mammalian feet, their specialization in form, and relative size, shape, and position, have constituted the chief data for both classification and phylogenetic series.

The teeth, as highly specialized organs, and as terminal parts of the individual organization, coming into most immediate contact with the outside elements of resistance to the life of the individual, are particularly sensitive expressions of the stages of evolution.

Any device of offence or defence, particularly when hardness and resistance to attrition are characteristics of its structure, becomes at once a mark of the effects of environment in inducing modifications, and of the stage of progress attained by the individuals in their evolution. Their resistance to destruction makes such parts most valuable records in the rocks of the history of organisms.

Laws Derived from the Study of the Teeth of Mammals by Osborne.—Professor H. F. Osborne, following the investigations of Rütemeyer and others, has recently written several instructive papers setting forth the laws to be observed in the history of the development of the teeth in mammals.

In a memoir (first read as the address of the vice-president of the section of Zoölogy, of the American Association for the Advancement of Science*) he narrates both concisely and admirably the laws expressed in the modification of the cusps or surface forms of the teeth of mammals.

Osborne shows how the tricuspid tooth is an evolution from a simple monocuspid tooth, which is the primitive type of tooth in all earlier vertebrates. He shows further that the multiple succession of teeth characteristic of reptiles is the primitive method of arrangement, and this, as is also the in-

* "The Rise of the Mammalia in North America," *Am. Jour. Sci.*, III., vol. XLVI., pp. 379–392 and 448–466.

definite number of teeth of the reptilian jaw, is a natural preliminary condition to the high specialization of the teeth, with particular form for each.

The selection and specialization seem to be brought about by the suppression of part of the multiple series, and the modification of the teeth retained in different parts of the jaw for special function.

In the primitive Marsupials and Insectivores, he observes, the regular reptilian succession was early interrupted, while in all the higher mammals the reptilian succession of two series was retained in the anterior part of the jaw. In the Edentates and whales retrogression takes place in fins as well as in teeth; it is the first set of teeth that persists, the second set being represented by a rudimental row of tooth-caps buried in the jaw.* He concludes that there is strong evidence that the stem mammals had a uniform number of each kind of teeth and a uniform dental formula; that homodontism is secondary, and was actually preceded in time by heterodontism in the mammalian dentition.

The ancestral formula for both Marsupials and Placentals, according to this author, is: incisors 4, canines and premolars 5, molars 4. By adopting Röse's suggestion that incisor 5 of the marsupials belongs with the second series of incisors, he supposes that Placentals have lost one incisor and one molar from the primitive formula. The paper is an important contribution to the interpretation of the method of evolution, and must be studied with care to be fully appreciated; the author's conclusions are quoted on page 324.

For the purposes of this treatise a sufficient number of illustrative cases has now been presented to show where the emphasis is placed by the facts of geological biology as to the true factors of evolution. A great many examples crowd themselves upon the attention which must be left for the student to investigate directly and in detail. The evidence to be derived from the study of living plants and animals is so vast, that a special treatise would be necessary to do justice

* p. 378.

to either, and the reader may find many admirable treatises giving account of this aspect of evolution.

Method and Purpose in the Selection of the Evidence here Set Forth.—The facts which have been selected in these chapters have been chosen for the purpose of ascertaining what the geological history of organisms has been.

Examples have been taken and analyzed to ascertain what has been the particular law of succession in particular cases where the evidence was full enough to be relied upon. If the interpretation of these selected cases has been correct, the principles discovered may be applied to other cases.

The facts have been examined for the purpose of learning (1) what the fossils indicate has been the order of succession in the initiation of different forms of organisms; (2) what relation this succession bears to the relative importance of the characters in the economy of the individual organism, as shown by the systematic classification of the Animal Kingdom; and (3) what have been the determining causes by which the multitudinous differences in organic structure have been brought about. The first consideration in their selection was that they should be from among those of which the most perfect record is preserved. The cases already cited in evidence are not selected because they are the most important examples, nor because they illustrate only the most important laws of evolution, but they are selected because they are the best examples to show what the geological records testify regarding the history of organisms.

Different Kinds of Evidence Borne by Living and Fossil Organisms.—Living organisms present the best evidence of the laws of ontogenetic development, because they furnish illustration of each stage in the development. A continuous series of the stages of development of a single organism is more satisfactory evidence of the essential nature of that development, than would be any number of detached exhibitions of sundry stages of development of different organisms.

So it is believed that the evidence borne by a series of fossils preserved in each stage of the geological record, of which specimens are well preserved and described from the first to the last, and which show the beginning, dominance, decrease, and extinction of the type they represent, is of the highest value as evidence of the actual order of evolution and of the general laws by which differentiation of form has taken place. And a few such cases far outweigh any number of detached specimens tied together by theoretical links.

Natural Selection seems Reasonable when Based alone upon the Study of Living Organisms. — When we observe living animals in competition—the vigorous ones living and the weaker dying, the strong overcoming and devouring the weak, the large and fewer in number making their daily food of the smaller and more abundantly produced; when we note how the places for the greatest abundance of individuals are determined by the presence of favorable conditions for obtaining food; and thus, in general, when we observe that animals as they are are actually adjusted, each to its own most favorable conditions of environment—it seems reasonable to draw the conclusion that the differences distinguishing one animal from another may have arisen as the result of better fitness for the struggle for existence on the part of those which survived and carried on the race.

Having once assumed that the law of evolution is a process in which the chief active determining force has been natural selection by the survival of the fittest, it was easy to find illustrations of adjustment of structure and function to the conditions of environment among fossil, as has been done among living, organisms.

Every Species of Organism that has Flourished in the Past the Fittest for its Place and Generation.—When, however, we stop one moment to consider the relations of organisms in the past to their own environment, it becomes evident that, at every particular stage in the geological history of organisms, the individuals then existing must have been as thoroughly well adapted to live under the conditions of their environment as the present inhabitants are adapted to live in their environment. Every organism that has lived on the earth has in some sense been the fittest to live in the particular time and conditions it occupied.

If environmental conditions (outside of organic environment) have determined the evolution of organisms, then we are obliged to assume a degree and amount of change in them of which the facts of geology give no evidence.

If the conditions which have changed with the geological ages have been the organisms themselves, and they have constituted the environment, then it becomes necessary to explain the more powerful contestants before their selecting agency can result in the survival of fitter races.

But leaving aside for the present the philosophical argument, the burden of these pages is to show what is in fact the testimony on these questions furnished by the organic history as found in the best-preserved parts of the record.

As previously explained, the records which are made at the place and time of the formation of the rocks are those which must on that account be the most perfect we can consult. The rocks bearing fossils are not wholly, but are in the large majority of cases, of marine origin. This determined the selection of the evidence from among marine animals. The animals of which the best records could be preserved in the rocks are those secreting hard parts—shells, or corals, or similar parts; hence the examples have been taken chiefly from the corals, the Mollusca, and Brachiopods.

The Geological Evidence does not Emphasize the Importance of Natural Selection as a Factor of Evolution.—What has already been said is sufficient to show that the emphasis of the testimony brought forward differs from the emphasis drawn by the embryologist, or by the student of living organisms, as to the relative prominence of the several factors in the evolutional history of organisms.

That which has seemed most conspicuous to the latter class of observers has been the intimate relationship existing between morphological difference and environmental conditions; paleontological facts point to the greater importance of the continuous and progressive process of differentiation and specialization of structure and function with the passage of geological time.

The law of natural selection, suggested to explain the evolution from the first point of view, calls for an extremely slow rate of modification, but uniform and continuous. The facts of the history itself point to the reality of rapid strides at critical points, with long periods of almost absolute cessation of progress; and suggest that the part played by what is called natural selection has determined rather the particular individuals and the place and time for advance steps, than, either the direction of the steps themselves, or the relative value of the particular modifications in relation to continuation of the race, which have taken place.

The study of the actual facts of the geological history of organisms points unmistakably to a course of evolution by descent, in which the progress attained by each succeeding form was a paramount condition of the origin of the next member of the race.

Objection may be taken to an argument based on so few examples. I think the force of this objection will be lessened when we bear in mind that the examples were selected primarily because of their fitness to testify upon the points in question, viz., the law of the history of organisms, the nature, the rate, and the order of modification of form, which organisms actually undergo in producing that divergence of specific forms observed at any particular stage of the history.

It may be said that the particular kinds of animals selected do not fairly represent the total life of the world. To this objection the reply may be made that a full quota of diversity of specific forms has been attained by the races examined, and the chief question before us is, How has that diversity arisen?

If the facts we have examined do not support the hypothesis that the chief factor in organic evolution is either external environment or natural selection, it is not on account of any lack of fitness to testify on this point, if it were true.

The facts examined—and we believe that fuller examination of other statistics both fossil and recent, will support the same conclusion—show that evolution is rather an intrinsic law of all organisms, and is to be discovered in the phenomena of variation, which appear to be constantly active, rather than in any accidental operations dependent upon the conditions of external environment.

The emphasis is placed upon the intrinsic rather than the extrinsic factors of evolution, as the actual determinants of the results attained by evolution in specific, generic, and the higher orders of differentiation.

A Statement of the Laws of Evolution Emphasized by Fossils.—The analysis of the facts regarding the order of succession and modification of organisms derived from this critical study of fossils suggests the following to have been some of the chief laws of the evolution by which the present conditions of the organic world have arisen:

(1st) An orderly succession in the geological history of organisms, which in the main has resulted in an increasing differentiation of structure and specialization of function with the progress of geological time. The general name for this process is *evolution*.

(2d) While the whole organism is concerned in this evotion, *certain parts* of an organism (or certain of the morphological characters) *exhibit the evolution more rapidly than do other parts* or characters.

(3d) When these characters are arranged in the order of relative rank of importance in the economy of the organism, *the characters of least importance* (the varietal and specific characters) *exhibit the evolution most constantly and persistently, but at a very slow rate*, chronologically considered.

(4th) The *characters of higher rank* (the branch, class, ordinal, and family characters) *were relatively more rapid in the expression of their initial evolution and thereafter were very constant in each successive race.*

(5th) These two tendencies are expressive of the two fundamental laws of evolution—variability and heredity. *Variability* is recognized as a common law of organism, according to which, in the ordinary process of generation, slight changes are continually taking place in the morphological features of the offspring as compared with the parent form.

Heredity is a common law of the organism, according to which a character once acquired in the parent tends, in the process of ordinary generation, to be repeated with increasing precision, and to result in the transmission of characters without change from generation to generation. The process of evolution is the combined result of the interaction of these two antagonistic laws of the organism.

(6th) The *mode of the evolution* consists in the acquirement of new characters by variation, and in the *acceleration* or the *retardation* of the development of characters already acquired.

(7th) The *cause of the evolution* is of a twofold nature—*extrinsic* and *intrinsic*.

In the first case, extrinsic evolution, the *direction* and *specific character* of the modifications appear to be *determined by the conditions of environment*—using that term in its broadest sense for all the outward conditions of life in which the individual organism finds itself after birth. *Adjustment* of the organism to the environment, *struggle for existence*, and *natural selection* are the terms under which extrinsic evolution is commonly defined.

The *intrinsic* cause of evolution acts previous to the individual birth, and it seems to be at the foundation of variability. The mode and manner of expression of this kind of evolution are more difficult to define than in the case of extrinsic evolution; but the facts of Paleontology clearly indicate that such a cause exists, prior to the morphological appearance of each individual and species.

(8th) In this discussion *classification* is recognized as an orderly and epitomized formulation of the facts already known regarding the extent and kind of differentiation actually attained in the evolution of the characters of organisms. The statistics of classification are therefore available for expressing, numerically, the relations existing between organic characters and time and place; and it is observed that the numerical relations of the different kinds of organisms to the time and the place of their appearance point with overwhelming force to the conclusion, that acquirement of morphological difference is co-ordinate with both the passage of geological time and the divergence of the conditions of external environment in which the organisms have lived.

CHAPTER XXI.

PHILOSOPHICAL CONCLUSIONS REGARDING THE CAUSES DETERMINING THE COURSE OF EVOLUTION.

What is the Philosophy of Evolution?—Statement of the Case. —In the foregoing chapters a few of the prominent facts regarding the history of organisms have been examined, and the primary conclusion from their study is that the method of acquirement of all that is characteristic of organisms has been evolutional. Evolution is a matter of fact in the description of the history of organisms; but there remain for consideration the questions, Why should organisms express a law of evolution? and, What are the immediate conditions determining the particular steps of evolutional history? and, finally, What is the rational philosophy of evolution?

The theory of natural selection may be so applied as to lead to the philosophical belief that difference in the conditions of environment is the primary cause of the differences expressed in the form and functions of organisms; and secondly, the theory of the unchangeableness of matter and the universal conservation of energy may be carried so far as to lead to the belief that in the matter of organism, under the names *germ plasm, biophors, pangenes, gemmules, physiological units*, or some other names resides the power and potency of all that is evolved in the course of the total history of organisms.

Are these beliefs incident to the proposition that evolution is a fact in nature, or is there a philosophy of evolution which more completely recognizes the whole body of facts in the case?

The Point of View.—If we were to discuss such a common topic as the weather, we would find that, although everybody has his notions as to the cause of the various changes in that very variable phenomenon, the likeness and differences in the theories advanced are determined primarily by the point of view in relation to land and water of their advocate. Land and water are sharply contrasted, natural and familiar phenomena; but the Bostonian is accustomed to look to the eastward for his ideal expanse of water, and for him the land extends from the solid *terra firma* upon which he walks for unmeasured miles to the westward. The man of St. Louis is familiar enough with land, but the ocean is a foreign thing to him; it does not come into his every-day reckoning. At San Francisco the Bostonian's notions are simply reversed; the point of view is totally different.

Residents of these three cities, unless they were to adjust their definitions to the points of view of their companions, could not talk about even the weather without constant misunderstanding.

The Act of Evolving as well as the Order of Events Included in the Discussion.—In the same way it may be said that some of the chief misunderstandings and differences of opinion regarding the problems of evolution are due to a failure to appreciate the differences of philosophical attitude from which the matter is viewed.

Evolution is concerned with two very distinct fields of human inquiry. On the one hand, *evolution* is the name for *the natural order of unfolding of the characters of organic beings that have lived on the earth;* on the other hand evolution is the name for our conception of *the mode of operation of the fundamental energy of the universe.* Thus it will be seen that the notion of God is as intimately involved in a discussion of evolution as is the notion of organism; in elaborating the definition of the one we consciously or unconsciously elaborate our definition of the other. We are obliged to consider the act of evolving as well as the results of the evolution.

The Course of the Discussion.—In the present discussion the reader has been led step by step from the detailed, statistical description of actually existing objects of nature, in their relations to time and space and to each other, through the consideration of their classification on the basis of order of arrangement, proportion, and intricate relationships of structure and function, up to a consideration of the scientific explanations proposed to account for them. We have passed from the promiscuous array of facts, through analysis and systematic classification, to the reasons for the classification and the interpretation of the meaning of it all in terms of force and cause.

So long as we deal only with sequence of forms, and consider only the relation of particular forms to particular places in the series, evolution is simply an analysis of the order of events. When we step one side or the other of this simple process of the narration and classification of facts and events, we leave the field of scientific observation and are dealing with the principles of causation. It is useless to disregard the philosophical side of the study of nature, and it is a mistaken notion to think that those who spend their time in measuring and recording phenomena have no need to consider the meaning of such terms as cause and effect which elude actual observation.

Darwin's Origin of Species Centres its Interest in the Search for Causes.—Darwinism is an attempt to find a cause for the differences in form and function observed in the organic world, and the search for this cause has aroused a world-wide interest in observing and recording the phenomena of nature; but the real stimulus inspiring all the investigations has been the expectation of discovering somehow the true cause of these things in some visible, tangible and describable form. Origins and creations have been said to be discovered in the search, but the calm philosopher knows full well that the origins described have been only apparent origins; they have not reached to the essence, or to a fundamental explanation of nature.

The Evolutional Idea of Creation.—It has been supposed by many that evolution is intrinsically antagonistic to, and has, in fact, replaced the creational conception of the origin of things in the world. In one respect this is partly true; the new view has fundamentally changed the conception of creation. Evolution has given us another notion of God. In the old conception God was an artificer making organisms out of inorganic matter directly, as one might build up a vessel of clay and then vivify it. The new conception of God as creator finds its concrete, empirical representation in the act of expressing a thought or purpose into the spoken word. Creation is the phenomenalizing of will, so sublimely described in that ancient formula, *In the beginning God spoke and it* (the whole phenomenal universe) *became*.

The origin of the universe is thus the becoming phenomenal of an eternal purpose; the only alternative is to deny all origin, and to assume that the phenomenal universe itself is eternal.

Evolution the Mode of Creation of Organic Beings.—And, as we have seen, the great distinction between organic and inorganic matter consists in the evolving of the organic characters in an appreciable and often very slow course of time; whereas the qualities of inorganic matter were originally committed to the particular matter, which has continued to exist from the beginning without change.

The slowness and continuity of the process of organic evolution is thus an evidence of the continual presence of creative energy in the world, and the permanence of qualities of inorganic matter is evidence of the ultimate distinctness between the created and the Creator. The human mind is utterly incapable of accounting for intrinsic differences in the universe except by conceiving of some mode of their origination, and we have not explained their origin by simply saying that they have evolved.

The change which the speculations of the last fifty years have wrought in the notion of creation has been a most important and radical one. There has been substituted for the old idea of an artificer constructing a machine out of materials, with the addition of his making his own materials out of nothing, the higher conception of the transformation of a conscious purpose into physical action—the visible expression of invisible will.

The Properties of Matter Coexistent with it, and either Eternal or Created.—The new notion of creation does not include the idea of the making of something out of nothing, but it does mean that what has existed already in one state of being (which we describe under the simile of purpose of the eternal mind) becomes expressed in another realm of existence (which we describe in terms of form and function of living matter).

When we define matter as being of various elemental kinds, their differences being expressed by their behavior under sundry conditions, and called properties or qualities, we proceed on the assumption that these properties are characteristic of the particular kind of matter, and have been from its first existence, so that there is no evolution: the properties are either eternal or were immediately created as they are.

In the case of organisms it does not free us from the same conclusion, if we liken their characters to the properties of matter, and imagine that there is some original endowment of differences which gradually finds expression by evolution.

If we attempt to treat the characters of organisms as if they were properties of matter, we are forced to imagine infinite and inconceivable ultimate units, like atoms, of which the original organic matter of ancestral organisms was composed, and it has been found necessary to endow these units with qualities of persistence and definition, of will and determination, of power over the environment in which they reside, and of judgment of the value of the to-be-attained morphological structure and functional activities of the organisms, which in the creational idea are ascribed to the will and mind of the Creator.

Any one who is not already prejudiced against the notion of God cannot fail to see in the theistic view of the Creator, in which eternal will and purpose constitute the powers and potencies back of phenomena, a more rational and satisfactory theory of the universe than the materialistic view in which the same powers and potencies, invisible and infinitesimal, are made to be the endowments of an infinite number of undying, determinant, organic units.

Evolution does not apply to the Mode of Becoming of Chemical or Physical Properties of Matter, but is the Distinctive Characteristic of Organisms.—In the case of chemical and physical properties, as related to particular material things and on the assumption that matter is not eternal, their creation can be considered only as having been immediate, since our whole science of physics and chemistry is based on the assumption that these properties persist without change.

But in the case of organisms their characters are constantly changing, and evolution as a theory is based upon the assumption of not only constant but progressive change. The origination of the organic characters was not done all at once, but evolution as the mode of creation of organisms has been more or less continuous throughout the geological ages. It is this continuation of the process of phenomenalizing that distinguishes the mode of creation in the organic realm from that in the lower realm of inorganic matter. Whatever is characteristic of organisms was not created at once in any remote beginning, but has been unfolded by degrees, and there is no reason for supposing that the process is not still going on. Such expressions as "effort," "growth force," "conscious endeavor, "reactions, "producing modification." "determination," "memory," etc., used in describing the phenomena of evolution, all express the notion of the pre-existence of some unphenomenal property, or power, or potency, which constitutes the cause of the particular characters which are acquired by organisms in the process of their evolution.

The Evolutional Idea an Enlargement of the Conception of God as Creator.—On the assumption that the ideas of creation and Creator are fundamental to a rational explanation of the universe—and such an assumption seems to be a logical necessity to account for any intrinsic heterogeneity—we observe that the effect of adding the idea of evolution to creation enlarges the conception of creation by making it a continuing process instead of an ancient act, and brings God into the midst of the present universe.

The purpose of the living God then becomes immanent by continuously phenomenalizing itself into living form. God thus becomes a living, present, active reality in the existing universe, and the course of the evolution of organisms becomes in a true sense the history of creation. This term "Schöpfungsgeschichte" was chosen by Haeckel for the title of his treatise on the laws of evolution, and in one of its closing chapters he acknowledged that there are only two ways of accounting for the original organisms—spontaneous generation or creation.* Both of these hypotheses are alike in recognizing that nothing in the visible universe is capable of accounting for the properties of living matter.

Evolution as an Account of the Course of the History of Creation a Gain upon the Older Idea of Arbitrary Creation, but not a Satisfactory Substitute for Creation.—Evolution as a theory of the mode of the orderly appearance of heterogeneity among organisms is a great gain upon the older theory of creation, which found no natural or regular method in the history, but only an arbitrary and unfathomable complexity and heterogeneity.

That this order of sequence is correlated with genetic succession, and is thus bound up with the organic nature of the evolving beings, is a most rational inference from the facts observed.

But evolution as a theory of origins, as an attempt to explain *why* things are as they are, as a philosophy of the cause of organic diversity, is an utterly inadequate substitute for creation. And we find the most zealous advocates of pure scientific observation unable entirely to avoid the inquiry Why are things as they are?

Consideration of Causation Indispensable to a Thoughtful Study of Nature.—In our studies we may for a time confine our attention to the "course of nature," entirely excluding all consideration of matters not pertaining strictly to definition and classification of the facts actually observed and measured; but sooner or later we must think, and when we think the question of cause, and the nature of the relation of cause and effect, inevitably arise.

A scientist, so ardent for the elimination of everything unscientific from science as Mr. Huxley, was not unconscious of something beyond, as is illustrated by the following quotations.

In the admirable study of the "crayfish" as a typical organism we find the following definition: "*The course of nature as it is, as it has been, and as it will be is the object of scientific inquiry; whatever lies beyond, above, or below this is outside science;*" but such a definition only follows the state-

* Vol. I. p. 345.

ment that "*the phenomena of nature are regarded as one continuous series of causes and effects, and the ultimate object of science is to trace out that series.*" . . .*

And in the same essay the remark is made that "*Under one aspect the result of the search after the rationale of animal structure thus set afoot is Teleology, or the doctrine of adaptation to purpose; under another aspect it is Physiology.*"

If we admit into the discussion of science the question as to the causal relation of one thing or event to another, the consideration of a supreme cause necessarily comes into the case. As is tersely phrased by Whewell: "*In contemplating the series of causes which are themselves the effects of other causes, we are necessarily led to assume a supreme cause in the order of causation, as we assume a first cause in the order of succession.*"†

Causes not Discovered by Observation, but Discerned by the Reasoning Mind.—In the scientific study of organisms it is possible to separate in our minds the act of observation from the act of the associating one observed fact with another as cause and effect. It is one thing, however, to observe, note, measure, define, and classify organisms and their structures and functions, and quite another thing to state that a particular structure and function is caused by a particular preceding structure and function or by any other preceding conditions of the world.

For instance, there can be no dispute that the heat of the sun, the various conditions of moisture, of air and soil, incident to the spring season, are the direct causes of the leafing out of the elm-trees on the street side; but it is far from the truth to say that these conditions of environment have had any causative agency whatever in producing the elm leaves, when the elm leaf is considered as differing from a maple leaf. The mere association of two phenomena together does not determine the one to be the cause of the other.

The fact that we are familiar with and understand the effects of heat and moisture, and do not understand the operation of the more hidden biological forces, does not influence

---

* "The Crayfish," p. 3.

† Nov. Org., III., x. § 7.

at all the decision that the sun, while it is the cause when we speak of the development of the leaf, is not the cause when we speak of the particular course of that development.

When we seek the cause of the changing of the characters of organisms in the course of geological history the same reasoning applies.

The fact that an infinitesimal part of the differences in the characters of organisms is an expression of adaptation to the immediate conditions of its local and temporal environment does not suffice to prove that the environment is the cause of the adjustment.

The determination of the true relations of cause and effect in nature is therefore not a matter of observation, but interpretation of cause is founded upon the philosophy we apply in the interpretation of the course of nature.

Ability to Adjust the Organization to Conditions of Environment a Chief Element in the Fitness for Survival.—It is undoubtedly true that the fittest do survive, but too much is made of the theory that fitness consists in precision of adjustment of organic structure to conditions of environment. If this were true the less variable would be more fit than the more variable, and the result of survival would be the cessation of variation; whereas it is probably much nearer the truth to say that fitness to survive is in almost direct proportion to the ability to vary.

Darwin did not find it essential to inquire why variation takes place: variation was assumed to be a common fact in the life of organisms, and it is one of the chief factors of evolution. But when we push the question, why has a particular variation arisen, become abundant, and been transmitted from generation to generation? we are forced to the conviction that the primal characteristic which distinguishes it from its unsurviving fellows is its greater capacity to modify its structure, function, and habits into fitness for the particular conditions of environment. It is the greater ability to adjust, not the closer adjustment of structure to environment, which constitutes the higher fitness to survive. An organism is the fittest to survive, not because it has less to oppose it, or less to overcome, not because the conditions of life are easier or more congenial to its particular condition, but because it has more of the essence of evolution in it.

The Philosophy of Evolution: a Summary.—It is this view of evolution which the geological history of organisms emphasizes. When we look back historically to the early geological ages, and not assuming that we have reached the beginning, but allowing that there may have been as long a stretch of time before the Cambrian as since, for organisms to evolve in,—when we compare the rate of initiation of characters of higher rank with the rate of initiation of varietal or specific rank,—we find it to be a striking fact that relatively the initiation of higher characters predominated in early times, and as time went on differentiation in each line was confined to characters of less and less taxonomic value; to use the oft-cited figure of a phylogenetic tree, all the main branches dichotomized near the roots of the tree, and as we advance chronologically toward the present the branching has been confined to secondary and tertiary limbs and terminal twigs. Although such a tree is used as a figure of the way in which differentiation has arisen, it seems never to have occurred to those adopting this analogy that all the branching of a tree is peripheral, at the very terminal twigs. The bifurcation of two contiguous twigs *becomes* the main crotch of the trunk only by the circumferential growth of the twig into a great limb; but does any one imagine that the difference between a Crustacean and a Pteropod was in any particular *of less taxonomic value* in the Cambrian time than it is now? or has the difference between two species of Silurian Rhynchonellas become of any greater significance by the continuous evolution of the Brachiopods up to recent time? No; natural selection only works at the adjustment of varietal modifications in making them permanent, or in dropping them out of the race; and the mere transmission of an insignificant character from parent to offspring for a million generations cannot in itself have the least effect in raising the economic importance of that character among the functions of its possessor. It is this view of the case which shows natural selection to be but one of the phenomena incident to evolution, and not the main factor in the case. The same force which is expressed in the appearance of the new variation in the first place is required to account for the appearance of the new generic, the new ordinal, or the new class character. This force has been distinguished as intrinsic evolution; it is expressed in variation itself, which is the chief factor assumed in the theory of natural selection. The nature of the force is expressed in the term *blastogenic* of Weismann, and in the term *centrifugal*, as used by Poulton; but whatever it is called the importance of the distinction lies in the fact that the selection, the preservation, or the transmission of a character does not account for its origin.

Evolution is thus seen to be a process that is primarily organic: it is expressed in the acquirement of new characters in the course of growth by living organisms; and we may as reasonably speak of evolution force, as of the growth force of the individual, or the force of gravitation. As the normal laws of growth of the individual are thwarted and diverted by external conditions, so undoubtedly a greater or less modification of the course of evolution has been produced by the conditions of environment.

When we attempt to explain the course of evolution by tracing it backward from the differentiated, adjusted organisms to their ancestors, it is natural to place great importance upon the fact of the accomplished adjustment of the individual to its particular environment; but when the point of view is reversed and the organism is traced from the earlier geological periods through the ages down to the present time, the conviction becomes impressed upon the student that environmental conditions are but the medium through which the organic evolution has been determinately ploughing its way.

Differentiation of form and function has been the expression of vitality, and environment is never exhausted. With the occupying of unexplored fields has come divergence and the appearance of new form and structure; progress has not been made in overcrowded fields by the survival of the fittest. The crowding of the field has led to division and co-ordination of labor. All die in due time, and thus end the struggle;

but they who could best adjust themselves or their actions to adverse conditions were the fittest while they lived, and it was they who diverged. Those expressing more strongly than their fellows the originative energy of life itself are the ones to push forward and furnish the surviving and persisting members of the race. The pioneers, the skirmishers in the front line, are those among whom appear the founders of new species and new races, as with men they are the makers of new nations and of higher civilization.

Thus evolution has been working in the midst of the races from the earliest recorded times; in each line it has been regularly progressive in its order, everywhere advancing as rapidly as the conditions already attained have rendered it possible.

The great facts attested by geology are that the grander and more radical divergences of structure were earliest attained; that, as time has advanced, in each line intrinsic evolution has been confined to the acquirement of less and less important characters: such facts emphasize with overwhelming force the conclusion that the march of the evolution has been the expression of a general law of organic nature, in which events have occurred in regular order, with a beginning, a normal order of succession, a limit to each stage, and in which the whole organic kingdom has been mutually correlated.

In closing, an illustration may be used to emphasize the real points at issue.

Suppose a handful of lead shot were placed in a blunderbuss, and the whole load discharged at a burglar climbing into my chamber window.

The individual shot, originally of globular form, would be found at the end of their journey of various shapes and in various positions. Some of them would have travelled till they expended their force and dropped to the ground in the distance comparatively unchanged; others would be slightly distorted by impact upon the soft clothing or flesh of the intruder; others would be flattened by meeting the resistance of bone; a few would be stamped with the shape of some brass button, surface of nail-head, or some other impenetrable substance against which they had struck. All the modifications of the separate shot, their particular stopping-places of rest, in fact every particular of the shape, condition, and position finally assumed by the shot, would in greater or less measure be the result of the influence upon them of the conditions of the environment.

The immediate conditions of environment in which each shot was found would appear to be a sufficient cause to explain the particular modification of that shot from its original simple globular condition. The exact repetition on the shot striking the brass button of its particular form would seem to be sufficient evidence to prove that the one cause of the form assumed was the adjustment of the shot to the conditions of its environment. The fact observed is the actual perfect adjustment of the lead pellicle to its conditions of environment; this adjustment is interpreted as an expression of equilibrium between the moving pellicle and the resisting environment, and the interpretation leads to the theory that the modification is the resultant of natural selection among the numerous forces expressed in the resisting obstructions to the once started shot. When we use results in this sense it is evident that the causes are various and of various values. There is the initial energy expressed in the properties of the explosive powder, the directive force expressed in the barrel of the gun which guides the explosion in one direction; and there is the aim of the gun made by the man shooting it, and even behind this is the mental direction of the muscular action. Each of these was a determining cause in bringing about the shape of the pellet, and in accounting for the distribution and shaping of the shot each was a cause of greater importance than the particular conditions of its place of final rest.

Although it is scientifically true and accurate to define the particular flesh, bone, button, or nail-head as determinants in bringing about the final result of the motion of the lead pellets through space, their actual and relative positions, and the shape they finally assume, these conditions of environment are but causes of diversion from the direction, position, and relative distribution which were determined before the environment was met with.

The reason why each individual pellet stopped exactly where it did is correctly defined as the result of its particular environment, but the reason why it got there is not so explained. So it is not difficult to understand that as long as we only microscopically examine the perfect adaptation of organic structure to the particular place it occupies in nature, the theory that species were originated by the action of the conditions of environment through natural selection and the survival of the fittest seems sufficient and apt. But when we consider what an immensely greater demand is made upon causative energy to account for variability, compared with that required to adjust to its environment an already living and varying organism, it becomes evident that evolution is a far greater matter than the result of natural selection.

To use the same illustration, we note that the fact, that the lead pellets are observed in the act of travelling through space, and finally stopping as they strike the resisting bodies, does not remove the necessity of assuming the initial explosion of the powder and the aim of the gun to account for their motions.

So were we to lengthen out the gyration of organic plastidules, or biophores, a million million years, continuously holding on to their original powers and potencies for all that time, we are not relieved in the least from the logical necessity of endowing them at the outset with the real directive energy which phenomenally expresses itself for the first time when the finally adjusted organism appears. And the increment to organic structure expressed by their final bursting into morphological reality, after travelling unobserved but potential through the organic matter of countless generations, is as much a result of creative energy as if a new species were to arise out of the dust of the earth.

INDEX.

Allen's Laboratory Exercises in Elementary Physics. By CHARLES R. ALLEN, Instructor in the New Bedford, Mass., High School. *Pupils' Edition:* x + 209 pp. 12mo. 80c., *net.* (*Teachers' Edition:* $1.00, *net.*)

Arthur, Barnes, and Coulter's Handbook of Plant Dissection. By J. C. ARTHUR, Professor in Purdue University, CHARLES R. BARNES, Professor in University of Wisconsin, and JOHN M. COULTER, President of Lake Forest University. xi + 256 pp. 12mo. $1.20, *net.*

Barker's Physics. See American Science Series.

Beal's Grasses of North America. For Farmers and Students. By W. J. BEAL, Professor in the Michigan Agricultural College. 2 vols. 8vo. Copiously illustrated. (*December*, 1895.)

Bessey's Botanies. See American Science Series.

Black and Carter's Natural History Lessons. By GEORGE ASHTON BLACK, Ph.D., and KATHLEEN CARTER. (For very young pupils.) x + 98 pp. 12mo. 50c., *net.*

Bumpus's Laboratory Course in Invertebrate Zoology. By HERMON C. BUMPUS, Professor in Brown University, Instructor at the Marine Biological Laboratory, Wood's Holl, Mass. *Revised.* vi + 157 pp. 12mo. $1.00, *net.*

Cairns's Quantitative Chemical Analysis. By FREDERICK A. CAIRNS. Revised and edited by Dr. E. WALLER. 279 pp. 8vo. $1.60, *net.*

Champlin's Young Folks' Astronomy. By JOHN D. CHAMPLIN, Jr., Editor of *Champlin's Young Folks' Cyclopædias.* Illustrated. vi + 236 pp. 16mo. 48c., *net.*

***Crozier's Dictionary of Botanical Terms.** By A. A. CROZIER. 202 pp. 8vo. $2.40, *net.*

Hackel's The True Grasses. Translated from "Die natürlichen Pflanzenfamilien" by F. LAMSON-SCRIBNER and EFFIE A. SOUTHWORTH. v + 228 pp. 8vo. $1.35, *net.*

Hall's First Lessons in Experimental Physics. For young beginners, with quantitative work for pupils and lecture-table experiments for teachers. By EDWIN H. HALL, Assistant Professor in Harvard College. viii + 120 pp. 12mo. 65c., *net.*

Hall and Bergen's Text-book of Physics. By EDWIN H. HALL, Assistant Professor of Physics in Harvard College, and JOSEPH Y. BERGEN, Jr., Junior Master in the English High School, Boston. xviii + 388 pp. 12mo. $1.25, *net.*

Postage 10 per cent additional on net books. Descriptive list free.

Hertwig's General Principles of Zoology. From the Third Edition of Dr. Richard Hertwig's *Lehrbuch der Zoologie.* Translated and edited by GEORGE WILTON FIELD, Professor in Brown University. About 200 pp. 8vo. (*October*, '95.)

Howell's Dissection of the Dog. As a Basis for the Study of Physiology. By W. H. HOWELL, Professor in the Johns Hopkins University. 100 pp. 8vo. $1.00, *net.*

Jackman's Nature Study for the Common Schools. By WILBUR S. JACKMAN, Teacher of Natural Science, Cook County Normal School, Chicago, Ill. x + 448 pp. 12mo. $1.20, *net.*

Kerner & Oliver's Natural History of Plants. From the German of ANTON KERNER VON MARILAUN, Professor of Botany in the University of Vienna, by F. W. OLIVER, Quain Professor of Botany in University College, London, with the assistance of MARIAN BUSH and MARY E. EWART. 4to. 2 vols., each bound in two parts. $7.50 a vol., *net.* (*Vol. I ready; Vol. II ready soon.*)

Macalister's Zoology of the Invertebrate and Vertebrate Animals. By ALEX. MACALISTER. Revised by A. S. PACKARD. 277 pp. 16mo. 80 cents, *net.*

MacDougal's Experimental Plant Physiology. On the basis of Oels' *Pflanzenphysiologische Versuche.* By D. T. MACDOUGAL, University of Minnesota. vi + 88 pp. 8vo. $1.00, *net.*

Macloskie's Elementary Botany. With Students' Guide to the Examination and Description of Plants. By GEORGE MACLOSKIE, D.Sc., LL.D. 373 pp. 12mo. $1.30, *net.*

McMurrich's Text-book of Invertebrate Morphology. By J. PLAYFAIR McMURRICH, M.A., Ph.D., Professor in the University of Cincinnati. vii + 661 pp. 8vo. $4.00, *net.*

McNab's Botany. Outlines of Morphology, Physiology, and Classification of Plants. By WILLIAM RAMSAY McNAB. Revised by Prof. C. E. BESSEY. 400 pp. 16mo. 80c., *net.*

Martin's The Human Body. See American Science Series.

***Merriam's Mammals of the Adirondack Region.** Northeastern New York. With an Introductory Chapter treating of the Location and Boundaries of the Region, its Geological History, Topography, Climate, General Features, Botany, and Faunal Position. By Dr. C. HART MERRIAM. 316 pp. 8vo. $3.50, *net.*

Newcomb & Holden's Astronomies. See American Science Series.

***Noel's Buz:** or, The Life and Adventures of a Honey Bee. By MAURICE NOEL. 134 pp. 12mo. $1.00.

Noyes's Elements of Qualitative Analysis. By WILLIAM A. NOYES, Professor in the Rose Polytechnic Institute. x + 91 pp. 8vo. 80c., *net.*

Postage 10 *per cent additional on net books. Descriptive list free.*

Packard's Entomology for Beginners. For the use of Young Folks, Fruit-growers, Farmers, and Gardeners. By A. S. PACKARD. xvi + 367 pp. 12mo. *Third Edition, Revised.* $1.40, *net.*

—— Guide to the Study of Insects. and a Treatise on those Injurious and Beneficial to Crops. For the use of Colleges, Farm-schools, and Agriculturists. By A. S. PACKARD. With fifteen plates and six hundred and seventy wood-cuts. *Ninth Edition.* xii + 715 pp. 8vo. $4.50, *net.*

—— Outlines of Comparative Embryology. By A. S. PACKARD. Copiously illustrated. 243 pp. 8vo. $2.00, *net.*

—— Zoologies. See American Science Series.

Remsen's Chemistries. See American Science Series.

Scudder's Butterflies. By SAMUEL H. SCUDDER. x + 322 pp. 12mo. $1.20, *net.*

—— Brief Guide to the Commoner Butterflies. By SAMUEL H. SCUDDER, author of "Butterflies of the Eastern United States and Canada," etc., etc. xi + 206 pp. 12mo. $1.00, *net.*

—— The Life of a Butterfly. A Chapter in Natural History for the General Reader. By SAMUEL H. SCUDDER. 186 pp. 16mo. 80c., *net.*

Sedgwick & Wilson's Biology. See American Science Series.

***Step's Plant Life.** By EDWARD STEP. Popular Papers on the Phenomena of Botany. 12mo. 148 Illustrations. $1.00, *net.*

Underwood's Our Native Ferns and their Allies. With Synoptical Descriptions of the American Pteridophyta North of Mexico. By LUCIEN M. UNDERWOOD, Professor in DePauw University. *Revised.* xii + 156 pp. 12mo. $1.00, *net.*

Williams's Elements of Crystallography. By GEORGE HUNTINGTON WILLIAMS, late Professor in the Johns Hopkins University. x + 270 pp. 12mo. *Revised and Enlarged.* $1.25, *net.*

Williams's Geological Biology. An Introduction to the Geological History of Organisms. By HENRY S. WILLIAMS, Professor of Geology in Yale College. 8vo. About 300 pp. *(Oct., '95.)*

Woodhull's First Course in Science. By JOHN F. WOODHULL, Professor in the Teachers' College, New York City.

I. Book of Experiments. xiv + 70 pp. 8vo. Paper. 50c., *net.*
II. Text-Book. xv + 133 pp. 12mo. Cloth. 65c., *net.*
III. Box of Apparatus. $2, *net (actual cost to the publishers).*

Zimmermann's Botanical Microtechnique. Translated by JAMES ELLIS HUMPHREY, S.C. xii + 296 pp. 8vo. $2.50, *net.*

Postage 10 per cent additional on net books. Descriptive list free.

HENRY HOLT & CO., 29 WEST 23D ST., NEW YORK.

September, 1895.

PSYCHOLOGY
ETHICS AND PHILOSOPHY
REFERENCE AND TEXT-BOOKS

PUBLISHED BY

HENRY HOLT & COMPANY

29 WEST 23D STREET, NEW YORK

*Books marked * are chiefly for reference and supplementary use, and to be found in Henry Holt & Co.'s List of Works in General Literature. For further particulars about books not so marked see Henry Holt & Co.'s Descriptive Educational Catalogue. Either list free on application. Postage on net books 10 per cent. additional.*

***Bain's John Stuart Mill.** A Criticism with Personal Recollections. By Prof. ALEXANDER BAIN of Aberdeen. 12mo. 214 pp. $1.00.

***—— James Mill.** A Biography. With portrait. 12mo. 498 pp. $2.00.

Baldwin's Handbook of Psychology. By Prof. JAMES MARK BALDWIN of Toronto. 2 vols. (sold separately). 8vo.
Vol. I. Senses and Intellect. 357 pp. $1.80, *net.*
Vol. II. Feeling and Will. 406 pp. $2.00, *net.*

—— Elements of Psychology. 372 pp. 12mo. $1.50, *net.*

Descartes: The Philosophy of Descartes. Selected and translated by Prof. H. A. P. TORREY of the University of Vermont. (*Sneath's Modern Philosophers.*) xii + 345 pp. 12mo. $1.50, *net.*

Falckenberg's Modern Philosophy from Nicolas of Cusa to the Present Time. By Prof. RICHARD FALCKENBERG of Erlangen. Translated with the author's co-operation by Prof. A. C. ARMSTRONG, Jr., of Wesleyan. 665 pp. 8vo. $3.50, *net.*

Hegel: Philosophy of Hegel. Translated extracts from Hegel's works, with an introduction by Prof. JOSIAH ROYCE of Harvard. (*Sneath's Modern Philosophers.*) (*In preparation.*)

***Hillebrand's German Thought.** From the Seven Years' War to Goethe's Death. Six Lectures delivered at the Royal Institution of Great Britain. By CARL HILLEBRAND. 12mo. 306 pp. $1.75.

***Holland's Rise of Intellectual Liberty. from Thales to Copernicus.** A History. By FREDERICK MAY HOLLAND. 8vo. 458 pp. $3.50.

Hume: The Philosophy of Hume. Extracts from the "Treatise of Human Nature." With an introduction by HERBERT A. AIKINS, Professor in Western Reserve University. 176 pp. 12mo. (*Sneath's Modern Philosophers.*) $1.00, *net.*

Hyde's Practical Ethics. By Pres. WM. DE WITT HYDE of Bowdoin. 12mo. 219 pp. 80 cents, *net.*

James' Principles of Psychology. (*American Science Series.*) *Advanced Course.* By Prof. WM. JAMES of Harvard. 2 vols. 8vo. 701 + 710 pp. $4.80, *net.*

—— Psychology. *Briefer Course.* 12mo. 491 pp. $1.60, *net.*

Jastrow's Chapters in Modern Psychology. By Prof. JOSEPH JASTROW of the University of Wisconsin. (*In preparation.*)

Kant: The Philosophy of Kant. Selected and translated by Prof. JOHN WATSON of Queen's College, Canada. x + 356 pp. 12mo. (*Sneath's Modern Philosophers.*) $1.75, *net.*

Locke: The Philosophy of Locke. Extracts from the "Essay concerning Human Understanding." Edited by Prof. JOHN E. RUSSELL of Williams College. 160 pp. 12mo. (*Sneath's Modern Philosophers.*) $1.00, *net.*

***Martineau's Essays, Philosoph'cal and Theological.** By JAMES MARTINEAU. 2 vols. 8vo. 428 + 438 pp. $5.00.

***Maude's The Foundation of Ethics.** By JOHN EDWARD MAUDE. Edited by Prof. WM. JAMES of Harvard. 12mo. 224 pp. $1.50.

***Mill's Three Essays on Religion, and Berkeley.** By JOHN STUART MILL. 8vo. 313 pp. $2.00.

***—— The Autobiography.** 8vo. 319 pp. $2.00.

***—— Dissertations and Discussions.** 5 vols. 8vo. $2.00 per vol.

***—— Examination of Sir William Hamilton's Philosophy.** 8vo. 354 pp. $2.75.

***—— Comte's Positive Philosophy.** 8vo. 182 pp. $1.25.

***Mill, John Stuart: His Life and Works.** Twelve sketches by HERBERT SPENCER, HENRY FAWCETT, FREDERIC HARRISON, and other distinguished authors. 16mo. 96 pp. $1.00.

***Nicholls' The Psychology of Time.** By HERBERT NICHOLLS, Fellow of Clark. 8vo. 140 pp. $1.50, *net.*

Paulsen's Introduction to Philosophy. By Prof. FRIEDRICH PAULSEN of Berlin. Translated with the author's sanction by Prof. FRANK THILLY of the University of Missouri. With a preface by Prof. WM. JAMES of Harvard. xxiv + 437 pp. 8vo. $3.50, *net.*

Reid: The Philosophy of Reid. The "Inquiry into the Human Mind on the Principles of Common Sense." Edited by Dr. E. HERSHEY SNEATH of Yale. viii + 368 pp. 12mo. (*Sneath's Modern Philosophers.*) $1.50, *net.*

Spinoza: The Philosophy of Spinoza. Parts I, II, V of the "Ethics," and Extracts from III, IV. Translated and edited by Prof. GEO. STUART FULLERTON of the University of Pennsylvania. *Second Edition, Enlarged.* vi + 358 pp. 12mo. (*Sneath's Modern Philosophers.*) $1.50, *net.*

Zeller's Outlines of the History of Greek Philosophy. By Dr. EDWARD ZELLER. Translated with the author's sanction by SARAH F. ALLEYNE and EVELYN ABBOTT. 12mo. 377 pp. $1.40, *net.*

Postage on net books 10 per cent additional. Descriptive list free.

HENRY HOLT & CO., 29 WEST 23D ST., NEW YORK.

September, 1895.

Translated by Chas. G. Leland. 75c.—Hertz's King René's Daughter. Translated by Theo. Martin. $1.25.—Lessing's Nathan the Wise. Translated by Ellen Frothingham. $1.50. —Lockhart's Ancient Spanish Ballads. $1.25.—Kalidasa's Shakuntala. Translated by Prof. A. H. Edgren. $1.50.

Lord's (A. E.) Days of Lamb and Coleridge. 12mo. $1.25.

Maine's (H. S.) Ancient Law. $3.50.—Early History of Institutions. $3.50.—Village Communities. $3.50.—Early Law and Custom. $3.50.—Popular Government. $2.75.—International Law. $2.75.—Life and Speeches. With portrait. $3.50.

Martin's (Arthur Shadwell) The Art of Parody. 16mo.

Martineau's (James) Essays. 2 vols. 8vo. $5.00.

Mill's (J. S.) Three Essays on Religion. $2.00.—Autobiography. $2.00.—Dissertations and Discussions. 5 vols. $10.00.—Representative Government. $2.00.—Sir William Hamilton's Philosophy. $2.75.—On Liberty and the Subjection of Women. $2.00. —Comte's Positive Philosophy. $1.25.

Morgan's (L. H.) Ancient Society. 8vo. $4.00.

Moscheles's (Ignatz) Recent Music and Musicians. 12mo. $2.00.

Pancoast's (H. S.) Introduction to English Literature. 16mo. $1.25 *net.*—Introduction to American Literature. 16mo.

Paulsen's (F.) Introduction to Philosophy. 8vo. $3.50, *net.*

Perry's (T. S.) Greek Literature. Richly illustrated. 8vo. $4.00.

Smith's (Chas. Jno.) Synonyms Discriminated. 12mo. $1.80, *net.*

Sumner's (W. G.) History of American Currency. 12mo. $3.00. Collected Essays. 8vo. $1.50.—Protectionism. 16mo. $1.00.

Symonds' (J. A.) Renaissance in Italy. 7 vols. $2.00 *each.*—Short History of the Renaissance. 1 vol. 12mo. $1.75.—Italian Byways. 12mo. $1.75.

Taine's (H. A.) Italy. 2 vols.—Notes on England.—Tour through the Pyrenees.—Notes on Paris.—On Intelligence. 2 vols.—Lectures on Art. 2 vols.—The Ancient Régime.—The French Revolution. 3 vols.—The Modern Régime. 2 vols.—History of English Literature. 2 vols. 12mo. $2.50 per vol.

Tylor's (E. B.) Primitive Culture. 2 vols. 8vo. $7.00.—Early History of Mankind. 8vo. $3.50.

Wagner's (Richard) Art Life and Theories. 12mo. $2.00.—Ring of the Nibelung (Explained and in part translated.) 12mo. $1.50.

Walker's (F. A.) Wages. $2.00.—Money. $2.00.—Money, Trade, and Industry. $1.25.

Wallace's (D. Mackenzie) Russia. With 2 maps. 8vo. $2.00.

Warren's (Prof. F. M.) History of The Novel, Previous to the Seventeenth Century (in Greece, Spain, Italy, Germany, England, and China). 12mo. $1.75.

List of General Literature free. Postage on NET *books 10% additional.*

HENRY HOLT & CO., 29 W. 23D ST., NEW YORK.

www.ingramcontent.com/pod-product-compliance
Lightning Source LLC
Chambersburg PA
CBHW021347210326
41599CB00011B/783

* 9 7 8 3 3 3 7 2 1 6 8 3 2 *